Carbon Dioxide Fixation and Reduction in Biological and Model Systems

Carbon Dioxide Fixation and Reduction in Biological and Model Systems

Proceedings of the Royal Swedish Academy of Sciences Nobel Symposium 1991

Edited by

CARL-IVAR BRÄNDÉN

and

GUNTER SCHNEIDER

Department of Molecular Biology,
Swedish University of Agricultural Sciences

Oxford New York Tokyo

OXFORD UNIVERSITY PRESS

1994

Oxford University Press, Walton Street, Oxford OX2 6DP
Oxford New York Toronto
Delhi Bombay Calcutta Madras Karachi
Kuala Lumpur Singapore Hong Kong Tokyo
Nairobi Dar es Salaam Cape Town
Melbourne Auckland Madrid
and associated companies in
Berlin Ibadan

Oxford is a trade mark of Oxford University Press

Published in the United States
by Oxford University Press Inc., New York

A catalogue record for this book is available from the British Library

Library of Congress Cataloging in Publication Data.
Carbon dioxide fixation and reduction in biological and model systems
: proceedings of the Royal Swedish Academy of Sciences nobel
symposium, 1991/ edited by Carl-Ivar Brändén and Gunter Schneider.
1. Carbon dioxide – Metabolism – Congresses. 2. Photosynthesis – Congresses. 3. Microbial metabolism – Congresses. 4. Carbon dioxide – Congresses. 5. Reduction (Chemistry) – Congresses. I. Brändén, Carl-Ivar. II Schneider, Gunter. III. Kungl. Svenska vetenskapsakademien.
QP535. C1C39 1994 574. 19'14–dc20 93–45258

ISBN 0 19 854782 X

Typeset by
EXPO Holdings Malaysia
Printed in Great Britain on acid-free paper by
St Edmundsbury Press, Bury St Edmunds, Suffolk

Contents

Contributors

Christian Amatore *Ecole Normale Supérieure, Département de Chimie, URA CNRS 1679, 24 Rue Lhomond, 75231 Paris Cedex 05, France.*

T. John Andrews *Plant Environment Biology Group, Research School of Biological Sciences, Australian National University, P O Box 475, Canberra 26017 Australia.*

Murray R. Badger *Research School of Biological Sciences, Australian National University, PO Box 475, Canberra, ACT 2601, Australia.*

Arno Behr *Henkel KGaA, Chemische Verfahrensentwicklung TVC, Abteilung Explorative Versuche, Geb. H40, D-40191 Düsseldorf, Germany.*

Christian Bruneau *Laboratoire de Chimie de Coordination Organique, URA CNRS 415, Campus de Beaulieu, Université de Rennes, 35042 Rennes, France.*

John R. Coleman *Department of Botany, University of Toronto, 25 Willcocks St, Toronto, Ontario, Canada M5S 3B2.*

Donald J. Darensbourg *Department of Chemistry, Texas A & M University, College Station, TX 77843, USA.*

Pierre H. Dixneuf *Laboratoire de Chimie de Coordination Organique, URA CNRS 415, Campus de Beaulieu, Université de Rennes, 35042 Rennes, France.*

Daryl L. Edmondson *Plant Environment Biology Group, Research School of Biological Sciences, Australian National University, PO Box 475, Canberra 2601, Australia.*

Giuseppe Filardo *Dipartimento di Ingegneria Chimica dei Processi e dei Materiali, Viale delle Scienze, 1–90128, Palermo, Italy.*

Masahiro Fujiwara *Osaka National Research Institute, AIST, 1–8–31, Midorigaoka, Ikeda, Osaka 563, Japan.*

Salvatore Gambino *Dipartimento di Ingegneria Chimica dei Processi e dei Materiali, Viale delle Scienze, 1–90128, Palermo, Italy.*

Michael Grätzel *Institut de Chimie Physique, Ecole Polytechnique Fédérale de Lausanne CH-1015 Lausanne, Switzerland.*

Steven Gutteridge *Central Research and Development Department, Experimental Station, Du Pont Company, Wilmington, DE19880–0402 USA.*

Mark R. Harpel *Protein Engineering Program, Biology Division, Oak Ridge National Laboratory, Oak Ridge, TN 37831, USA.*

Fred C. Hartman *Protein Engineering Program, Biology Division, Oak Ridge National Laboratory, Oak Ridge, TN 37831, USA.*

Anny Jutand *Ecole Normale Supérieure, Département de Chimie, URA CNRS 1679, 24 Rue Lhomond, 75231 Paris Cedex 05, France.*

Heather J. Kane *Plant Environment Biology Group, Research School of Biological Sciences, Australian National University, PO Box 475, Canberra 2601, Australia.*

Ronald Kluger *Lash Miller Laboratories, Department of Chemistry, University of Toronto, Toronto, Canada M5S 1A1.*

Bernhard Kräutler *Institute of Organic Chemistry, University of Innsbruck, A-6020 Innsbruck, Austria.*

Frank W. Larimer *Protein Engineering Program, Biology Division, Oak Ridge National Laboratory, Oak Ridge, TN 37831, USA.*

Eva H. Lee *Protein Engineering Program, Biology Division, Oak Ridge National Laboratory, Oak Ridge, TN 37831, USA.*

Thomas J. Meyer *Department of Chemistry, University of North Carolina Chapel Hill, NC 27514 USA.*

Matthew K Morell *Plant Environment Biology Group, Research School of Biological Sciences, Australian National University, PO Box 475, Canberra 2601, Australia.*

Richard J. Mural *Protein Engineering Program, Biology Division, Oak Ridge National Laboratory, Oak Ridge, TN 37831, USA.*

Merete F. Nielsen *Ecole Normale Supérieure, Département de Chimie, URA CNRS 1679, 24 Rue Lhomond, 75231 Paris Cedex 05, France.*

Marion H. O'Leary *Department of Biochemistry, University of Nebraska, Lincoln, NE 68583, USA.*

Kalanethee Paul *Plant Environment Biology Group, Research School of Biological Sciences, Australian National University, P O Box 475, Canberra 2601, Australia.*

G.D. Price *Research School of Biological Sciences, Australian National University, PO Box 475, Canberra, ACT 2601, Australia.*

Gabriel A. Quinlan *Plant Environment Biology Group, Research School of Biological Sciences, Australian National University, P O Box 475, Canberra 2601, Australia.*

Vicente Rubio *Instituto de Investigaciones Citológicas, Fundación Valenciana de Investigaciones Biomedicas (Centro Asociado de CSIC), Amadeo de Saboya, 4, 46010, Valencia. Spain.*

Gunter Schneider *Department of Molecular Biology, Uppsala Biomedical Center, PO Box 590, S–751 24, Uppsala, Sweden.*

Giuseppe Silvestri *Dipartimento di Ingegneria Chimica dei Processi e dei Materiali, Viale delle Scienze, 1–90128, Palermo, Italy.*

Harry B. Smith *Protein Engineering Program, Biology Division, Oak Ridge National Laboratory, Oak Ridge, TN 37831, USA.*

Thomas S. Soper *Protein Engineering Program, Biology Division, Oak Ridge National Laboratory, Oak Ridge, TN 37831, USA.*

Yoshie Souma *Osaka National Research Institute, AIST, 1–8–31, Midorigaoka, Ikeda, Osaka 563, Japan.*

Belinda Tsao *Lash Miller Laboratories, Department of Chemistry, University of Toronto, Toronto, Canada M5S 1A1.*

Günter Wächtershäuser *Munich Am Tal 29, 8000 2, Germany.*

Alain Wasserfallen *Department of Microbiology, University of Illinois, Urbana, IL 61801, USA.*

Ralph S. Wolfe *Department of Microbiology, University of Illinois, Urbana, IL 61801, USA.*

J-W. Yu *Research School of Biological Sciences, Australian National University, PO Box 475, Canberra, ACT 2601, Australia.*

1

Photosynthetic carbon dioxide fixation: structural and functional aspects of ribulose bisphosphate carboxylase/oxygenase

Gunter Schneider

Introduction

Every year, about 10^{11} tons of CO_2 are incorporated into the biosphere by the photosynthetic activities of plants and microorganisms. Photosynthetic CO_2 fixation proceeds through the incorporation of CO_2 into ribulose bisphosphate (RuBP) which is then split into two molecules of phosphoglycerate. Part of the phosphoglycerate is used to synthesize energy-rich compounds such as starch or fatty acids. The remaining phosphoglycerate is used to regenerate RuBP, the primary acceptor of CO_2, in a cyclic series of reactions, the reductive pentose phosphate or Calvin cycle.

The initial step in the fixation of carbon dioxide, the addition of CO_2 to RuBP and formation of two molecules of phosphoglycerate, is catalysed by the enzyme ribulose-1,5-biphosphate carboxylase/oxygenase (Rubisco). The enzyme also catalyses the initial step in photorespiration in which oxygen instead of carbon dioxide is added to RuBP, yielding one molecule of phosphoglycerate and one molecule of phosphoglycolate. The latter is metabolized in the glycolate pathway, where it is ultimately converted to CO_2 and the energy is dissipated as heat. In C3 plants (most of the major crop plants in moderate climate zones belong to this class), photorespiration significantly reduces the efficiency of photosynthetic carbon dioxide fixation, and consequently crop productivity, by up to 50 per cent.

The dual function of Rubisco, catalysing the primary steps in both photosynthetic carbon dioxide fixation and photorespiration, makes it a challenging target for attempts to improve the efficiency of photosynthesis. Recombinant DNA techniques provide a promising tool to modify the carboxylase/oxygenase ratio by genetic engineering. A detailed understanding of Rubisco catalysis for both carboxylation and oxygenation reactions in structural and chemical terms is needed for a rational attempt to modify the enzyme's substrate specificity. Biochemical, genetic, and structural studies have provided a rather detailed picture of the mechanism of the carboxylation reaction (for recent review see Andrews and Lorimer 1987; Gutteridge 1990; Brändén *et al.* 1991; Schneider *et al.* 1992). In the following, some of the chemical and structural information will be summarized and the structural data to biochemical and genetic studies correlated.

Chemical events during carbon dioxide fixation

Overall carboxylation reaction

Carboxylation of RuBP is a complicated reaction which involves a series of events and a number of intermediates. The overall reaction can be divided into a number of individual steps (Fig. 1.1). The first step in this mechanism is the enolization of RuBP, resulting in 2,3-enediolate. Carboxylation at the nucleophilic centre at C2 creates the six-carbon intermediate 2-carboxy-3-keto-arabinitol-1,5-bisphosphate (3-keto-CABP), which undergoes hydration to the *gem*-diol form. An analogue of this intermediate, reduced at the C3 position, 2-carboxy-arbinitol-1,5-bisphosphate (CABP), is a tightly binding inhibitor of the enzyme (Fig. 1.2). Deprotonation of the *gem*-diol at the O3 atom initiates carbon–carbon cleavage which results in one molecule of phosphoglycerate and the C2 carbanion form of another phosphoglycerate molecule. The carbanion is then stereospecifically protonated at C2 yielding the second phosphoglycerate molecule. The release of products completes the catalytic cycle.

Partial reactions and reaction intermediates

Enolization of bound RuBP is considered to be the very first step in the catalytic cycle. In fact, this step is common to both the carboxylation and oxygenation reactions. Enolization, which is initiated by abstraction of the C3 proton of the substrate, leads to formation of the 2,3-enediolate of RuBP as the first intermediate during turnover. The exchange of the proton at C3 of RuBP can be fol-

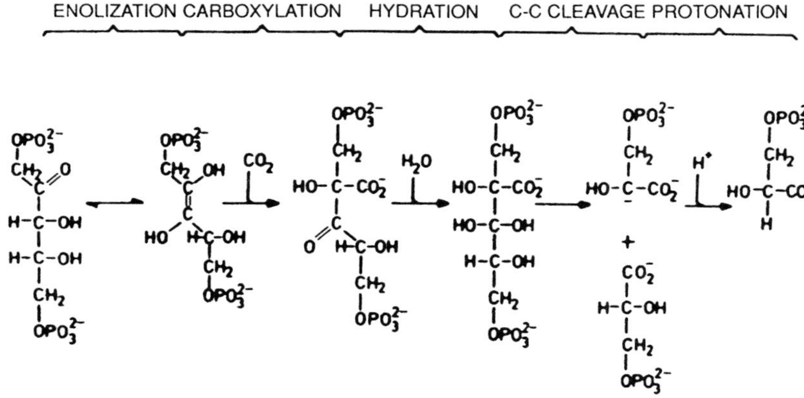

ENOLIZATION CARBOXYLATION HYDRATION C-C CLEAVAGE PROTONATION

Fig. 1.1 Chemical steps during the carboxylation reaction.

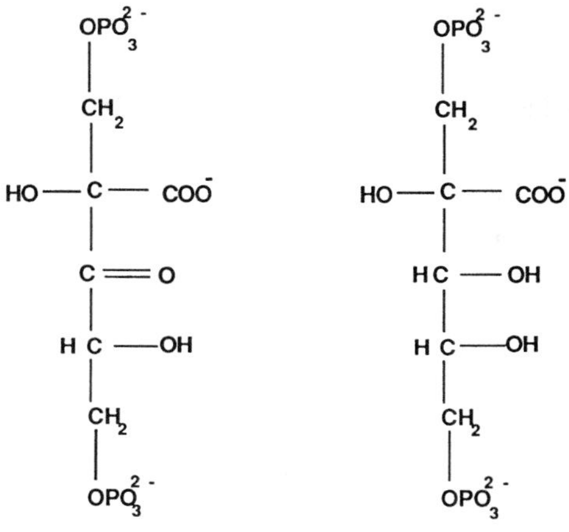

2-carboxy-3-keto-arabinitol 1,5 bis-phosphate

2-carboxy-arabinitol 1,5 bisphosphate

Fig. 1.2 The reaction intermediate 2-carboxy-3-keto-arabinitol-1,5-bisphosphate and an analogue, the strong binding inhibitor 2-carboxy-arabinitol-1,5-bisphosphate.

lowed by the 'wash in' of solvent ^3H into C3 of RuBP and the 'wash out' of [3-^3H]RuBP into the solvent (Sue and Knowles 1978; Sue and Knowles 1982). In this way, the enolization reaction can be studied independently of the overall carboxylation reaction.

It is at the stage of the 2,3-enediolate that the reaction proceeds either towards carboxylation or oxygenation. In both reactions, the gaseous substrate, CO_2 or O_2 respectively, react with the C2 carbon atom of RuBP. Electrophilic attack of CO_2 on the C2 carbon atom yields the six-carbon intermediate, 3-keto-CABP. One of the unique features of the carboxylation reaction is that the intermediate 3-keto-CABP can be isolated and that it is surprisingly stable (Pierce *et al.* 1986a). By acid quenching of the reaction, sacrificing large amounts of enzyme, the intermediate can be isolated and stored at –80 °C for months. At room temperature, 3-keto-CABP decarboxylates with a half-time of approximately 1 h.

After acid quenching of the carboxylation reaction, the intermediate can be trapped by borohydride reduction to the corresponding 2-carboxypentitol bisphosphate (Schloss and Lorimer 1982). The borohydride reduction of the intermediate occurs only in free solution not when bound to the enzyme. This is probably due to the fact that the six-carbon intermediate exists on the enzyme predominantly as the hydrated C3 *gem*-diol which cannot be reduced by borohydride (Lorimer *et al.* 1987).

On the enzyme, the intermediate 3-keto-CABP exists predominantly in the *gem*-diol form, resulting from the addition of a water or hydroxide molecule to the C3 carbon atom. The O3 oxygen of RuBP is completely retained during carboxylation (Lorimer 1978; Sue and Knowles 1978). This means that the hydration step is either kinetically reversible and/or stereospecific.

The isolated 3-keto-CABP can be added back to the enzyme and is then hydrolysed to products. The availability of the six-carbon intermediate provides another partial reaction to probe the functional defects of site-directed mutants. Mutant Rubisco's which are deficient in the overall carboxylation reaction might be able to catalyse the hydrolysis of the six-carbon intermediate to products. One can thus distinguish between mutant deficient in the enolization reaction, verified by the 'wash in' or 'wash out' experiments, or mutants deficient in one of the subsequent steps of catalysis.

The carbon–carbon bond cleavage of the *gem*-diol results in the formation of one molecule of phosphoglycerate and one molecule of the C2 carbanion of phosphoglycerate. However, no firm experimental evidence for the existence of the C2 carbanion as an intermediate has been obtained so far. Stereochemical considerations (Andrews and Lorimer 1987) suggest that the enzymic base donating the proton to the C2 carbonion is different from the groups involved in proton–proton transfer during the earlier enolization–carboxylation steps of catalysis.

The enzyme, ribulose-1,5-bisphosphate carboxylase

Rubisco from higher plants, algae, and most photosynthetic microorganisms is a multisubunit complex built up of eight large (Mw 54 kDa) and eight small (Mw 14 kDa) subunits. The catalytic activities for both the carboxylation and oxygenation reaction reside on the large subunit. The primary structure of the large subunits of higher plants and algeal carboxylases studied so far exhibit a high degree of amino acid homology, in the range of 70–90 per cent (Andrews and Lorimer 1987).

In contrast to these L_8S_8 type carboxylases, the enzyme from the photosynthetic bacterium *Rhodosprillium rubrum* differs considerably in primary and quaternary structure. This carboxylase is only a dimer of large subunits and lacks the small subunits (Tabita and McFadden 1974; Schloss *et al.* 1979). The overall amino acid homology to the large subunit of higher plant carboxylases is approximately 28 per cent.

Irrespective of the type of Rubisco, the enzyme has to be activated in order to become catalytically competent. Activation involves the reaction of a CO_2 molecule with the ϵ-amino group of a lysine residue under formation of a carbamate (Lorimer 1981). The CO_2 molecule forming the carbamate is different from the CO_2 molecule that is incorporated into RuBP during catalysis. The carbamylated enzyme then forms the active ternary complex with a Mg(II) ion (Lorimer 1979).

Overall structure of the enzyme

A number of crystallographic studies have focused on Rubisco, in its activated and non activated form, with and without bound ligands, and structural information is now available for both the L_2 and the L_8S_8 type of the enzyme (Table 1.1). Since the catalytic activities reside on the L chain of the enzyme, only the three dimensional structure of this chain will be considered in the following. A more detailed description of the structure for the L_8S_8 enzyme can be found elsewhere (Andersson *et al.* 1989; Chapman *et al.* 1988; Knight *et al.* 1989,1990), as well as a comparison of the L_2 and L_8S_8 enzyme with a discussion of the function of the small subunit (Schneider *et al.* 1990*b*)

The large subunit is divided in two domains, one smaller N-terminal domain linked to a C-terminal domain which has an eight-stranded barrel type structure (Schneider *et al.* 1986). The domain arrangement and the secondary structure of these domains is shown in Fig. 1.3. The two subunits interact tightly to form the functional L_2 Rubisco molecule of *R. rubrum* (Fig. 1.4). The core of this binding area consists of interactions between the C-terminal domains around a local twofold axis. In addition, two regions from the N-terminal domain of one subunit interact with regions from the C-terminal domain of the second subunit (Fig. 1.5). These subunit interactions are of functional importance as some of the residues involved occur in or close to the active site region. Each active site of the dimer is thus built up from residues of both subunits.

The corresponding functional dimer of large subunits occurs as part of the L_8S_8 Rubisco molecule from spinach (Andersson *et al.* 1989) and from tobacco (Chapman *et al.* 1987). In the higher plant type enzyme, four such dimers are arranged around a fourfold axis, building up the L_8 core of the molecule.

The active site of Rubisco is located at the carboxy ends of the eight β strands in the barrel. The site is shaped like a funnel and is mainly formed by the eight

Table 1.1 Crystal structures of Rubisco

Source	Species	Resolution (Å)	Reference
R. rubrum	Native	1.7	Schneider *et al.* 1986,1990*b*
R. rubrum	Enzyme – phosphoglycerate	2.9	Lundqvist and Schneider 1989*a*
R. rubrum	Enzyme – CABP	2.6	Lundqvist and Schneider 1989*b*
R. rubrum	Enzyme – Mg(II) – CO_2	2.6	Lundqvist and Schneider 1991*a*
R. rubrum	Enzyme – Mg(II) – CO_2 – RuBP	2.6	Lundqvist and Schneider 1991*b*
R. rubrum	D193N	2.6	Söderlind *et al.* 1992
Spinach	Enzyme – Mg(II) – CO_2 – CABP	2.4	Knight *et al* . 1990
Tobacco	Native	2.8	Chapman *et al.* 1987, 1988

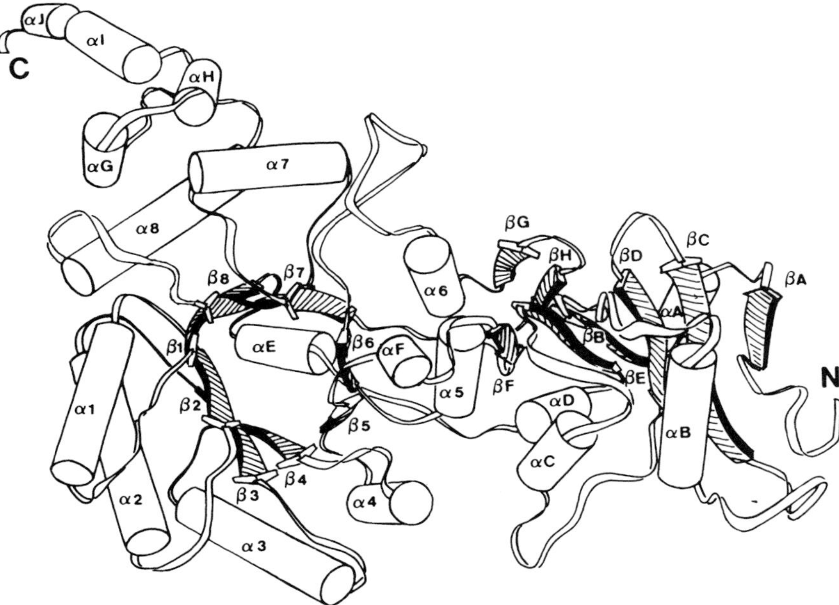

Fig. 1.3 Schematic view of the subunit of Rubisco from *R. rubrum*. The secondary structural elements are indicated (cylinders represent α helices and arrows represent β strands).

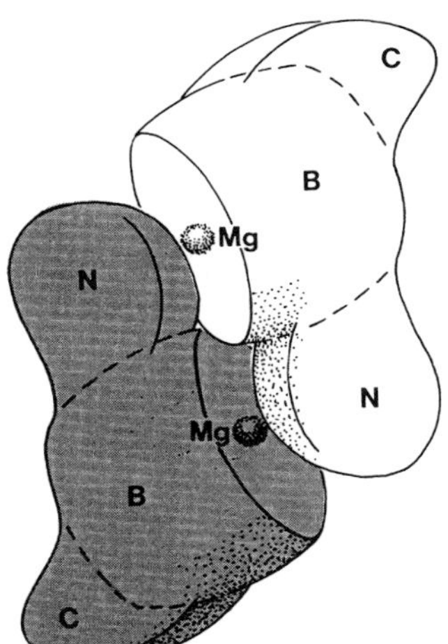

Fig. 1.4 Subunit arrangement in the L_2 dimer of Rubisco. The location of the active sites is indicated by the position of the active site Mg(II) ion. One active site is located between the C-terminal domain from one subunit (B) and the N-terminal domain (N) from the other subunit. The distance between the two active site metals is 36Å (Drawing by U. Uhlin.)

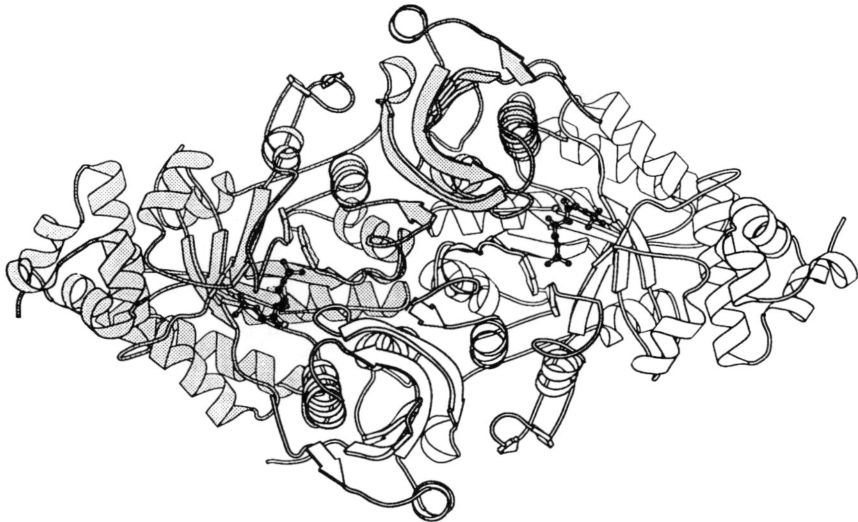

Fig. 1.5 View of the Rubisco dimer from *Rhodospirillum rubrum*. One of the L subunits is shown in grey. Bound RuBP at the active site is included in the figure.

loop regions that connect the eight β strands with the corresponding helices in the barrel domain (Fig. 1.6). The N-terminal domain from the second subunit in the L_2 dimer covers part of the top of the active site. In particular, two loop regions of this domain provide residues to the active site.

The activation process

The amino acid residue of central importance for activation is Lys201*. Lys201 is the last residue in β-strand 2 of the α/β barrel and is located at the bottom of the active site. The side chain forms hydrogen bonds to the main chain oxygen of Asn202 and a water molecule. The side chain of residues Asp203, His294, Leu266, and Ile173 are in van der Waals distance to the Lys side chain. The addition of a CO_2 molecule to the ϵ-amino group of Lys201 results in the formation of a carbamate. One of the carbamate oxygen atoms is ligated to the Mg(II) ion. Other protein ligands of the metal ion are the side chain of Asp203 and Glu204. By binding Mg(II) the active site becomes poised to properly bind and orient the substrate. In the activated ternary complex of Rubisco from *R. rubrum*, at least one water molecule could be identified within the first coordination sphere of the metal. The activation process thus changes a positive charge, located at a central position in the active site to a negative charge, which can now accommodate a positively charged metal ion.

* (All amino acid sequence numbering is based on the unified numbering system suggested in Schneider *et al.* (1991*a*).

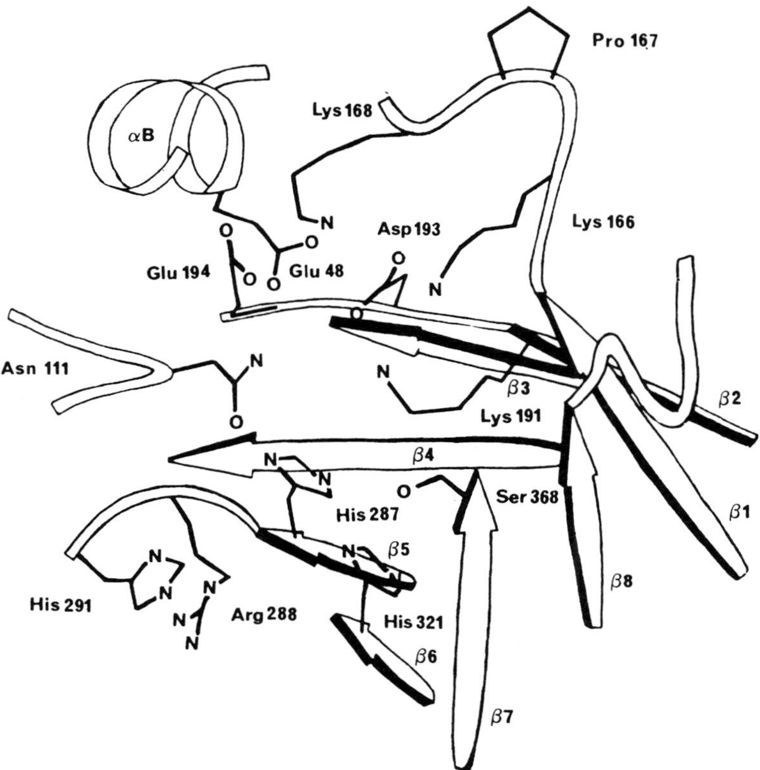

Fig. 1.6 Schematic view of the active site of Rubisco from *R. rubrum*. The eight β strands of the α/β barrel are shown as arrows. The side chains of conserved amino acids are included (the amino acid sequence numbering is that for the enzyme from *R. rubrum*).

Binding of phosphorylated compounds to the active site

The structure of two complexes of phosphorylated sugar compounds with non-activated Rubisco have been determined (Lundqvist and Schneider 1989*a*,*b*). The binding of the product, phosphoglycerate, revealed the location of the active site and one of the phosphate-binding sites. The phosphate group is bound close to loops 5 and 6 with residues Arg295, His327, and main-chain nitrogen atoms of residues 328 and 329 interacting with the phosphate group. Furthermore, the side chain of Ser379 from loop 7 forms a hydrogen bond to the bridging oxygen atom of the phosphate group. The carboxyl group of phosphoglycerate interacts with the side chain of Lys201.

Nonactivated Rubisco binds the inhibitor CABP across the active site, with one of the phosphate groups bound at the binding site close to loops 5 and 6 described above. The second phosphate binding site is located on the other side

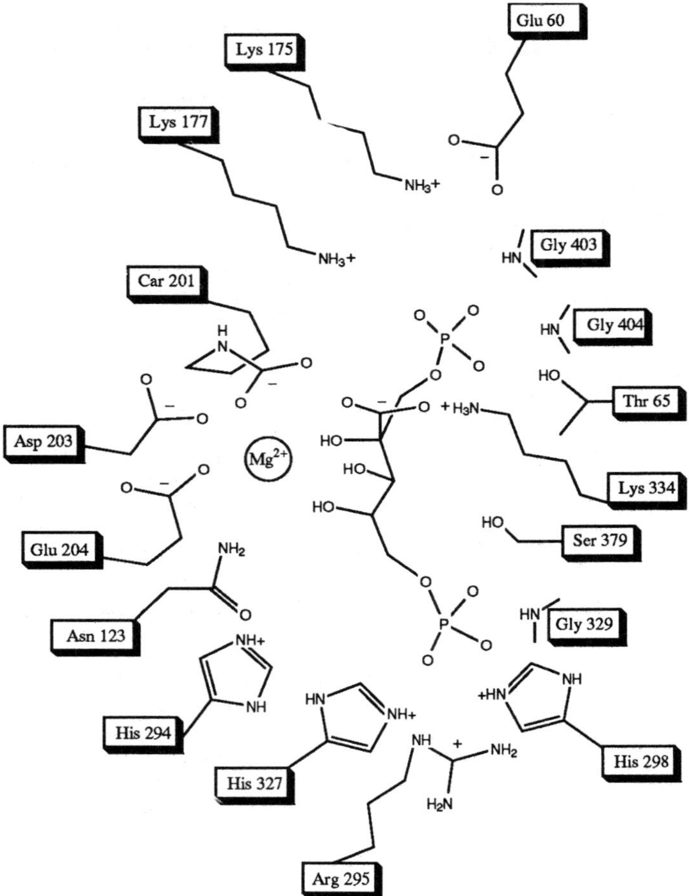

Fig. 1.7 Schematic view of the inhibitor CABP bound at the active site of activated spinach Rubisco.

of the barrel, close to loop 8. The phosphate group interacts with the main-chain nitrogen of the conserved residue 403 which is located at the N-terminal end of a very short helix. Other interactions to the enzyme are through hydrogen bonds and salt bridges with Lys175, Lys334, and Asn123.

The crystal structure analysis of the quaternary complex of spinach Rubisco with bound inhibitor CABP (Andersson *et al.* 1989; Knight *et al.* 1990) provided the first evidence of the binding of substrate analogues to the active site of the activated enzyme. Studies of the activated complex of tobacco Rubisco with CABP (Curmi *et al.* 1992) confirmed these results and demonstrated the high similarity of the active sites of the two L_8S_8 enzymes. The reaction intermediate analogue binds in an extended conformation at the active site (Fig. 1.7). The two phosphate groups bind at the two phosphate sites of the α/β barrel. The P1 phosphate interacts with the main-chain nitrogens of residues 403 and 404 of the short helix in loop 8 as well as with the main-chain nitrogen of residue 381 of loop 7. A hydrogen bond to the side chain of Thr65 and two salt links, involving Lys175 and Lys334 also contribute to the interactions at the phosphate binding site (Knight *et al.* 1990). The P2 phosphate binding site is formed by residues Arg295, His327, and Ser379. One of the oxygen atoms of the 2-carboxyl group which simulates the substrate CO_2 after the carboxylation event is a ligand to the Mg(II) ion. The second oxygen forms a hydrogen bond to the side chain of the conserved Lys334. Two of the hydroxyl oxygen atoms of CABP are additional ligands to Mg(II). The remaining hydroxyl group forms a hydrogen bond to the side chain of the invariant residue Ser379. Other polar interactions of the analogue with the groups on the enzyme are contacts to Asn123, Thr173, the main-chain nitrogen of residue 380, and the carbamate at 201.

A comparison of the crystal structures of the complexes of activated and non-activated Rubisco with CABP shows that the metal ion has a profound influence on the mode of inhibitor binding (Lundqvist and Schneider 1989*b*). In the complex with nonactivated enzyme, the inhibitor binds 'the other way around' at the active site, with the phosphate groups interchanged at their respective binding sites, as compared to the quaternary complex of activated spinach Rubisco with CABP (Knight *et al.* 1990). It seems that the Mg(II) ion plays an important role in orienting the analogue, and by inference the substrate, into the proper, catalytic competent binding mode at the active site.

In activated L_8S_8 Rubisco with bound CABP, the side chain of Lys334 is close enough to the reaction intermediate analogue to form a hydrogen bond to the 2-carboxy group of CABP.

The binding of the substrate, RuBP to activated crystals of Rubisco from *R. rubrum* has been studied to 2.6 Å resolution (Lundqvist and Schneider 1991*b*). The substrate binds in a rather bent conformation at the active site, with one of the phosphate groups at the phosphate binding site close to loops 5 and 6 (Fig. 8). The second phosphate interacts with residue 403 from the short helix in loop 8 and the side chain of Lys175. Due to the different position of loops 7 and 8 and helix $\alpha 8$ in the L_2 Rubisco (Schneider *et al.* 1990*a*), the second phosphate binding site is shifted in relation to its position in the L_8S_8 Rubisco. As a consequence of this species-dependent conformational difference, the substrate is bound in a more bent fashion to L_2 Rubisco than in the active site of L_8S_8 Rubisco.

The C2 oxygen atom and the C3 hydroxyl group are ligands to the Mg(II) ion. This leads, after abstraction of the C3 proton, to the formation of the *cis*-enediol. Due to the limited resolution of the electron density maps, the conformation at the C3 and C4 atom of RuBP cannot be established with confidence. The refinement of the structure of the quaternary complex of spinach Rubisco with CABP to 1.7 Å resolution is expected to settle this question.

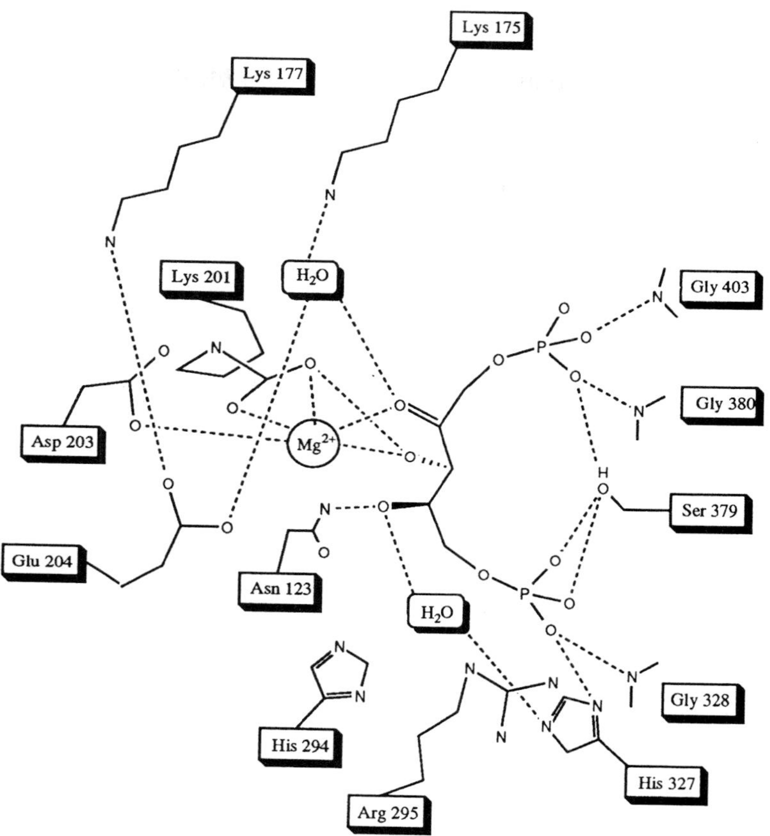

Fig. 1.8 Schematic view of the bound substrate, RuBP at the active site of activated Rubisco.

Conformational changes during catalysis

Comparisons of the nonactivated forms of Rubisco from *R. rubrum* (Schneider *et al.* 1990*b*) and tobacco (Chapman *et al.* 1988) with the quaternary CABP complexes of activated Rubisco from spinach (Knight *et al.* 1990) and tobacco (Curmi *et al.* 1992) reveal important conformational differences around the active site. In the activated quaternary complexes of both spinach and tobacco, loop 6 of the α/β barrel is folded over the CABP molecule so that the invariant residue Lys334 interacts with one of the oxygen atoms of the carboxyl group of CABP. In addition this residue participates in salt bridges to both the phosphate group P1 and Glu60 in the N-terminal domain of the other subunit of the dimer. This loop has a different conformation and position in the nonactivated tobacco enzyme which is reflected in a difference in the $C\alpha$ positions of residue 335 of

12.3 Å. In the nonactivated *R. rubrum* enzyme this loop is disordered (residues 330–341) and not visible in the electron density maps at 1.7 Å resolution.

The C terminus from residue 460 and onwards extends towards the active site and folds over loop 6 in the activated ternary complexes. In the nonactivated tobacco enzyme some of these residues are folded differently and the remaining ones, after residue 468, are disordered. The conformational change of these C-terminal residues are in all probability correlated to the movement of loop 6.

Thr65, which is involved in binding the phosphate group P1, is part of a loop region in the N-terminal domain which is close to the active site in the quaternary complexes. This loop is disordered in nonactivated Rubisco both from *R. rubrum* (residues 66 to 76) and tobacco (residues 64 to 68). The N-terminal residues 9 to 21 fold over this loop in the quaternary complexes but are disordered in the nonactivated tobacco enzyme. Presumably the ordering of both loops 64 to 68 and the N terminus are correlated. Conformational changes of the N and C terminal are likely to be different in L_2 Rubisco, since both the N terminus and the C terminus are quite different in sequence as well as in structure in *R. rubrum* Rubisco.

In conclusion, the following structural changes are observed when going from the nonactivated enzyme to quaternary complexes which simulate the active site after CO_2 addition. Two loop regions, one from each domain fold over the reaction intermediate analogue and provide side chains which are directly

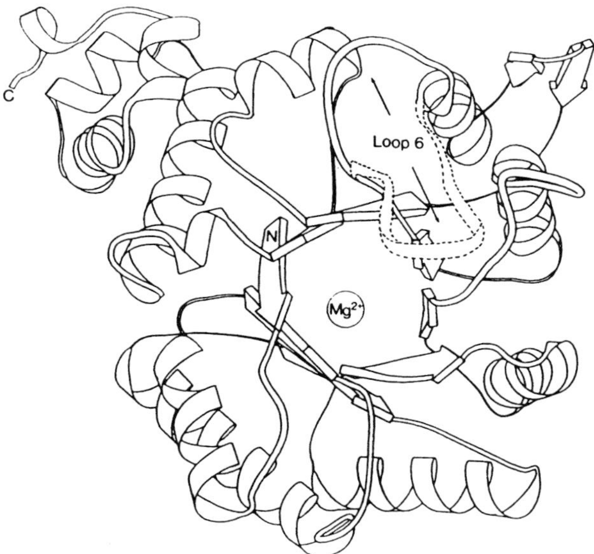

Fig. 1.9 Schematic view of the C-terminal domain of Rubisco, illustrating parts of the conformational changes during binding of the inhibitor CABP. The location of the active site is indicated by the position of the Mg(II) ion. The different conformations of loop 6 of the α/β barrel are shown. Solid lines: 'open' conformation; dashed lines: 'closed' conformation.

involved in binding the analogue (Fig. 1.9). The N and C terminii are coupled to these changes so that they in turn fold over the loop regions, locking them in position. One effect of these structural changes is to shield the reaction intermediate analogue almost completely from solution, a second effect is to bring binding residues into the active site.

Structure and function

Activation

Activation involves the condensation of CO_2 with an unprotonated Lys201 (Lorimer 1981) to form the carbamate. An explanation for the high reactivity of Lys201 as compared to other lysine side chains has been suggested based on examination of the electrostatic field at the active site of the enzyme (Lu *et al.* 1992). It is found that Lys201 is in an area of high positive potential making it prone to release a proton and react with CO_2.

Mutation of any of the residues involved in metal-ion binding result in a catalytically inactive enzyme (Estelle *et al.* 1985; Lorimer *et al.* 1987). Replacement of Lys201 by a glutamic acid side chain to mimic the carbamate, resulted in an enzyme devoid of activity, corroborating the important of a correctly positioned metal ion.

Although no large conformational changes occur upon activation of the enzyme, the microenvironment of the active site is greatly changed which can explain the large difference in physical and chemical properties between the nonactivated and activated forms of the enzyme.

Enolization

The initial step in enolization of the substrate RuBP is abstraction of the C3 proton. Much effort has been spent to identify the enzymic base involved in this step. Based on biochemical and mutagenesis data, Lys175 has been suggested as the base abstracting the C3 proton in the first step of the enolization reaction (Hartman *et al.* 1987; Lorimer and Hartman 1988). The evidence for this conclusion seemed compelling. Both affinity labelling and crystallographic studies showed that Lys175 is located at the active site. Based on biochemical data, it was concluded that Lys175 has an unusually low pK_a value of 7.8 (Hartman *et al.* 1985) which agrees well with the pK_a of 7.5 determined form the pH dependency of the deuterium isotope effect for proton abstraction from C3 of the substrate (Van Dyke and Schloss 1986). Site-directed mutagenesis of this residue, replacing the lysine residue by glycine, alanine, serine, glutamine, arginine, cysteine, or histidine results in mutant proteins severely deficient in carboxylase activity (Hartman *et al.* 1987). The Gly mutant is able to undergo activation and to bind CABP, but is unable to catalyse the enolization reaction(Lorimer and Hartman 1988). In the crystal structures of activated Rubisco with RuBP or

CABP, the Nϵ-nitrogen atom of the side chain is about 6 Å away from the C3 atom of the substrate. Model-building experiments cannot bring this side chain close enough for proton abstraction without destroying the conformation of loop 1. This would in turn influence the interactions at the subunit–subunit interface and such a conformational change seems unlikely. In fact, crystallographic studies have revealed conformational difference between L_2 and L_8S_8 Rubisco and between the nonactivated and quaternary complex of the enzyme. However, no differences in the conformation of loop 1 in any of the crystal structure determined so far have been observed. Based on the crystal structure of the complex of Rubisco from *R. rubrum* with RuBP it has been suggested that the Lys175 side chain acts not as a base but as an acid and polarizes the C2 oxygen bond, thus facilitating the abstraction of he C3 proton (Lundqvist and Schneider 1991*b*). Dependent on whether the resulting enediolate would be in the *cis* or *trans* conformation, the P1 phosphate group, His294, a water molecule or the carbamate was suggested as possible proton abstractors. But clearly, this issue remains to be settled and more experiments are required.

Carboxylation

The active site Mg(II) ion plays a crucial role in the mechanism of carboxylation by Rubisco (Lorimer *et al.* 1987; Knight *et al.* 1990; Lundqvist and Schneider 1991*b*). The profound influence of the metal on the overall reaction and the partitioning between carboxylation and oxygenation can be seen in enzyme species where the Mg(II) ion has been replaced by Co(II) (Robison *et al.* 1979, Christeller 1981) or Cu(II) (Brändén *et al.* 1984). The Co(II) substituted enzyme is not active as a carboxylase, but can catalyse the oxygenation reaction. The 2,3 enediolate, which is necessary for CO_2 addition, is stabilized by coordination of its hydroxyl oxygen to Mg(II). Similar 2,3 enediolates are unstable in free solution and undergo β elimination of the C1 phosphate group (Paech *et al.* 1978; Jaworowski *et al.* 1984). The transition state for CO_2 addition is in all probability stabilized by coordination of one of the carboxyl oxygen atoms to Mg(II), since such a bond is present in the complex with the reaction intermediate analogue, CABP. There is no kinetic evidence for the formation of a Michaelis complex with CO_2 prior to the chemical reaction (Pierce *et al.* 1986*b*). However, the structure of the active site of Rubisco is compatible with such a complex. The Mg(II) ion in the quaternary complex with RuBP is accessible from solution and could form a transient Mg–CO_2 complex with the CO_2 group in a suitable position for the subsequent chemical reaction with the C2 atom of RuBP.

The carboxyl group of the transition state adduct is in addition stabilized by interactions with the positive side chain of Lys334, as deduced from the structure of the quaternary CABP complex. This residue is located at the tip of loop 6 which undergoes a major conformational change during the reaction from an open to a closed from (Fig. 1.9). Mutant Rubisco molecules in which Lys334 has been changed to other residues are severely deficient in carboxylase activity and do not form strong complexes with CABP (Soper *et al.* 1988).

A Rubisco mutant *in vivo* from *Chlamydomonas reinhardtii* has been isolated where a residue at the base of loop 6, Val331, was replaced by Ala (Chen and Spreitzer 1989). These mutant Rubisco molecules show a reduction of the CO_2/O_2 specificity of 37 per cent in favour of oxygenation. The equivalent residue in the spinach enzyme, Val331, is at the rim of the hydrophobic pocket between β-strand 6 and α-helix 6 of the α/β barrel. The side chain of Val 331 interacts directly with the side chain of Thr342 in this pocket. The mutation Val331 to Ala would be expected to create a hole in this hydrophobic pocket which could alter either the position or the conformation of loop 6 and hence change details of the interaction between Lys334 and transition state. Interestingly, an intragenic suppressor mutation has been identified in *Chlamydomonas reinhardtii* (Chen and Spreitzer 1989) that increases the CO_2/O_2 specificity of the mutant enzyme by 33 per cent. In this suppressor mutant Thr342 has been changed to Ile. The changes in this double mutant, Val331 to Ala and Thr342 to Ile have thus at least partially restored the hydrophobic pocket and hence the role of Lys334 in differentially stabilizing the transition states for carboxylation and oxygenation.

Lys334 interacts not only with the carboxyl group of CABP in the quaternary complex but also with the side chain of Glu60 (Fig. 1.7). Replacement of Glu60 by the shorter side chain of Asp gives a drastic reduction in catalytic rate (Mural *et al.* 1990). The effects of a longer side chain have been explored by a combination of mutation and chemical modification (Smith *et al.* 1990). Glu60 was first mutated to Cys which produced a catalytically inert mutant. Catalytic activity was restored by treating the mutant with iodacetate. The cys side chain becomes carboxymethylated and an acidic side chain is produced that contains an extra sulphur atom compared to the Glu side chain. The catalytic rate of this Rubisco molecule has decreased tenfold compared to wild-type Rubisco and the CO_2/O_2 specificity has changed by a factor of five in favour of oxygenation. This is the first engineered mutant with a change in specificity and as such demonstrates the possibility to alter the carboxylation/oxygenation ratio by rational design.

Future prospects

In the first phase of protein engineering of Rubisco, the combination of site-directed mutagenesis and protein crystallography has been used to probe the function of certain key amino acid residues in the catalytic cycle. Encouraging in this respect is the fact that already at this stage, mutants with a changed specificity ratio have been obtained. These mutants are of special importance, since they identify features of the enzyme involved in partitioning between the carboxylase and oxygenase reaction. In a second phase, this knowledge will now be extended by modifying the active site by changing nonconserved residue and/or conserved residues not within immediate contact with the substrate or reaction intermediates.

Since the specificity amongst enzymes from different species varies to a certain extent, useful information might also be retrieved from a detailed comparison of the amino acid sequences and three-dimensional structures of enzymes with different specificity factors. Of utmost importance is also a careful biochemical analysis of the oxygenation reaction, since detailed chemical knowledge on that reaction is missing.

Acknowledgements

This work was supported by grants from the Swedish research councils NFR and SJFR.

References

Andersson, I., Knight, S., Schneider, G., Lindqvist, Y., Lundqvist. T., Brändén, C.-I., and Lorimer, G. (1989). *Nature*, **337**, 229–34.

Andrews, T.J. and Lorimer, G.H. (1987). *The biochemistry of plants*, Vol. 10 (ed. M.D. Hatch), pp. 131–218. Academic Press, Orlando, FL.

Brändén, C.-I., Lindqvist, Y., and Schneider, G. (1991). *Acta Crystallogr. B*, **47**, 824–35.

Brändén, R., Nilson, T., and Styring, S. (1984). *Biochemistry*, **23**, 4373–7.

Chapman, M., Suh, S.W., Cascio, D., Smith, W.W., and Eisenberg, D. (1987). *Nature*, **329**, 354–6.

Chapman, M.S., Suh, S.W., Curmi, P.M.G., Cascio, D., Smith, W.W., and Eisenberg. D. (1988). *Science*, **24**, 71–4.

Chen, X. and Spreitzer, R.J. (1989). *J. Biol. Chem.*, **264**, 3051–3.

Christeller, J.T. (1981). *Biochem. J.*, **193**, 839–44.

Curmi, P.M.G., Schreuder, H., Cascio, D., Sweet, R., and Eisenberg, D. (1992). *J. Biol. Chem.*, **267**, 16980–9.

Estelle, M., Hanks, J., McIntosh, L., and Somerville, C. (1985). *J. Biol. Chem.*, **260**, 9523–6.

Gutteridge, S. (1990). *Biochem. Biophys. Acta*, **1015**, 1–14.

Hartman, F.C., Soper, T.S., Niyogi, S.K., Mural, R.J., Foote, R.S., and Mitra, S. *et al.* (1987). *J. Biol. Chem.*, **262**, 3496–501.

Hartman, F.C., Milanez, S., and Lee, E.H. (1985). *J. Biol. Chem.*, **260**, 13 968–75.

Jaworowski, A., Hartman, F.C., and Rose, I.A. (1984). *J. Biol. Chem.*, **59**, 6783–89.

Knight, S., Andersson, I., and Brändén, C.-I. (1990). *J. Mol. Biol.*, **215**, 113–60.

Knight, S., Andersson, I., and Brändén, C.-I. (1989). *Science*, **244**, 702–5.

Lorimer, G.H. (1981). *Biochemistry*, **20**, 1236–40.

Lorimer, G.H. (1979). *J. Biol. Chem.*, **254**, 5599–601.

Lorimer, G.H. (1978). *Eur. J. Biochem.*, **89**, 43–50.

Lorimer, G.H. and Hartman, F. C. (1988). *J. Biol. Chem.*, **263**, 6468–71.

Lorimer, G.H., Gutteridge, S., and Madden. M. W. (1987). In *Plant and molecular biology* (ed. D. von Wettstein and N.-H. Chua), pp. 21–31. Plenum Press, New York.

Lu, G., Lindqvist, Y., and Schneider, G. (1992). *Proteins: Structure, Function, Genetics*, **12**, 117–27.

Lundqvist, T. and Schneider, G. (1991*a*). *Biochemistry*, **30**, 904–8.

Lundqvist, T. and Schneider, G. (1991*b*). *J. Biol. Chem.*, **266**, 12604–11.

Lundqvist, T. and Schneider, G. (1989*a*). *J. Biol. Chem.*, **264**, 3643–6.

Lundqvist, T. and Schneider, G. (1989*b*). *J. Biol. Chem.*, **264**, 7078–83.

Mural., R.J., Soper, T.S., Larimer, F.W., and Hartman, F.C. (1990) *J. Biol. Chem.*, **265**, 6501–5.

Paech, C., Pierce, J., McCurry, S.D., and Tolbert, N.E. (1978). *Biochem. Biophys. Res. Commun.*, **83**, 1084–92.

Pierce, J., Andrews, T.J., and Lorimer, G.H. (1986*a*). *J. Biol. Chem.*, **261**, 10 248–56.

Pierce, J., Lorimer, G. H., and Reddy, G.S. (1986*b*). *Biochemistry*, **19**, 1636–44.

Robison, P.D., Martin, M.N., and Tabita, F.R. (1979). *Biochemistry*, **18**, 4453–8.

Saver, B.G. and Knowles, J.R. (1982). *Biochemistry*, **21**, 5398–403.

Schloss, J.V., and Lorimer, G.H. (1982). *J. Biol. Chem.*, **257**, 4691–4.

Schloss, J.V., Phares, E.F., Long, M.V., Norton, I.L., Stringer, C.D., and Hartman, F.C. (1979). *J. Bacteriol.*, **137**, 490–501.

Schneider, G., Lindqvist, Y., Brändén, C.-I., and Lorimer, G. (1986). *EMBO J.*, **5**, 3409–15.

Schneider, G., Knight, S., Andersson, I., Brändén, C.-I., Lindqvist, Y., and Lundqvist, T. (1990*a*). *EMBO J.* **9**, 2045–50.

Schneider, G., Lindqvist, Y., and Lundqvist, T. (1990*b*). *J. Mol. Biol.*, 211, 989–1008.

Schneider, G., Lindqvist, Y., and Brändén, C.-I., (1992). *Ann. Rev. Biophys. Biomol. Struct.*, **21**, 119–43.

Smith, H.B., Larimer, F.W., and Hartman, F.C. (1990). *J. Biol. Chem.*, **265**, 1243–45.

Soper, T.S., Mural, R.J., Larimer, F.W., Lee, E.H., Machanoff, R., and Hartman, F.C. (1988). *Prot. Eng.*, **2**, 39–44.

Sue, J.R. and Knowles, J.R. (1978). *Biochemistry*, **17**, 4041–4.

Sue, J.R. and Knowles, J.R. (1982). *Biochemistry*, **21**, 5404–10.

Söderlind, E., Schneider, G., and Gutteridge, S. (1992). *Eur. J. Biochem.* **206**, 729–35.

Tabita, F.R. and McFadden, B.A (1974). *J. Biol. Chem.*, **249**, 3459–64.

Van Dyke, D. and Schloss, J.V. (1986). *Biochemistry*, **25**, 5145–56.

2

Partial reactions and chemical rescue of site-directed mutants of Rubisco as mechanistic probes

Mark R. Harpel, Frank W. Larimer, Eva H. Lee, Richard J. Mural, Harry B. Smith, Thomas S. Soper, and Fred C. Hartman

Introduction

Given the current state of knowledge of the reaction pathways catalysed by Rubisco* (Jaworowski and Rose 1985; Pierce *et al.* 1986*a*; Schloss 1990) and the elucidation of the three-dimensional structure of several different forms of the enzyme (Chapman *et al.* 1988; Andersson *et al.* 1989; Knight *et al.* 1990; Schneider *et al.* 1990; Lundqvist and Schneider 1991), site-directed mutagenesis offers the potential to decipher catalytic roles of active-site residues and to unravel the functional significance of various structural elements. Especially intriguing are intersubunit, electrostatic interactions at the active site between Glu48[+] and Lys168 of the nonactivated (noncarbamylated) enzyme (Schneider *et al.* 1990) and between Glu48 and Lys329 of the activated (carbamylated) enzyme (Knight *et al.* 1990).

The active sites of Rubisco are created by interacting domains from adjacent subunits (Larimer *et al.* 1987; Chapman *et al.* 1988; Andersson *et al.* 1989). The large COOH-terminal domain comprises an eight-stranded α/β barrel and contains most active-site residues, including Lys168 and Lys329. The smaller NH_2-terminal domain from the adjacent subunit extends partially across the top of the barrel and includes active-site residues Glu48 and Asn111. Lys329 is located within the flexible loop 6 of the α/β barrel; in the nonactivated form of the enzyme, this loop is so mobile that it cannot be seen in the electron density map (Schneider *et al.* 1986). However, in the activated form of the enzyme with a reaction-intermediate analogue (CABP) bound, loop 6 folds over the top of the barrel and becomes immobilized, in part, by electrostatic interactions between Lys329 and Glu48 of the adjacent subunit and between Lys329 and the carboxylate of the bound analogue (Fig. 2.1). One might readily imagine that loop 6 and the NH_2-terminal segment of the active site, due to their positioning, are involved in controlling ligand access to the active site and in precluding dissociation of reaction intermediates from the active site.

1* Abbreviations: Rubisco, D-ribulose-1,5-bisphosphate carboxylase/oxygenase; RuBP, D-ribulose-1,5-bisphosphate; XuBP, D-xylulose-1,5-bisphosphate; CABP, 2-carboxy-D-arabinitol-1,5-bisphosphate; PGA, 3-phospho-D-glycerate.

[+] Numbers refers to residue positions in the carboxylase from *Rhodospirillum rubrum*.

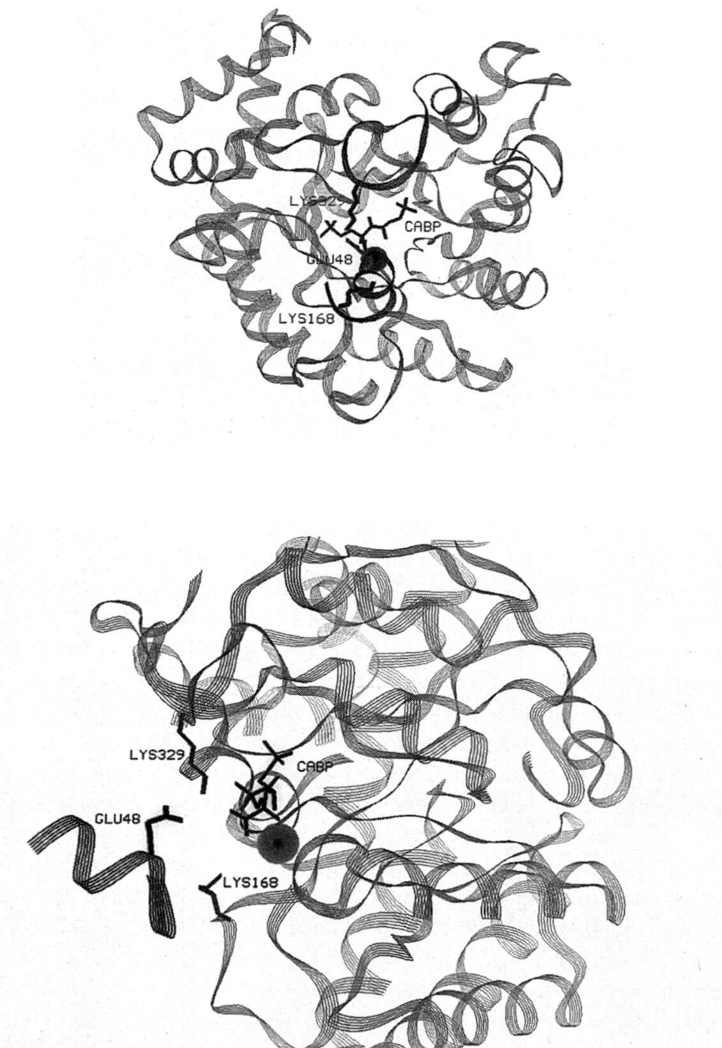

Fig. 2.1 Computer graphic generated model of the Rubisco active site. *Top panel*, view from top of barrel; *bottom panel*, view from side of barrel with the top on the left. The only displayed portion of the adjacent subunit is a segment of the NH_2-terminal domain that is an integral part of the active site. Loop 6 of the α/β barrel (residues 326–337), the NH_2-terminal segment (residues 43–52), bound CABP, Mg^{2+} and side chains of Glu48, Lys168, and Lys329 are the darker parts of the displayed structure. Atomic coordinates for the model (derived for the spinach enzyme) were kindly provided by Professor Carl Brändén and Gunter Schneider of the Swedish University of Agricultural Sciences, Uppsala, Sweden.

In this paper, we describe two approaches to address the roles of electrostatic interactions at the active site and the roles of the participant residues: (1) characterization of pertinent site-directed mutants, including their abilities to catalyse partial reactions and (2) subtle alteration of the active-site microenvironment by manipulation of these proteins with exogenous reagents.

Application of partial reactions

The overall carboxylation or oxygenation of RuBP as catalysed by Rubisco consists of discrete partial reactions illustrated in Fig. 2.2. Because an active-site residue will not necessarily be involved in all catalytic steps, site-directed mutants (devoid of overall activity due to altered active sites) may retain competence in one or more of the partial reactions. Realization of this expectation provides an avenue for discerning the particular step(s) facilitated by a given active-site residue.

Application of chemical rescue

Despite its revolutionary impact on enzymology, site-directed mutagenesis (Zoller and Smith 1983), as a means for altering structure, is generally restricted

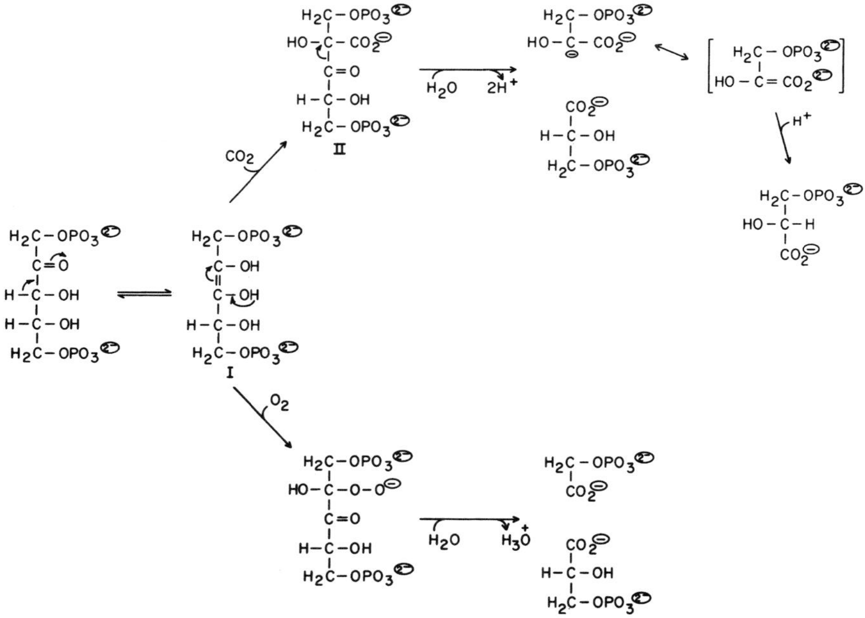

Fig. 2.2 Reaction pathways for the carboxylase and oxygenase reactions.

to the 20 amino acids normally occurring in proteins. Thus, reliance on homologous series of compounds to establish structure-reactivity correlations, a hallmark of mechanistic studies with non-enzymic catalysts, has not been possible with enzymes. This limitation is partially overcome by the demonstration that an enzyme, crippled because of an active-site substitution, can be partially rehabilitated ('rescued') merely by the addition of exogenous organic compounds that mimic the missing side chain (Toney and Kirsch 1989). For example, the virtually inactive K258A* mutant of aspartate aminotransferase is stimulated by primary amines; the degree of stimulation, after correcting for steric effects, correlates with the pK_a of the amine in accordance with the Brønsted relationship. These observations provide strong, direct evidence that the ϵ-amino group of Lys258 is the catalytic base that abstracts the α-carbon proton from the aldimine intermediate as postulated from earlier crystallographic studies (Kirsch *et al.* 1984). In other instances, the catalytic group lacking in a deficient enzyme can be imported into an active site by a properly designed substrate. Thus, the generally inactive subtilisin mutant H64A (His64 is the general base in the well-characterized catalytic triad of serine proteases) exhibits activity against peptides containing appropriately positioned histidyl residues (Carter and Wells 1987). Similarly, the H166G mutant of hexose-1-phosphate uridyltransferase utilizes uridine 5′-(phosphoimidazolate) as substrate, whereby the uridinyl phosphate moiety is transferred to a hexose-1-phosphate with release of imidazole (Kim *et al.* 1990).

Chemical rescue of deficient site-directed mutants can also be achieved through covalent chemical modification, thereby expanding the diversity and subtlety of structural changes that can be effected through mutagenesis. Examples include substitution of lysyl with aminoethylcysteinyl residues (net replacement of the γ-methylene group with a sulphur atom) (Smith and Hartman 1988; Planas and Kirsch 1991), substitution of glutamyl with carboxymethylcysteinyl residues (net insertion of a sulphur atom between the β- and γ-methylene groups with lengthening of the side chain by ~1 Å) (Lukac and Collier 1988; Smith *et al.* 1990), and substitution of arginyl with homoarginyl residues (net insertion of a methylene group with lengthening of the side chain by ~1 Å) (Beyer *et al.* 1987; Engler *et al.* 1992) (Fig. 2.3).

General strategies for the characterization of mutant proteins

Rubisco from most photosynthetic organisms consists of eight large (L) (53 000 dalton) and eight small (S) (14 000 dalton) subunits; hence, two gene products are necessary for the *in vivo* formation of a catalytically competent enzyme. Despite stringent requirements for coexpression of two distinct genes and proper

* The single-letter code is used to describe mutant proteins. The first letter denotes the amino acid present in the wild-type enzyme at the numbered position. The final letter denotes the amino acid present at the corresponding position in the mutant protein.

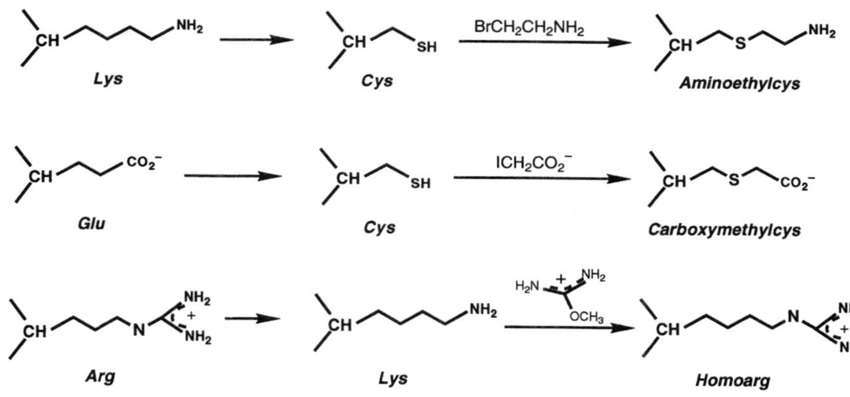

Fig. 2.3 Potential manipulations of protein structure by site-directed mutagenesis in combination with chemical modification.

assembly of their dissimilar encoded polypeptides in a foreign host, expression system for L_8S_8 enzymes have been described, thereby enabling application of site-directed mutagenesis (Voordouw *et al.* 1987; McFadden and Small 1988; Fitchen *et al.* 1990; Newman and Gutteridge 1990; Lee *et al.* 1991). However, the functionally analogous enzyme from the purple, non-sulphur bacterium *Rhodospirillum rubrum*, a homodimer (50.5 kDa subunit) altogether lacking small subunits, has been the target of most mutagenesis studies. Species invariance of active-site residues and homologous three-dimensional structures justify extrapolation of mechanistic conclusions from the L_2 protein to the L_8S_8 form. The original clone of the *R. rubrum* carboxylase gene was expressed as a fusion protein (Somerville and Somerville 1984), but all the mutant proteins generated in our laboratory were derived from a reconstruction that encodes authentic, wild-type enzyme (Larimer *et al.* 1986). Mutant proteins were constructed by either the single primer extension method utilizing an appropriate single-stranded M13 vector (Zoller and Smith 1983) or by a technique applied directly to an expression plasmid (Childs *et al.* 1985: Mural and Foote 1986). Several strains of *E. coli* (JM107, DH1 or MV1190) were used as expression hosts; mutant proteins were purified to hear homogeneity from whole-cell extracts by various chromatographic procedures.

Because Lys329, Lys168, and Glu48 are active-site residues in contact with bound CABP (Fig. 2.4) and are furthermore engaged in electrostatic interactions at subunit–subunit interfaces, drastic effects of amino acid substitution are anticipated. A general challenge is to ascertain if catalytic deficiencies of mutant proteins result from improper folding of polypeptide, failure of subunits to associate, inability to undergo carbamylation as necessary for activation, failure to bind substrates, or absence of a group that participates directly in catalysis.

If mutation of a single codon leads to substantially lowered yields of the protein, improper folding of the polypeptide and consequential proteolysis is

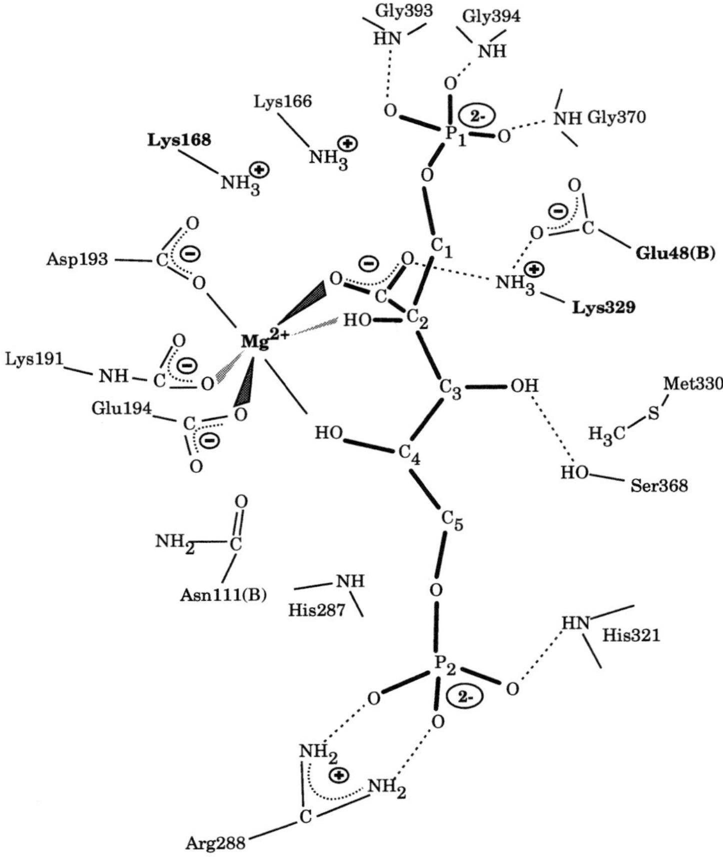

Fig. 2.4 Active-site residues of Rubisco in the immediate vicinity of bound CABP as determined by X-ray crystallography (Andersson *et al.* 1989; Knight *et al.* 1990; Lundqvist *et al.* 1991). 'B' after E48 and N111 denotes their location in the adjacent subunit. Adapted from the schematic of the spinach enzyme illustrated in Andersson *et al.* (1989).

probable. With mutants of *R. rubrum* Rubisco, distinguishing monomers (indicative of difficulties in subunit association) from dimers is readily accomplished by gel permeation chromatography or polyacrylamide gel electrophoresis under nondenaturing conditions.

The reaction-intermediate analogue, CABP, offers a convenient and powerful tool for deducing whether an inactive mutant can nevertheless form a carbamate and bind phosphorylated ligands. Only the activated form of the wild-type enzyme binds the analogue with sufficient tenacity to form a stable quaternary complex composed of equimolar amounts of enzyme subunit, CO_2, Mg^{2+}, and CABP. This complex is readily isolated by gel filtration (Miziorko and Sealy

1980). Its stability is emphasized by a consideration of the exchange rates between bound ligand and excess free ligand added to the complex: $t_{1/2}$ (*R. rubrum* enzyme) = 20 h (Smith *et al.* 1988); $t_{1/2}$ (plant enzyme) = 530 days (Schloss 1988). Because tight binding of each of the three ligands is dependent on the presence of the other two, mere demonstration of complex formation proves competence in activation chemistry and binding of phosphorylated ligands (RuBP). Failure to form a stable complex does not equate with failure of the mutant protein to recognize phosphoesters, because the tight binding of CABP to form the exchange-inert complex is a two-step process – rapid, reversible binding followed by a slow conformational change that 'locks' the ligands in place (Pierce *et al.* 1980). A mutant protein might bind CABP but not undergo the subsequent 'conformational change'. Based on crystallographic analyses (Knight *et al.* 1990), a major component of this 'conformational change', would be immobilization of loop 6 across the top of the α/β barrel.

Any catalytically deficient mutant (carrying a single amino acid substitution) that assembles properly, forms a carbamate, and binds substrate is deduced to be deficient because of the absence of a key residue. However, the possibility that the detrimental consequences are due to a subtle, localized reorientation of catalytic groups cannot be rigorously excluded without high-resolution X-ray crystallographic analysis. This shortcoming can be ameliorated by dissection of the carboxylase pathway into its component parts provided that mutant proteins devoid of overall carboxylation activity are competent in catalysing one or more of the discrete partial reactions illustrated in Fig. 2.2.

Formation of the enediolate of RuBP (compound I in Fig. 2.2), the first step in overall catalysis, is readily assayed on the basis of exchange of solvent protons with the C3 proton of substrate (Saver and Knowles 1982; Sue and Knowles 1982):

$$\text{Enz–B:}\searrow \begin{array}{c}\text{R}\\|\\\text{C=O}\\|\\\text{H–C–OH}\\|\\\text{R}'\end{array} \rightleftharpoons \begin{array}{c}\text{E–BH}\\ \text{T}_2\text{O}\downarrow\uparrow\text{H}_2\text{O}\\ \text{E–BT}\end{array} \quad \begin{array}{c}\text{R}\quad\text{O}^-\\ \diagdown\ \diagup\\ \text{C}\\ \|\\ \text{C}\\ \diagup\ \diagdown\\ \text{R}'\quad\text{OH}\end{array} \rightleftharpoons \text{Enz–B:}\searrow \begin{array}{c}\text{R}\\|\\\text{C=O}\\|\\\text{T–C–OH}\\|\\\text{R}'\end{array}$$

The enediolate is occasionally (about 1 per 400 molecules of RuBP processed) protonated on the wrong face of the plane, resulting in the formation of XuBP (the C3 epimer of RuBP) (Edmondson *et al.* 1990). This epimerase activity of Rubisco is but one of several side reactions that can be useful in characterizing deficiencies of mutant proteins. The XuBP formed is quantified by cleavage with aldolase, which acts only sluggishly on RuBP.

The six-carbon reaction intermediate (2-carboxy-3-ketoarabinitol-P_2, compound II in Fig. 2.2) is sufficiently stable to permit its isolation and partial purification; synthesis entails rapid acid quench after mixing essentially equimolar amounts of RuBP and the carboxylase in the presence of $^{14}CO_2$. Availability of the intermediate (labelled at the carboxyl group) allows its conversion to product (3-phosphoglycerate) and its decomposition to be monitored independently of overall carboxylation activity (Pierce *et al.* 1986*b*). Product formation is observed as an increase in acid-stable radioactivity, and decomposition is equated with a decrease in radioactivity that can be stabilized by borohydride.

$$\begin{array}{c} H_2C{-}OPO_3^{2-} \\ | \\ C{-}OH \\ \| \\ C{-}OH \\ | \\ H{-}C{-}OH \\ | \\ H_2C{-}OPO_3^{2-} \end{array} \xleftarrow{\;^{14}CO_2\;} \begin{array}{c} H_2C{-}OPO_3^{2-} \\ | \\ HO{-}C{-}^{14}CO_2^{-} \\ | \\ C{=}O \\ | \\ H{-}C{-}OH \\ | \\ H_2C{-}OPO_3^{2-} \end{array} \xrightarrow[H_2O \;\; 2H^+]{} 2\begin{array}{c} ^{14}CO_2^{-} \\ | \\ H{-}C{-}OH \\ | \\ H_2C{-}OPO_3^{2-} \end{array}$$

Another side reaction, discovered by Andrews and Kane (1991), permits the status of the final step in catalysis, protonation of the C2 *aci*-acid of PGA, to be examined with mutants that retain any carboxylase activity. This side reaction entails β elimination of the phosphate group from the intermediate to form pyruvate; elimination of phosphate rather than proper protonation occurs about once in every one hundred turnovers as catalysed by wild-type enzyme:

$$\begin{array}{c} H_2C{-}OPO_3^{2-} \\ | \\ HO{-}C{-}COO^{-} \\ \ominus \end{array} \begin{array}{l} \xrightarrow{H^+} \begin{array}{c} COO^{-} \\ | \\ H{-}C{-}OH \\ | \\ H_2C{-}OPO_3^{2-} \end{array} \\ \xrightarrow[P_i]{} \begin{array}{c} COO^{-} \\ | \\ C{=}O \\ | \\ CH_3 \end{array} \end{array}$$

Properties of K168Q, K329Q, and E48Q mutants

The consequences of substituting Glu48, Lys168, or Lys329 by glutamine are summarized in Table 2.1 (Hartman *et al.* 1987; Soper *et al.* 1988; Hartman and Lee 1989; Mural *et al.* 1990). The selection of glutamine was dictated by the desire to remove the potential for electrostatic interactions, while retaining a side chain with similar polarity and geometry. Each mutant was isolated in high yield as a stable dimer; hence, elimination of either a Glu48–Lys168 or Glu48–Lys329 salt bridge is not disruptive of normal subunit–subunit interactions. In contrast,

Table 2.1 General properties of mutant proteins

Protein	Subunit structure	Carboxylase activity (% wild type)	Quaternary complex (exchange time, h)	Enolization activity	
				Specific activity (% wild type)	K_m (μM RuBP)
Wild type	dimer	100	20	100	10
E48Q	dimer	0.5	<0.5	10	100
K168Q	dimer	0.01	n.d*	2	200
K329Q	dimer	<0.01	n.d*	5	200

* Complex not detected.

these amino acid replacements severely impair carboxylase activity. K329Q lacks detectable activity, the threshold of which is 0.01 per cent of wild type; and K168Q exhibits only the threshold level of activity. The impact of placing glutamine at position 48 is less severe, but even then the mutant's carboxylase activity is diminished more than a hundredfold relative to wild type. These losses of activity are not due merely to charge neutralization, because the E48D, K168R, and K329R mutants are defective like their glutamine counterparts.

The E48Q mutant forms a sufficiently stable quaternary complex (Enz•CO_2•Mg^{2+}•CABP) to permit its isolation by gel filtration; however, when challenged with exogenous CABP, the observed rate of ligand exchange is much faster with the mutant than with wild-type enzyme. Despite these weakened interactions, complex formation *per se* demonstrates that Glu48 is not required for the CO_2/Mg^{2+}-dependent activation or for binding of phosphorylated ligands. Complexes with the K329Q and K168Q mutants could not be isolated, compatible with either lack of complex formation or the formation of complexes which are simply too labile to survive gel filtration.

Each mutant displays substantial activity in the enolization reaction, the first step in overall catalysis. Thus, Glu48, Lys168, and Lys329 can be eliminated as candidates for the base that accepts the C3 proton from RuBP. Rather, each of the three residues must play a crucial role at some stage of the reaction coordinate beyond the initial enolization. The observed catalytic competence of these mutants, even in a partial reaction, provides dramatic evidence of intact subunit structures, normal activation chemistry, and substrate binding. The binding of RuBP may be somewhat perturbed, however, based on the ten to twentyfold increases in K_m values. In the cases of the K168Q and K329Q mutants, the retention of enolization activity also provides indirect evidence that these proteins can form quaternary complexes with CABP, but ones that are insufficiently stable to detect by gel filtration. This logic and the observed CO_2/Mg^{2+} dependence of enolization by the mutants imply that Lys168 and Lys329, like Glu48, are not involved in the activation process.

Despite the simplicity of the proton exchange experiments to monitor enolization, they are revealing. Retention of enolization activity by position 329 mutant proteins, lacking in detectable carboxylase activity, shows that this step can proceed independently of others in the overall pathway. Failure of CO_2 or O_2 to react with the enediol generated by position 329 mutant proteins suggests that neither carboxylation nor oxygenation is spontaneous with the wild-type enzyme but require direct intervention by amino acid side chains.

K329G (identical to K329Q in terms of general properties shown in Table 2.1) was also evaluated as a catalyst for turnover of the six-carbon, carboxylated reaction intermediate (Lorimer *et al.* 1993). Despite its total lack of detectable carboxylase activity, K329G readily converts the intermediate to PGA (Fig. 2.5). Thus, Lys329 is not needed for enolization nor for processing of six-carbon, carboxylated intermediate; by deduction, it must be required for reaction of gaseous substrate with enediolate. This conclusion is entirely consistent with the location of the ϵ-amino group of Lys329 in the enzyme•CABP complex as seen by crystallography (Knight *et al.* 1990).

A caveat in assigning the role of an active-site residue based on efficacy of mutant proteins to catalyse partial reactions is their interdependence, rather than clean compartmentalization as illustrated on paper. Although a number of mutant proteins, lacking significant carboxylase activity, are competent in one or more partial reactions, invariably the observed rates are depressed relative to wild type. For example, the K329G mutant does catalyse both enolization of

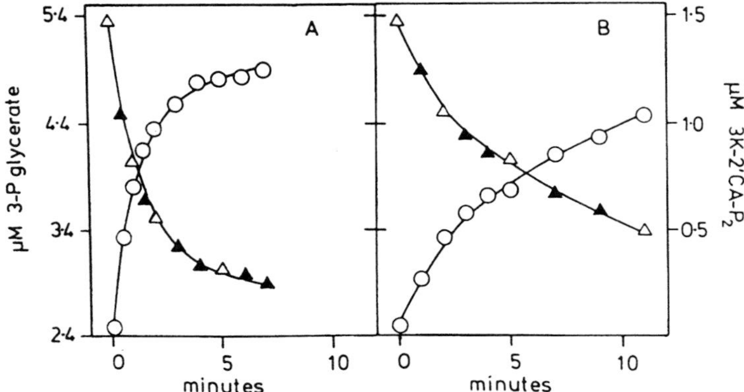

Fig. 2.5 Hydrolysis of the six-carbon reaction intermediate (3K-2′CA-P_2) by (A) the wild-type carboxylase and (B) the K329G mutant enzyme. PGA (circles) and intermediate (triangles) were determined by the difference method of Pierce *et al.* (1986*b*). Empty symbols represent experimentally determined values; shaded symbols represent calculated values for the concentration of intermediate assuming a stoichiometry of 2 moles of phosphoglycerate formed per mole of intermediate consumed. These experiments were carried out in collaboration with Dr George Lorimer of Du Pont.

RuBP and turnover of the six-carbon reaction intermediate but at rates which are ~10 per cent of wild type. The assignment of Lys329 to the carboxylation step is then based on greater disruption of this step compared to other ones. Parenthetically, the k_{cat} for turnover of six-carbon reaction intermediate by wild-type enzyme is only 3 per cent of the k_{cat} for overall carboxylation of RuBP; a plausible explanation for this disparity is a requirement for a slow conformational change in processing of the intermediate (Lorimer and Hartman 1988). An alternative view is that the isolated intermediate (a free ketone) in not kinetically competent, and only its hydrated form exists on the normal reaction pathway (Cleland 1990).

The carboxylase-catalysed reaction proceeds by a Theorell–Chance type of kinetic mechanism: ordered addition and enolization of RuBP followed by bimolecular reaction of the enediol with gaseous substrate (Pierce *et al.* 1986*a,b*; Van Dyk and Schloss 1986). Evidence in favour of this ordered, sequential reaction includes NMR, direct binding assays, isotope trapping, and kinetic analyses indicating that the gaseous substrates do not bind to free enzyme or to enzyme•RuBP. Earlier speculation that the universal oxygenase activity of Rubiscos reflects an unavoidable consequences of the inherent reactivity of the enediol is certainly compatible with the elucidated kinetic mechanism (Lorimer and Andrews 1973). Carboxylation or oxygenation of the enediol could then be viewed as nonenzymic. However, the enolization of RuBP as catalysed by position 329 mutant proteins, without concomitant carboxylation or oxygenation, argues that the enzyme plays an active role in facilitating the addition of gaseous substrate to enediol. Remembering that presteady-state kinetics detect a rate-determining step between enolization and carboxylation of enediol (Schloss 1990), it is reasonable to invoke Lys329 in polarization of enediol and development of the nucleophilic centre at C2. Alternatively, this residue could stabilize a transition state leading to the carboxylated intermediate.

The propensity of E48Q to catalyse the side reactions leading to XuBP and pyruvate were also determined. As seen in Table 2.2, the rates of XuBP and PGA formation are the same; i.e with this mutant, misprotonation of the enediolate intermediate is just as likely to occur as its reaction with CO_2. In stark contrast, XuBP formation during the turnover of RuBP by the wild-type *R. rubrum* enzyme could not be detected under our assay conditions. One might argue that the longer resident time of the enediolate at the active site of the mutant, due to very slow conversion to PGA, provides more opportunity for misprotonation and accounts for the increased XuBP/PGA ratio relative to wild-type enzyme. However, K329C, which catalyses enediolate formation but not the production of PGA, does not convert [3-^{3}H]RuBP to XuBP during the time period necessary for complete detritiation. In comparison to wild-type enzyme, E48Q also gives rise to an elevated formation of pyruvate relative to PGA. Greater accessibility of the active site to solvent, due to perturbing the NH_2-terminal domain and eliminating the Glu48–Lys329 intersubunit link, could account for the relative enhancement in both XuBP and pyruvate production.

Table 2.2 Comparison of wild type and E48Q activities

	Specific activity (units mg^{-1})	
	wt	E48Q (% wt)
CO_2 fixation (PGA formation)	4.8	0.02 (0.5%)
Enolization (Detritiation)	3.1	0.35 (10%)
Epimerization (XuBP formation)	n.d.†	0.02
Pyruvate formation	0.025	0.003 (1.2%)

*When measured from carboxylation, enolization, or pyruvate formation, the K_m for RuBP is ~120 μM with the mutant enzyme and ~10 μM with wt.

† XuBP formation was not detected.

Manipulation of active-site microenvironment by chemical rescue

As noted earlier in this article, chemical rescue broadens the horizons of site-directed mutagenesis by allowing subtle and systematic alteration of protein

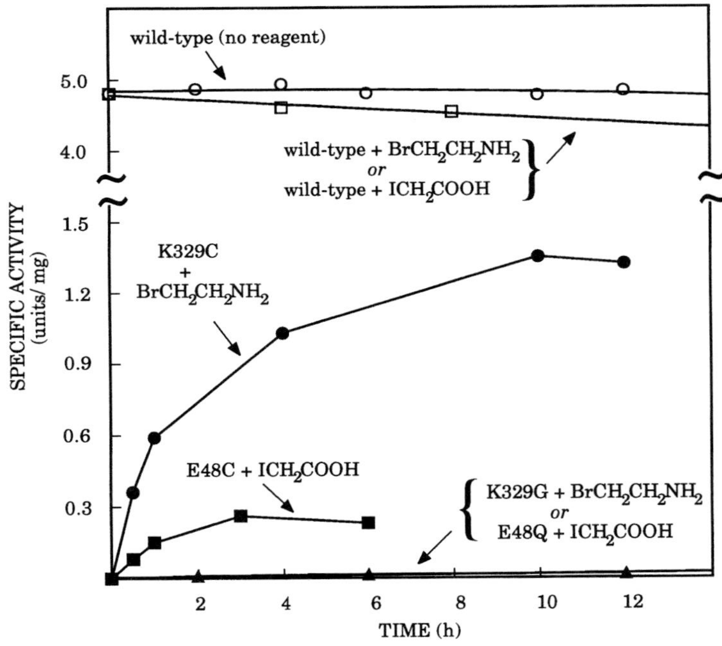

Fig. 2.6 Covalent chemical rescue of K329C by aminoethylation with 100 mM 2-bromoethylamine and chemical rescue of E48C by carboxymethylation with 100 mM iodoacetate. Modifications were carried out at pH 8.0 in the presence of 50 mM $NaHCO_3$ and 10 mM $MgCl_2$.

structure that would not be possible by genetics or chemistry individually. Successful examples of covalent chemical rescue of catalytically inactive K329C and E48C by aminoethylation (Smith and Hartman 1988) and carboxymethylation (Smith *et al.* 1990), respectively, are illustrated in Fig. 2.6. Treatment of K329C with 2-bromoethylamine partially restores enzyme activity, presumably as a consequence of selective aminoethylation of the introduced thiol group. Amino acid analyses, isoelectric focusing under denaturing conditions, slow inactivation of the wild-type carboxylase by bromoethylamine, and the failure of bromoethylamine to restore activity to the corresponding glycyl mutant protein support this interpretation. The observed facile, selective aminoethylation is consistent with an active-site microenvironment similar to that of the native enzyme, in which the ϵ-amino group of Lys329 exhibits unusually high acidity and nucleophilicity. The catalytic constant (k_{cat}) of this novel carboxylase, which contains a sulphur atom in place of a specific lysyl γ-methylene group, is 2.5-fold lower than that of the wild-type enzyme. This detrimental effect by such a modest structural change underscores the stringent requirement for a lysyl side chain at position 329. In contrast, the aminoethylated mutant protein exhibits K_m values that are unperturbed relative to those for the wild-type enzyme. Clearly, major reduction in k_{cat} with unaltered K_m values argue for a direct role of Lys329 in catalysis, as concluded from characterization of K329Q in catalysis of partial reactions.

When E48C is treated with iodoacetate, 4–6 per cent of the original wild-type activity is restored. Isoelectric focusing under denaturing conditions, refractiveness of E48Q to the reagent, and slight sensitivity of wild-type enzyme to the reagent all indicate that rescue is due to selective alkylation of the Cys48. These results emphasize the necessity for a carboxylate at position 48 and the sensitivity of catalytic activity to precise positioning of that carboxylate. A 1-Å lengthening of the glutamyl side chain is reasonably well tolerated. However, a 1-Å shortening is as detrimental as removal, because E48D is no more active than E48C (~0.05 per cent of wild-type activity).

Position 329 mutants (for example K329A) are also amenable to noncovalent chemical rescue by aliphatic amines (Fig. 2.7), similarly to the K258A mutant of aspartate aminotransferase (Toney and Kirsch 1989). For example, at 80 mM ethylamine, the rate of CO_2 fixation is ~0.5 per cent of wild-type and the K_m for RuBP is elevated some thirtyfold. The system is saturable with respect to amine concentration, the apparent K_d for ethylamine being 120 mM. In addition, amine-rescued K329A forms a detectable complex with CABP. Given the mobility of loop 6, the effectiveness of an exogenous amine in stabilizing the catalytically competent conformation of the protein while concomitantly fulfilling the functionality of a lysyl side chain normally occupying position 329 is rather remarkable. Presumably, loop 6 of K329A, even in the absence of amine, can adopt the conformation necessary for catalysis; but without the lysyl side chain, reaction of gaseous substrate with the enediolate of RuBP cannot occur. The role of the lysyl side chain appears to transcend that of merely providing a positive charge, because the K329R mutant is devoid of carboxylase activity.

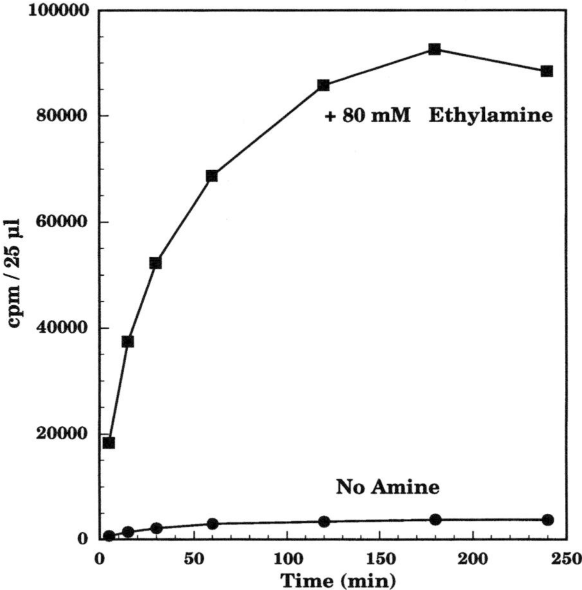

Fig. 2.7 Noncovalent chemical rescue of K329A by 80 mM ethylamine in the presence of 24 mM $NaHCO_3$ and 10 mM $MgCl_2$ at pH 8.0. The plateau in fixed $^{14}CO_2$ reflects exhaustion of RuBP.

Specificity for gaseous substrates

The kinetic characteristics of the enzyme which determine relative rate of carboxylation and oxygenation (ν_c/ν_0) can best be appreciated as a ratio of the catalytic efficiencies (V_{max}/K_m) for the two activities, $(V_c/K_c)/(V_0/K_0)$. This latter term defines the substrate specificity factor (Laing *et al.* 1975), denoted as the constant τ, and readily allows ν_c/ν_0 to be expressed as a function of the relative concentrations of the two gaseous substrates: $\nu_c/\nu_0 = \tau \bullet ([CO_2]/[O_2])$. When [1-^{3}H]RuBP is used as substrate, the distribution of tritium between phosphoglycolate (derived from the oxygenation pathway) and PGA (derived from the carboxylation pathway) is a direct measure of flux between the two pathways (ν_c/ν_0). A chromatographic profile of an assay mixture with wild-type enzyme, from which its τ value can be calculated, is shown in Fig. 2.8.

Outstanding questions concerning the dual activities of Rubisco include the identity of those structural features which account for partial discrimination against oxygen and prospects for altering the carboxylase/oxygenase ratio by manipulating the microenvironment of the active site. To judge the potential of chemical rescue in addressing these issues, we have determined the specificity

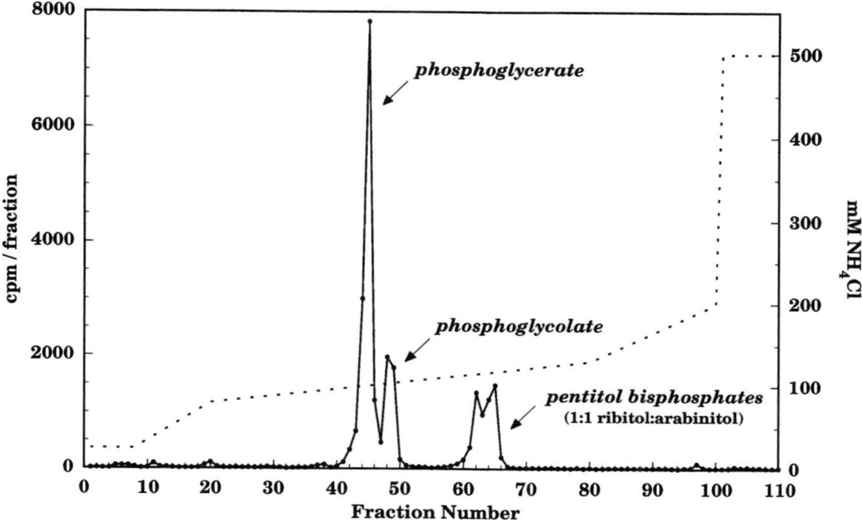

Fig. 2.8 Chromatography of a wild-type assay mixture on MonoQ with a FPLC unit (Pharmacia). [1-^{3}H]RuBP was used as substrate in the presence of 7 mM HCO_3^-, 255 μM O_2, and 10 mM $MgCl_2$. Prior to chromatography, the reaction was terminated by reduction of the remaining RuBP with sodium borohydride to yield an epimeric mixture of pentitol bisphosphates.

Table 2.3 Specificity factor (τ) of various carboxylases

Protein	τ
Wild type	10
E48Q	0.3
E48CmCys	2
K329A rescued	6

factor for some of the parental and manipulated proteins under discussion (Table 2.3), The τ value for the *R. rubrum* enzyme decreases dramatically upon substitution of Glu48 by Gln. Reinsertion of a carboxylate at this position by carboxymethylation of E48C, which restore ~5 per cent of the wild-type carboxylase activity, elevates the τ value more than sixfold. In the case of K329A rescued by ethylamine, the τ value decreases about twofold relative to wild-type enzyme. These observations show that residues in loop 6 and in the NH_2-terminal segment of the active site exert large influences in partioning between the carboxylation and oxygenation pathways catalysed by Rubisco. We suggest that both of these regions are attractive targets for further microsurgery in pursuit of thorough understanding of specificity determinants.

Conclusions

1. Intersubunit electrostatic interactions at the active site of Rubisco are not necessary for maintenance of dimeric structure.
2. The side chains of Glu48, Lys168, and Lys329 that comprise these salt linkages serve key roles in catalysis; they are not required for enzyme activation or for binding of substrates.
3. These three side chains all preferentially facilitate catalytic steps beyond the initial enolization of RuBP.
4. The ϵ-amino group of Lys329 facilitates, directly or indirectly, the reaction of gaseous substrates with the enediolate of RuBP.
5. Both the ϵ-amino group of Lys329 and the γ-carboxylate of Glu48 influence the relative reactivity of the enediolate toward CO_2 and O_2.
6. The γ-carboxylate of Glu48 and/or proper positioning of the NH_2-terminal segment across the top of the barrel domain of the active site participate in shielding the active site from solvent and in retaining reaction intermediates at the active site during catalytic turnover.

Acknowledgements

The authors' research was supported by the Office of Health and Environmental Research, US Department of Energy under contract DE-AC05-84OR21400 with Martin Marietta Energy Systems, Inc. The authors are especially indebted to Professors Carl Brändén and Gunter Schneider of the Swedish University of Agricultural Sciences at Uppsala, Sweden, for providing atomic coordinates for Rubisco prior to publication.

References

Andrews, T.J. and Kane, H.J. (1991). *J. Biol. Chem.* **267**, 9447–52.

Anderson, I., Knight, S., Schneider, G., Lindqvist, Y., Lundqvist. T., and Brändén, C.-I. *et al.*, (1989). *Nature,* **337**, 229–34.

Beyer, W.F., Jr., Fridovich, I., Mullenbach, G.T., and Hallewell, R. (1987). *J. Biol. Chem.*, **262**, 11 182–7.

Carter, P. and Wells, J.A. (1987). *Science,* 237, 394–9.

Chapman, M.S., Suh, S.W., Curmi, P.M.G., Cascio, D., Smith, W.W., and Eisenberg. D.S. (1988). *Science,* **241**, 71–4.

Childs, J., Villanueba, K., Barrick, D, Schneider, T.D., Stormo, G.D., and Gold, L. *et al.* (1985). *Sequence specificity in transcription and translation* (ed. R. Calendar and L. Gold), pp. 341–50. Alan R. Liss, New York.

Cleland, W.W. (1990). *Biochemistry,* **29,** 3194–7.

Edmondson, D.L., Kane, H.J., and Andrews, T.J. (1990). *FEBS Lett.* **260,** 62–6.

Engler, D.A., Campion, S.R., Hauser, M.R., Cook, J.S., and Niyogi, S.K. (1992). *J. Biol. Chem.*, **267**, 2274–81.

Fitchen, J.H., Knight, S., Andersson, I., Brändén, C.-I, and McIntosh, L. (1990). *Proc. Nat. Acad. Sci.* USA, **87**, 5768–72.

Hartman, F.C. and Lee, E.H. (1989). *J. Biol. Chem.*, **264**, 11 784–9.

Hartman, F.C., Larimer, F.W., Mural, R.J., Machanoff, R., and Soper, T.S. (1987). *Biochem. Biophys. Res. Commun.*, **145**, 1158–63.

Jaworowski, A. and Rose, I.A. (1985). *J. Biol. Chem.*, **260**, 944–8.

Kim, J., Ruzicka, F., and Frey, P.A. (1990). *Biochemistry,* **29**, 10 590–3.

Kirsch, J.F., Eichele, G., Ford, G.C., Vincent, M.G., Jasonius, J.N., and Gehring. H. *et al.* (1984). *J. Mol. Biol.*, **174**, 497–525.

Knight, S., Anderson, I., and Brändén, C.-I. (1990). *J. Mol. Biol.*, **215**, 113–60.

Laing, W.A., Ogren. W.L., and Hageman, R.H. (1975). *Biochemistry*, **14**, 2269–75.

Larimer, F.W., Machanoff, R., and Hartman, F.C. (1986). *Gene*, **41**, 113–20.

Larimer, F.W., Lee, E.H., Mural. R.J., Soper, T.S., and Hartman, F.C. (1987). *J. Biol. Chem.,* **262**, 15 327–9.

Lee, B., Berka, R.M., and Tabita, F.R. (1991). *J. Biol. Chem.*, **266**, 7417–22.

Lorimer, G.H. Chen, Y.R., and Hartman, F.C. (1993). *Biochemistry* , 9018–24.

Lorimer, G.H. and Hartman, F.C. (1988). *J. Biol. Chem.,* **263**, 6468–71.

Lorimer, G.H. and Andrews, T.J. (1973). *Nature*, **243**, 359–60.

Lukac, M. and Collier, R.J. (1988). *J. Biol. Chem.*, **263**, 6146–9.

Lundqvist, T. and Schneider, G. (1991). *J. Biol. Chem.*, **266**, 12 604–11.

McFadden, B.A. and Small, C.L. (1988). *Photosynthesis Res.*, **18**, 245–60.

Miziorko, H.M. and Sealy, R.C. (1980). *Biochemistry*, **19,** 1167–71.

Mural, R.J. and Foote, R.S. (1986). *DNA*, **5**, 84.

Mural, R.J., Soper T.S., Larimer, F.W., and Hartman, F.C. (1990). *J. Biol. Chem.*, **265**, 6501–5.

Newman, J. and Gutteridge, S. (1990). *J. Biol. Chem.*, **265**, 15 154–9.

Pierce, J., Tolbert, N.E., and Barker, R. (1980). *Biochemistry,* **19**, 934–42.

Pierce, J., Lorimer, G.H., and Reddy, G.S. (1986*a*). *Biochemistry*, **25**, 1636–44.

Pierce, J., Andrews, T.J., and Lorimer, G.H. (1986*b*). *J. Biol. Chem.*, **261**, 10 248–56.

Planas, A. and Kirsch, J.F. (1991). *Biochemistry*, **30**, 8268–76.

Saver, B.G. and Knowles, J.R. (1982). *Biochemistry*, **21**, 5398–403.

Schloss, J.V. (1990). In *The proceedings of NATO ASI on enzymatic and model carboxylation and reduction reactions for carbon dioxide utilization* (ed. M. Aresta and J.V. Schloss), pp. 321–45. Kluwer, Netherlands.

Schloss, J.V (1988). *J. Biol. Chem.*, **263**, 4145–4150.

Schneider, G., Lindqvist, Y., and Lundqvist, T. (1990). *J. Mol. Biol.*, **211**, 989–1008.

Schneider, G., Lindqvist, Y., Brändén, C.-I., and Lorimer, G. (1986). *EMBO J.* **5**, 3409–15.

Smith, H.B. and Hartman, F.C. (1988). *J. Biol. Chem.*, **263**, 4921–5.

Smith, H.B., Larimer, F.W., and Hartman, F.C. (1990). *J. Biol. Chem.*, **265**, 1243–45.

Smith, H.B., Larimer, F.W., and Hartman F.C. (1988). *Biochem. Biophys. Res. Commun.*, **152**, 579–84.

Somerville, C.R. and Somerville, S.C. (1984). *Mol. Gen. Genetics*, **193**, 214–19.

Soper, T.S., Mural, R.J., Larimer, F.W., Lee, E.H., Machanoff, R., and Hartman, F.C. (1988). *Prot. Eng.*, **2**, 39–44.

Sue, J.M. and Knowles, J.R. (1982). *Biochemisty*, **21**, 5404–10.

Toney, M.D. and Kirsch, J.F. (1989). *Science*, **243**, 1485–8.

Van Dyk, D.E. and Schloss, J.V. (1986). *Biochemistry*, **25**, 5145–56.

Voordouw, G., de Vries, P.A., van den Berg, W.A.M., and de Clerk, E.P.J. (1987). *Eur. J. Biochem.*, **163**, 591–8.

Zoller, M.J. and Smith, M. (1983). *Methods Enzymol.*, **100**, 468–500.

3

Elements of the structure of ribulose-P_2 carboxylase/oxygenase affecting substrate partitioning

Steven Gutteridge

Introduction

A combination of mutagenesis and crystallographic analysis is being used to unravel the processes leading to fixation of atmospheric CO_2 by photosynthetic organisms. The first steps in this complicated chemistry are due to one protein entity, ribulose-P_2 carboxylase/oxygenase (Rubisco) that mediates at least five distinct chemical reactions commencing with binding of ribulose-P_2 to 3P-glycerate release (for recent reviews see Hartman 1992; Gutteridge 1990; Andrews and Lorimer 1987). An analysis of the fate of the atoms of substrate molecules as they proceed to product indicates that the intermediates generated during catalysis will have C centres with quite different geometries. Since all the reactions occur at a common site, then the enzyme must be flexible enough to retain each of the intermediates until the first stable product, 3P-glycerate is generated. For example, Fig. 3.1 shows the partial reactions leading to P-glycerate formation and the stereochemistry of the intermediates. Planar configurations of C2 and C3 centres of the bisphosphate alternate with tetrahedral conformers as the reaction proceeds to completion.

It is not surprising, therefore, to discover that some elements comprising the active site are flexible (Schneider *et al.* 1986; Knight *et al.* 1990). However, from the point of view of defining the reaction chemistry in terms of the function of individual amino acids using mutagenesis, is proving difficult (particularly ascribing altered activity) to either specific function or to changes which have arisen simply from structural modifications. Because elements of the structure are flexible, then amino acids thought to be involved in one specific part of active site chemistry, may in fact play a significant role in other partial reactions. It is becoming clear that the detailed analysis afforded by crystallography will be required to understand the influence of mutations, particularly on those mobile elements involved in the mechanism (Schneider and Lindqvist 1992).

Carboxylation is not the only fate of the enediol intermediate of ribulose-P_2. The presence of molecular oxygen results in the consumption of the intermediate and the formation of P-glycolate and only one molecule of P-glycerate. At present ambient concentrations of both O_2 and CO_2 there is a significant loss of

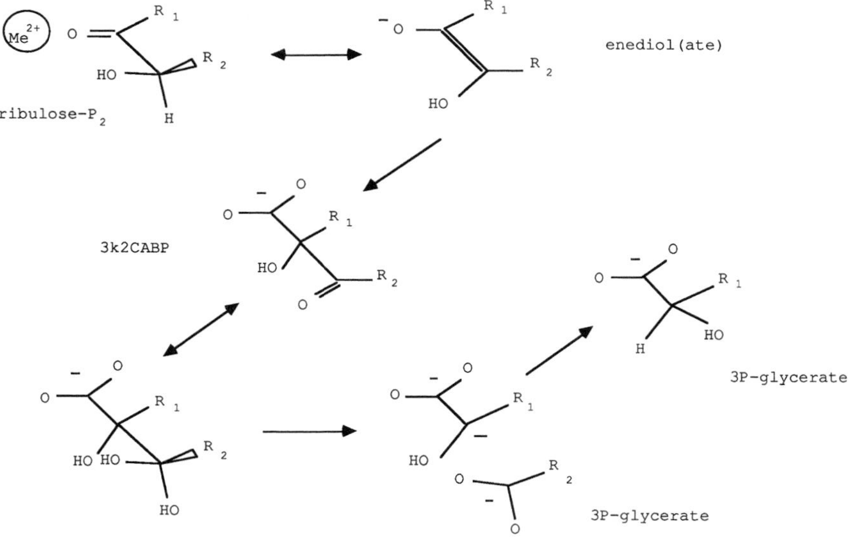

Fig. 3.1 The five partial reactions of ribulose-P_2 carboxylation catalysed by rubisco. The first step is proton abstraction producing an enediol intermediate that is carboxylated by gaseous CO_2 to form 3keto-2carboxy arabinitol bisphosphate (3k2CABP). Hydrolytic cleavage of the gemdiol at the enzyme active site leads to the formation of two molecules of 3P-glycerate.

ribulose-P_2 by oxygenation. The absence of any C gain due to this reaction in many major crop plants, combined with the loss of CO_2 and energy required to recycle the P-glycolate to a useable form, imposes a significant limitation on the efficiency of photosynthesis. Unfortunately the steps that lead to P-glycolate production are not well defined and have derived from those known to operate for CO_2 fixation (Lorimer 1981). A major fascination of the protein engineer is therefore to predict and introduce mutations that perturb these two competing reactions differentially. The results described below indicate that changes to the sequence of loop 6 fall into this category.

One region of the protein that has been altered by site-specific changes and now being analysed crystallographically is that of contributing ligands involved in metal binding. Magnesium binds at the active site following carbamylation of one specific Lys residue (K201) in loop 2 of the C-terminal domain of the L subunit. Two other amino acids of the same loop, an Asp203 and Glu residue (204) also coordinate to the metal (Lundqvist and Schneider 1991). The active site Mg^{2+} plays a central role at all stages of catalysis (Gutteridge and Gatenby 1987). Other metal ions that replace Mg^{2+} and support catalysis, in nearly all cases alter the relative specificities of carboxylation and oxygenation (Christeller 1981). Thus mutations at these positions were intended to reveal details of the importance of these amino acids in modulating specificity of the two reactions

(Lorimer *et al.* 1988). Mutations at those amino acids known to be within the first coordination sphere of the metal, produce inactive enzyme. However, analysis of individual steps of catalysis indicate that some of the partial reactions are modified differentially. These data were interpreted to suggest that coordination around the metal changes at different steps of catalysis (Gutteridge *et al.* 1988). Thus the ligand identity around the metal with ribulose-P_2 bound is quite distinct from the coordination that accommodates the 2′-carboxyl-3-keto-arabinitol-bisphosphate intermediate of carboxylation (or its analogue, 2′-carboxy-arabinitol-bisphosphate) at the site.

Flexible elements of the structure

At least two loops of the protein that contain active site amino acids are known to be mobile and must occupy different positions as the reactions proceed. One of the loops is in a region of the N-terminal domain of the L subunit that composes part of the active site (see section on mutations outside loop 6). The other is a loop of the C-terminal barrel domain of the L subunit that must assume either an open or closed conformation relative to the site. This loop is the sixth of eight that occupy positions above the surface of the β-barrel core (Schneider *et al.* 1986) and are variously involved in metal coordination, activation, and substrate binding.

In the structure of the inactive carboxylase from *Rhodospirillum rubrum* the peptide backbone cannot be discerned for loop 6, yet becomes well defined in the structure of the higher plant enzyme with bound intermediate analogue, 2CABP (Andersson *et al.* 1989). In this structure (shown in Fig. 3.2) the loop and more specifically the Lys residue (K334) at its apex interacts directly in concert with the metal, with the two 'O' groups of the 2′-carboxyl of the inhibitor. The amount of the active site exposed in this closed conformation is minimal. Therefore it is reasonable to propose that during catalysis the capture of substrate CO_2 with bound enzyme enediol involves the closure of the loop over the active site. However, the ability of bisphosphate substrate to penetrate to the site and for product to be released requires the loop to move to a more open configuration.

It is tempting to propose that the movement of loop 6 and K334 to the closed position imprisons a water molecule with the enediol and CO_2 around the coordinated metal. This results in formation of the hydrated six-carbon intermediate, 2′-carboxy-3-keto-arabinitol-P_2 (3k2CABP) which rapidly decomposes, with appropriate protonation, to the two P-glycerate product molecules. The result of a metal-bound water being consumed and the loss of binding affinity due to the breakage of the C2–C3 bond of the six-carbon intermediate, allows the enzyme to relax back to the resting state and the cycle to restart.

Apart from the crystallographic evidence that indicates the ϵ-amino group of K334 in loop 6 stabilizes at least the carboxyl intermediate of carboxylation, there is now supporting data as a result of mutagenesis. Hartman and Lee (1989)

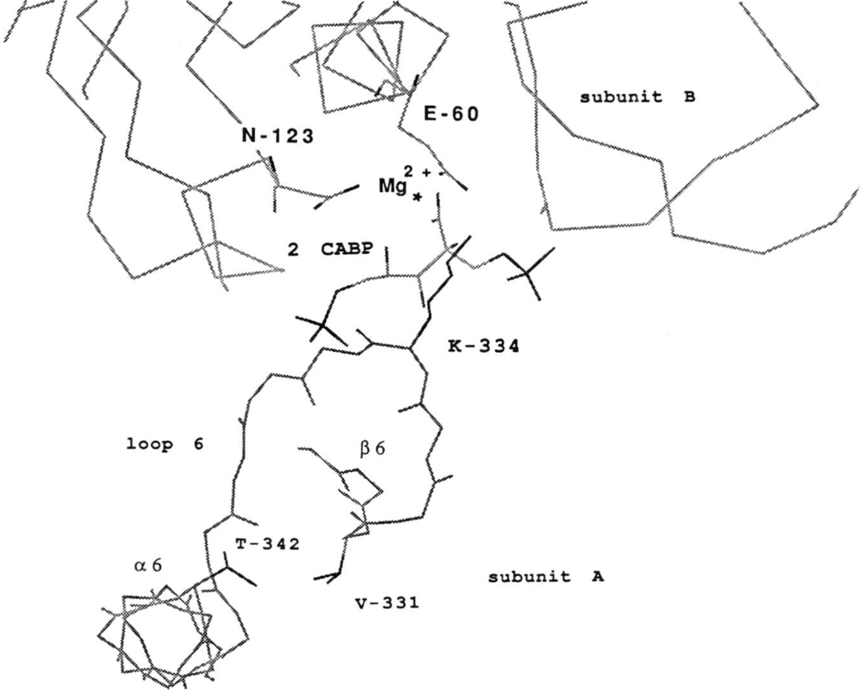

Fig. 3.2 The organization of the active site of Rubisco and the relative positions of those amino acids subjected to site specific mutations. Note that another amino acid of the N-terminal domain, Glu (E) 60 that also affects relative specificity (Hartman 1992) has been included.

mutated the active-site Lys residue to demonstrate the essential role of this amino acid in catalysis, yet show that the mutant retained the ability to catalyse the first reaction, enediol formation. They did not report any change in relative specificity for these mutants.

The first evidence that changes to loop 6 residues influence the partitioning of the two reactions was obtained with a high CO_2 requiring mutant of *Chlamydomonas reinhardii* (Chen and Spreitzer 1989). The chemically induced mutation was localized in the L subunit of the enzyme at position 331 where a Val residue was replaced by Ala. This mutant was only 7 per cent active as wild type and the authors reported a 37 per cent decrease in relative specificity. A second mutant that exhibited reversion back to wild-type activities was not due simply to replacement of the Ala with Val, rather the effects of the first change were effectively suppressed by a Thr at 342 changing to an Ile. This double mutant had 33 per cent turnover and a relative specificity 84 per cent of wild type.

Although both V331 and T342 are not directly involved in the composition of the active site, the crystal structure of the higher plant enzyme indicates how these mutations may cause these changes. The position of V331 is toward the N-terminal end of loop 6, whereas T342 is in the first turn of helix 6 of the barrel. However, both side chains of these residues are the only two that form a hydrophobic pocket at the base of the loop and therefore must influence the geometry of the loop. It is quite consistent with this interaction that loss of hydrophobic packing density due to a Val changing to Ala can be readily compensated for by increasing the side chain at 342. A similar combination of mutations have evolved naturally (Anderson and Caton 1987). The ability to reverse the effects of the first mutation, particularly the loss of carboxylation efficiency, by compensatory changes at an interacting residue is worth investigating by site-specific mutagenesis. There may be a chance to alter relative specificities in a positive direction with an appropriate combination of residues in these two positions.

A comparison of the primary sequences around loop 6 of different hexadecameric carboxylases indicate that this is one of the few regions of the C-terminal barrel that is low in homology. Residues 338–341 at the C-terminal end of the loop show quite wide variation. Furthermore it is also a region of the protein that interacts with other elements of the L subunit that must affect its mobility during catalysis. Therefore it is clearly a region of structure worth investigating by mutagenesis.

Finally, it has been established that the reaction mechanism of the enzyme involves firstly formation of a 2,3-enediol intermediate of the bisphosphate which then reacts with either CO_2 or O_2 at the active site (Gutteridge *et al.* 1984). There is no evidence that either gaseous substrate binds to the enzyme in a Michaelis complex prior to reaction with the enediol (Pierce *et al.* 1986*a*). The role of K334 is intimately involved in stabilization of the catalytic intermediate immediately following reaction with CO_2 and it is the position occupied by this residue relative to those other groups at the active site, such as Mg^{2+} ion that determines the partitioning of the enediol. If the oxygenase activity of the enzyme involves the same amino acids and partial reactions as carboxylation, then altering K334 should have no effect on the fate of the bisphosphate substrate.

Plasmid constructs and mutagenesis

The expectation that functional modifications resulting from mutagenesis will require the resolution of crystallographic analysis to explain the structural basis for the activity changes, immediately imposes a constraint when constructing expression systems. Enough recombinant material must be produced and purified to obtain at least one quality crystal for diffraction purposes. For the *Synechococcus* carboxylase the nature of the crystal packing is such that over 90° of data must be collected to compile a useable data set (Newman and

Gutteridge 1989). Normally, at least five crystals must be grown to withstand the intensity of the radiation from a synchrotron source long enough to ensure data of adequate resolution is obtained. Some of the mutants are less robust than the wild-type enzyme which means that even more crystals are required for a complete analysis.

The construct shown in Fig. 3.3 has produced consistently high yields of intact, fully assembled, active wild-type carboxylase. It is based on a high-copy number pUC plasmid containing the genes for the L and S subunit of *Synechococcus* PCC6301 inserted into the *Pst*1 site of the multiple cloning site

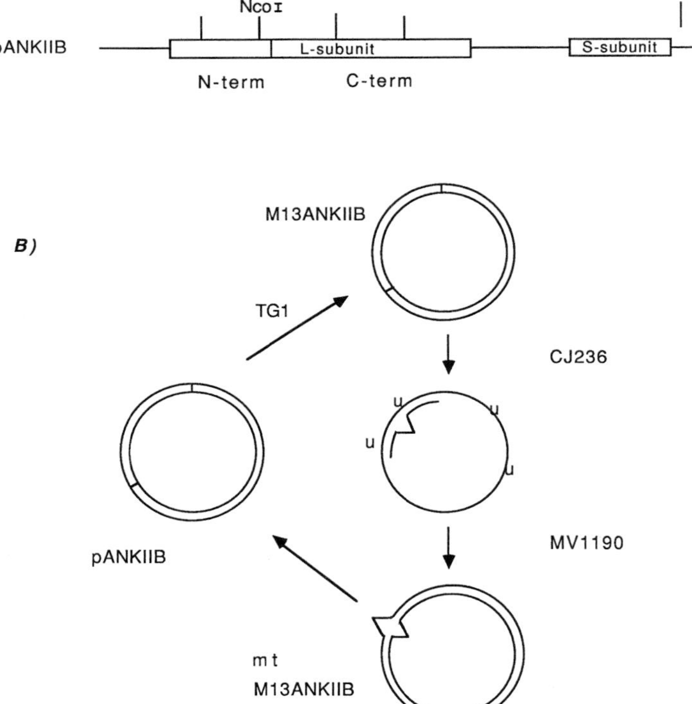

Fig. 3.3 The relative positions of restrictions sites in pANKIIB and the strategy used to introduce the mutations into the L-subunit of Rubisco. In A, the positions of the L- and S-subunit genes are shown and some of the unique restrictions sites used to remove regions of the protein for mutagenesis. The strategy for site-specific mutations (B) involves polymerase extension of oligos harbouring the mutation using a uracil-enriched single-stranded template of M13ANKIIB. The template is selected against using MV1190 strain of *E.coli* resulting in the mutant form of M13ANKIIB predominating. The mutated fragment, usually verified by restriction analysis is removed and reintroduced into the expression vector, pANKIIB for production of mutant enzyme.

and under the control of the *lac* promoter. The initial versions pSynRES (Gutteridge *et al.* 1986) and pAn92 (Kettleborough *et al.* 1988) have been through a number of rounds of mutagenesis leading to the present vectors, pANK1 and pANKIIB that have useful unique restriction sites located at various points throughout the gene (Newman and Gutteridge 1990; Ranty *et al.* 1991).

The positions of these sites (see Fig. 3.3(a)) are not as well placed as those in, for example, the constructs expressing *Rhodospirullum rubrum* carboxylase (Lorimer *et al.* 1988) which has precluded using cassette mutagenesis to alter sequences in the L or S subunits of the hexadecamer. However, with the development of mutant enrichment techniques using single-stranded based systems has meant that there is little delay in producing a desired mutation. In this particular case mutagenesis involves a shuttle type approach where the two subunit gene from pANKIIB has been inserted into the replicative form of M13mp18 to produce M13ANKIIB Fig. 3.3(b)). A single stranded version of this is isolated from phage generated in a bacterial host with *dut*, *ung* double mutation that replaces some of the thymine bases with uracil. A synthetic oligonucleotide harbouring the mutation is annealed to the uracil-containing M13 strand and the partial duplex is extended with polymerase in the presence of all four deoxynucleotides.

Transformation of a wild-type strain results in degradation of uracil containing parental strands and enrichment of the mutant strand. Mutated RF M13ANKIIB is used to isolate the mutant region of the gene on fragments about 400 base pairs in length, exploiting the unique restriction sites. This fragment is then used to replace the same segment in the wild-type expression vector. Usually the synthetic oligonucleotide can be organized to remove a restriction site in the gene so that restriction pattern analysis of double stranded M13 is adequate to confirm successful mutagenesis. Once the fragment is inserted into the expression vector, the gene is sequenced through the two restriction sites used to transfer the fragment. Many of the mutants generated with this system were obtained with about 30–60 per cent success. Fig. 3.3(a) shows the position of the *NsiI–BstXI* fragment used to insert the mutations into loop 6.

Production and isolation of mutant enzyme

Although expression of the protein is under *lac* promoter control the high copy number of the vector is in abundance relative to repressor and thus no IPTG is required for induction. The protein is expressed constitutively when the bacterium is grown in rich media. An interesting yet little understood response is that expression is much higher when the cells are starved of oxygen during log phase (Newman and Gutteridge 1990). Some of these requirements have been circumvented by omitting glycerol as the C source, so that both high protein yield and cell densities can be achieved, but it still remains that we do not enjoy full control of expression with these constructs (Gutteridge 1991). Fortunately most of the mutants produced using these plasmids retain the ability to generate acceptable levels of enzyme often exceeding 15 per cent of the soluble cell contents. This means that purification to an extent that yields usable crystals is not onerous.

A further advantage of a high-yielding recombinant system apart from the demands of the crystallographer, is that detailed functional and spectroscopic analyses are feasible. Although a mutant may only turnover 10^3 times slower than wild type, this rate is still significant relative to zero (see Chapter 2) yet to complete a full analysis of the effect of such a mutation requires tens of milligrams. Such yields are readily obtained from overnight growth in a few litres of culture.

The purification of the protein requires that the soluble contents of the bacterium after lysis be separated from the cell debris. The large quantities of plasmid DNA accompanying the isolation are degraded with DNAse to smaller fragments that can be washed from the proteins by filtration over semipermeable membranes (Pierce and Gutteridge 1985). The resulting washed and concentrated retentate chromatographs well using strong anion exchange columns developed with linear KCl or NaCl gradients at pH 7.6. In those extracts that still show significant contamination with other soluble proteins, further chromatography is possible on the same columns but this time using phosphate buffers. The mutants are finally desalted and concentrated using semipermeable membranes and stored as 16 per cent glycerol solutions at –80°C.

Analysis of mutant functionality

The mutant enzymes are roughly grouped into two sets; those that retain a substantial amount of overall activity and those that have essentially no detectable activity. The former group are subjected to a detailed analysis of changes to kinetic parameters, for example substrate affinities and particularly the relative rates of carboxylation and oxygenation. With the group that has less activity, the investigations are more directed to the partial reactions of catalysis and activation (Lorimer *et al.* 1988). The absence of catalysis can mean simply that structural alterations due to the mutation have, for example, destroyed the ability of the enzyme to coordinate Mg^{2+}. Assuming the mutation has not seriously interfered with bisphosphate binding then evidence for both activation of the enzyme and changes to bisphosphate affinity can be studied by looking for formation of a stable enzyme quaternary complex with the tight-binding inhibitor 2CABP.

In the presence of CO_2 and Mg^{2+}, wild-type hexadecameric Rubisco binds 2CABP almost irreversibly. This is determined most simply by following the exchange of labelled inhibitor from the complex challenged with a large excess of cold inhibitor with time. More dramatic changes to structure such as the inability of the subunit to fold correctly often come to light through electrophoretic or chromatographic analysis of the relative amounts of L and S subunits and their behaviour in nondenaturing conditions.

If the mutant clearly forms a quaternary complex with 2CABP, however tenuous, then the partial steps of carboxylation might be investigated. The reactions of the enzyme require firstly, the formation of the enediolate form of ribulose-P_2 and this step can be analysed with [3-^{3}H] substrate. The ability of mutants to catalyse this first step is observed as washout of the label from the substrate (Andrews and Lorimer 1985). Absence of the initial reaction does not necessarily mean that the enzyme is completely inactive particularly if there is affinity for 2CABP binding. The isolated intermediate of carboxylation, 3k2CABP can be added to the enzyme and the extent to which it is hydrolysed to 3P-glycerate can be determined (Lorimer *et al.* 1988).

Those species that have some activity are also analysed in terms of how they partition ribulose-P_2. Relative specificity measurements of the partitioning of the bisphosphate between carboxylation and oxygenation is achieved by using [1-^{14}C] labelled substrate and chromatographically separating the labelled (dephosphorylated) products of the two reactions, glycolate and glycerate (Andrews and Kane 1990). The following results were obtained for the mutants described above.

Mutations affecting loop 6

The mutations that were introduced to alter the composition of loop 6 and adjacent segments are shown in Fig. 3.4.

V331 and T342

Valine at position 331 was replaced with amino acids that have smaller and larger hydrophobic side chains. Apart from V331 all of the mutants were active, although with altered catalytic specifities as shown in Table 3.1. With a

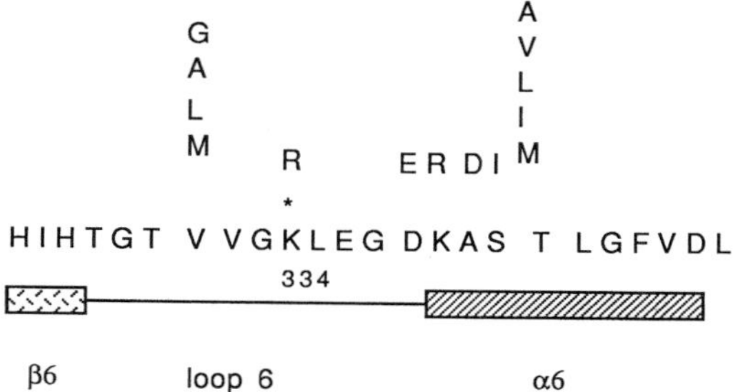

Fig. 3.4 The location of the mutations introduced into the loop 6 region of *Synechococcus* rubisco. Lys (K) 334 resides at the apex of the loop and is involved in stabilization of the carboxylated intermediate of the catalytic reaction (see Fig. 2).

Table 3.1 Effects of loop 6 mutations on the kinetic parameters of Rubisco

	Activity (%)	K_m (mM) ribulose-P_2	Relative specificity
V331 (wt)	100	0.03	56 ± 2
G	0.5	0.1	20 ± 4
A	5.0	0.05	30 ± 3
L	5.0	0.5	49 ± 3
M	0.5	1.6	16 ± 4
T342			
A	0.2	n.d.	n.d.
V	40	0.07	50 ± 2
L	40	0.20	56 ± 2
I	30	0.30	50 ± 2
M	6	0.13	56 ± 3

smaller or no side chain (Ala or Gly) overall catalysis suffers dramatically, and both mutants have substantially reduced relative specificity compared to wild type. Interestingly, the smaller amino acids have little impact on either CO_2 or bisphospate substrate affinity.

The larger amino acids cause a decrease in nearly all the parameters. Affinity for ribulose-P_2 is significantly weaker by at least a factor of ten and relative specificities have also suffered declining by 70 per cent.

Replacement of Thr at position 342 was tolerated much more than mutations at 331. Significant loss of catalysis was obtained with the residues that had the smallest and largest side chains, Ala and Met. Both had reduced turnover of 0.2 per cent and 6 per cent respectively, compared to wild type. The relative specificity declined by only a modest amount and less than with mutations at 331.

One reaction investigated in some detail was the rate of formation of enediol using [3-^{3}H] ribulose-P_2. If the role of loop 6 is simply to position the Lys at 334 to stabilize the 2′-carboxyl group of carboxylation, then an altered geometry of the loop should not effect enediol formation. Table 3.2 indicates that this is not the case. Even with changes that do not interfere with substrate affinities, for example V331A or V331G the rates of tritium exchange approximate the rate of overall catalysis. Clearly, enediol formation is apparently tightly coupled to substrate consumption.

Mutations at K334

All of the mutations at this position caused a dramatic loss of catalytic turnover, at least based on carboxylation. This was not unexpected for the Cys and Met mutations where there is no longer an amino group of a Lys to stabilize a reaction intermediate. The K334M mutant was completely inactive. However, replacement of Lys with another basic amino acid, Arg did restore some

Table 3.2 Changes in quaternary complex stability and enediol formation

	[3-^{3}H] exchange	2CABP (h) exchange
V331	100	> 168
A	0.6	1.0
G	0.3	1.5
L	0.2	10.0
T342 A	<0.2	0.5
L	40.0	2.1
M	8.0	0.8
K334 R	12.0	2.0
C	5.0	<0.5
M	3.0	<0.5

carboxylase activity 0.5 per cent of wild type. More significantly, the mutant exhibited high rates of oxygenation. The relative specificity of this enzyme was therefore the lowest yet reported for Rubisco with Mg^{2+} bound at the active site. Nevertheless, the mutant responded as expected to increased amounts of CO_2 indicating that this substrate still acts as a competitive inhibitor of the oxygenase reaction. Based on initial rate data, the affinity for CO_2 is less by at least a factor of two although ribulose-P_2 binding is essentially unchanged.

The ability of K334R to catalyse oxygenation at the expense of carboxylation suggests that the loss of Lys side chain has compromised stabilization of the 2′-carboxyl intermediate at the active site. Therefore the stability of the quaternary complex of all the 334 mutants was compared to that of the wild type. All three mutants bound a stoichiometric amount of the inhibitor to the activated form of the protein. However, after addition of excess unlabelled inhibitor to the Cys and Met complexes, exchange occurred within the time required to gel filter excess inhibitor from the solutions. The quaternary complex with the Arg mutant was more stable with a $t_{1/2}$ for the exchange of 2 h relative to wild type; a loss of at least two orders of magnitude in stabilization.

These values can be compared to those other Rubisco species with distinct relative specificities. This indicates that binding affinity of 2CABP is not a direct measure of carboxylation efficiency, although there is clearly a trend indicating that less efficient carboxylation is reflected in less 2CABP binding affinity.

The organization of the active site of Rubisco with bound 2CABP particularly within a limited sphere of the 2′carboxyl group shows contributions from more than just the ϵ-amino group of K334. Not only does one of the 'O' groups of the carboxyl form an ionic interaction with K334, but the other is ligated to the Mg^{2+} ion. Within this sphere are also two side chains from amino acids of the N-terminal domain of the second L-subunit, N123 and E60 (see Fig. 3.2). The γ-carboxyl of the latter residue is also close to the amino group of K334 and may influence the interaction with the 2′carboxyl. Thus, assuming this is the state of the active site immediately following reaction between enediol and CO_2, then the charges around the captured substrate are finely poised with the CO_2 polarized between the positive charge of the Lys amino group and the Mg^{2+} ion. Replacement of the Lys side chain with the more basic and bulky Arg clearly disturbs this balance enough that carboxylation of the enediol is almost completely suppressed. The more opportunistic substrate, oxygen therefore must generate an intermediate (presumably hydroperoxide) that is not as stringent in its choice of groups for stabilization and oxygenation proceeds almost undiminished.

Nevertheless, the intermediate of oxygenation requires some stabilization other than the metal because the K334C and K334M mutants do not have appreciable oxygenase rates (<0.1%), although both readily catalyse the exchange of the C3 proton of the bisphosphate substrate. Both of these mutants indicate that the exchange process and thus enediol formation can be uncoupled from overall catalysis. They catalyse the exchange reaction at rates that are at least 10 times faster than, for example, the 331 mutants, suggesting that the position that loop 6 adopts relative to the other active site elements is important for enediol formation. The smaller side chains of the 334 mutants allow the loop to adopt a conformation more like that of wild type than does the disruption of the hydrophobic pocket formed by amino acids at position 331 at the base of the loop.

Fortunately, the K334 mutants all crystallize with the same space group as the wild-type enzyme and thus a detailed description of the active site will be forthcoming.

Chimeric constructs

The response of the enzyme to mutations in loop 6 indicate that the amino acids that compose this segment are fundamentally involved in the fate of the substrates, presumably through their direct involvement in intermediate stabilization or through altering the geometry and thus mobility of the loop. Fig. 3.2 shows the organization of this region of the L subunit relative to the active-site-bound inhibitor and metal ion. Although, the residues that occupy positions 338–341 are somewhat distant from the active site they do interact with other elements of protein structure. The absence of homology in this segment suggested that replacement with the sequence found in more efficient carboxylases might impart a chimera with increased relative specificity.

Table 3.3 indicates that there was little effect on the kinetic parameters associated with carboxylation of such a chimera compared with wild-type enzyme, although a slight decrease in overall turnover was observed. Furthermore, the relative specificity increase detected for the mutant was determined to be insignificant.

Mutations outside loop 6

N123

This residue is conserved throughout all Rubisco species including the dimeric form II. This amino acid resides in a loop region of the N-terminal domain of the

Table 3.3 Kinetic parameters and quaternary complex stability of Rubisco mutants

	Activity (%)	K_m (mM) ribulose-P_2	Relative specificity	2CABP (h) exchange
ERDI	60	0.08	58	>168
K334R	0.5	0.10	0.3	2.0
N123	100	0.02	20	12.0
V	6.0	0.02	3.0	12.0

L subunit that contributes to the active site. The amino side chain is part of the coordination sphere of the metal ion. However, as catalysis proceeds and metal coordination adjusts, N123 is one of the ligands that may be replaced within the first shell.

There is now ample evidence to conclude that the enzyme bound metal is involved at many, if not all stages of catalysis. Many other metals can replace the Mg^{2+} at the active site, but not all support catalysis (Christeller 1981). With some transition metals, for example Mn^{2+} and Co^{2+}, activity is characterized by a decrease in relative specificity as oxygenation is favoured at the expense of carboxylation. The question remained whether this loss of carboxylation efficiency is due to altered coordination through metal preference compared with Mg^{2+}, or the nature of the chemistry involving metal orbitals. The former might be investigated by changing the nature of the ligands to the metal.

Other mutations of residues known to be involved with metal binding have interfered with activation of the enzyme and thus their effects on function were difficult to interpret without a detailed analysis of the partial reactions. Because the amide group of N123 was considered to have only a relatively weak interaction with the metal, it was expected that modification would be less disruptive. In this case, the mutations were most readily introduced into *R. rubrum* Rubisco using cassette mutagenesis (Lorimer *et al.* 1988).

Replacement of the Asn at position 123 with Asp or Glu produced only inactive enzyme unable to activate with CO_2 and Mg^{2+}. Based on carboxylase activity, a mutant with Val in place of Asn also showed little reaction. However, the oxygen electrode indicated that nearly all the oxygenase activity of the wild-type enzyme was retained. As a result, the relative specificity of the mutant is a factor of six less than the wild-type enzyme, yet overall catalysis based on oxygenase activity is about 33 per cent of wild type. The relative specificities of the K334R and N123V mutants are compared in the plot shown in Fig. 3.5 (Jordan and Ogren 1981) along with the wild-type forms of recombinant enzymes and spinach Rubisco.

Because the Val replacement generated a mutant with significantly lower relative specificity with little interference to overall turnover as found for the K334R mutant of *Synechococcus* Rubisco, it was of interest to compare the stability of the quaternary complexes. Interestingly, the complex with N123V

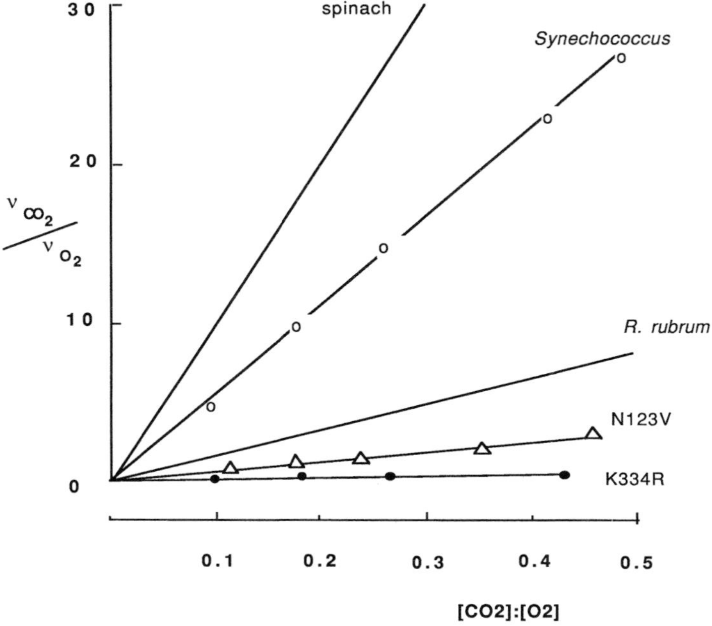

Fig. 3.5 A specificity plot of various species of Rubisco. The relative specificity of the enzymes is computed from the slope of the line, the steeper the slope the more ribulose-P_2 is committed to carboxylation. The relative specificity of the two mutants, N123V and K334R indicate the large shift toward oxygenation caused by the sequence alteration.

has almost the same stability as the wild-type enzyme and is thus six times more stable than K334R.

One other use was made of this mutant. Changes in relative specificity of wild-type Rubisco due to replacement of Mg^{2+} with other metals might be due to their electronic structure and/or their ligand preference at the active site, i.e. they occupy a slightly different binding site than Mg^{2+}. The similar binding affinity of 2CABP in both wild type and mutant suggests that the ^{31}P chemical shifts of the inhibitor might be used to report on the nature of the metal site. Table 3.4 indicates that although the mutation has dramatically altered the ^{31}P shifts of 2CABP with bound Mg^{2+}, in fact the two phosphates are almost resolved, the Ni complex shows little change in P chemical shift. This would suggest that indeed the two metals have different coordination geometry in the quaternary complexes.

The basis of relative specificity differences

Interpreting the differences in relative specificity of some of the mutants described in this chapter in terms of quaternary complex stability must be viewed with caution. K334R supports the proposal in that 2CABP readily dissociates from the quaternary complex with Mg^{2+} and CO_2, whereas N123V does not. The proposal would also suggest that generating a mutant with increased affinity for 2CABP might translate into an enzyme that has increased carboxylation efficiency, i.e. to a first approximation, the mutation will have increased the affinity of the enzyme for the transition state.

One such mutant exists, at least of *R. rubrum* carboxylase. With Lys 201, i.e. the residue involved in carbamylation, replaced by Cys, the enzyme forms a quaternary complex more stable than wild type by at least a factor of 5. This does not translate into increased carboxylation efficiency. In fact, the mutant is unable to catalyse any of the reactions that should occur with an appropriately bound metal. Superficially, the reason might be that the absence of the essential carbamate at the active site blocks the first step of catalysis. Thus the ability of 2CABP to form a stable complex is that the binding affinity of the inhibitor so favours the quaternary complex with CO_2 and metal, that the equilibrium is all toward formation. However, this does not reflect the situation with substrate. No such complex was detectable with ribulose-P_2 in the presence of CO_2 and Mg^{2+}, suggesting that with the lower affinity for substrate and the absence of carbamate at the active site (Gutteridge and Lorimer, unpublished), then little, if any, active ternary complex with Mg^{2+} exists long enough for ribulose-P_2 to bind.

Evidence that this might be the case was obtained from an analysis of the rate of formation of the complex with mutant compared to wild-type enzyme. It was clear that the K201C mutant quaternary complex takes at least 60 min to form in the presence of optimal CO_2 and Mg^{2+} compared to the almost instantaneous rate obtained with wild-type Rubisco.

Natural mutants of Rubisco

The present data base available of Rubisco relative specificities is woefully small covering at most, some four dozen species of higher plant enzyme. Yet even within that limited sample, differences of carboxylation efficiency have been reported. It is likely that a distribution of relative specificities exists amongst higher plants of which spinach is equally unlikely to be the best example. Indeed two other species are known to have superior enzyme, i.e. sunflower and wheat (Parry *et al.* 1989); both with at least a 16 per cent higher relative specificity compared with spinach or tobacco Rubisco. It is within the bounds of statistical probability that enzymes with even higher partitioning await discovery.

Acknowledgements

The author wishes to convey his appreciation for continuing discussions about the structure of Rubisco with the crystallography group at the Biomedical

Center, Uppsala, and George Lorimer for input concerning enzyme mechanism. Dan Rhoades and Chuck Herrmann provided technical support.

References

Anderson, K. and Caton, J. (1987). *J. Bacteriol.*, **169**, 4547–58.

Andersson, I., Knight, S., Schneider, G., Lindqvist, Y., Lundqvist, T., and Brändén, C.-I. *et al.* (1989). *Nature*, **337**, 229–34.

Andrews, T.J. and Kane, H. (1991). *Plant Physiol.*

Andrews, T.J. and Lorimer, G.H. (1985). *J. Biol. Chem.*, **260**, 4632–6.

Andrews, T.J. and Lorimer, G.H. (1987). In *The biochemistry of plants*, Vol. 10, pp. 131–218. Academic Press, New York.

Chen, Z. and Spreitzer, R.J. (1989). *J. Biol. Chem.*, **264**, 3053–5.

Christeller, J.T. (1981). *Biochem. J.*, **193**, 839–44.

Gutteridge, S. (1990). *Biochim. Biophys. Acta,* **1015**, 1–14.

Gutteridge, S. (1991) *J. Biol. Chem.*, **266**, 7359–62.

Gutteridge, S. and Gatenby, A.A. (1987). In *Oxford surveys of plant molecular and cell biology*, (ed. B.J. Mifluir), pp. 95–135. Oxford University Press, Oxford.

Gutteridge, S., Parry, M., Schmidt, C.N.G., and Feeney, J. (1984). *FEBS Lett.*, **170**, 355–9.

Gutteridge, S., Phillips, A.L., Kettleborough, K., Parry, M.A.J., and Keys, A.J. (1986). *Phil. Trans. R. Soc. Lond.*, **B313**, 433–45.

Gutteridge, S., Pierce, J., and Lorimer, G.H. (1988). *Plant Physiol. Biochem.*, **26**, 675–82.

Hartman, F.C. (1992). In *Plant protein engineering*, (ed. P.R. Shewry and S. Gutteridge), pp. 61–92. Cambridge University Press, Cambridge.

Hartman, F.C. and Lee, E. (1989). *J. Biol. Chem.*, **246**, 11 784–9.

Jordan, D.B. and Ogren, W.H. (1981) *Nature*, **291**, 513–15.

Kettleborough, K., Parry, M.A.J., Burton, S., Gutteridge, S., Keys, A.J., and Phillips, A.L. (1987). *Eur. J. Biochem.*, **170**, 335–42.

Knight, S., Anderson, I., and Brändén, C.-I. (1990). *J. Mol. Biol.*, **215**, 113–60.

Lorimer, G.H. (1981). *Ann. Rev. Plant Physiol.*, **32**, 349–83.

Lorimer, G.H., Madden, M., and Gutteridge, S. (1988). In *NATO proceedings of plant molecular biology* (ed. D. Von Wettstein and N.-H. Chua), pp. 21–31, Plenum Press, New York.

Lundqvist, T. and Schneider, G. (1991). *Biochemistry*, **30**, 904–8.

Newman, J. and Gutteridge, S. (1990). *J. Biol. Chem.*, **265**, 15 154–9.

Parry, M.A., Keys, A.J., and Gutteridge, S. (1989). *J. Exp. Botany*, **40**, 317–20.

Pierce, J. and Gutteridge, S. (1985). *Appl. Environ. Microbiol.*, **49**, 1094–1100.

Pierce, J., Andrews, T.J., and Lorimer, G.H. (1986*a*). *J. Biol. Chem.*, **261**, 10 248–56.

Ranty, B., Lorimer, G., and Gutteridge, S. (1991). *Eur. J. Biochem.*, **200**, 353–8.

Schneider, G. and Lindqvist, Y. (1992). In *Plant protein engineering* (ed. P.R. Shewry and S. Gutteridge), pp. 42–58. Cambridge University Press, Cambridge.

Schneider, G., Lindqvist, Y., Brändén, C.-I., and Lorimer, G.H. (1986). *EMBO J.*, **5**, 3409–15.

4

Learning from Rubisco's mistakes

T. John Andrews, Matthew K. Morell, Heather J. Kane, Kalanethee Paul, Gabriel A. Quinlan, and Daryl L. Edmondson

Introduction

A uniquely important, yet imperfect, carboxylase

The reaction catalysed by Rubisco (D-ribulose-1,5-bisphosphate (ribulose-P_2) carboxylase/oxygenase) is distinguished from other biological CO_2-fixing reactions in having been chosen by nature as the means of acquiring nearly all of its carbon (for a review, see Andrews and Lorimer (1987)). Therefore, we have to ask ourselves not only how Rubisco functions but also how it is better than other potential CO_2-acquiring mechanisms. Because we believe that, if nature had ever found a better mechanism for fixing CO_2, then we would see that mechanism operating in photosynthetic organisms today. So, although most of this chapter is devoted to an analysis of Rubisco's many imperfections, it is important to emphasize at the outset that, imperfect though it is, Rubisco is still the best mechanism that has evolved for fixing CO_2 under present atmospheric conditions.

Rubisco's imperfections illustrate that enzymes, in their courses of evolutionary refinement, have not only to maximize the rate at which they catalyse the transformation of substrates into products but also, and equally importantly, they have to minimize the rate at which they catalyse unwanted reactions to which the reactive intermediates that they use might be prone. Rubisco seems to fall short on both counts. It is a very slow carboxylase and it also catalyses at least four other reactions for which no useful purpose can be seen. Some of these side reactions have deleterious physiological consequences and they all require ancillary mechanisms for disposal of their products. The side reactions are also quite informative. They provide clues about Rubisco's catalytic chemistry that would be difficult to obtain by other means. Furthermore, by analysing the effects of mutations on these side reactions, we can learn about features in the active site which serve to suppress them.

Activation by carbamylation

For Rubisco to become catalytically competent, a specific lysine residue in its active site must be carbamylated so that the catalytically essential magnesium ion may bind (Fig. 4.1) (Andrews and Lorimer 1987). This lysine residue is number 201 in the sequence of the large subunit of spinach Rubisco. (We will use the spinach numbering throughout (see Andrews and Lorimer (1987) for an alignment with the sequences of Rubiscos for other species).

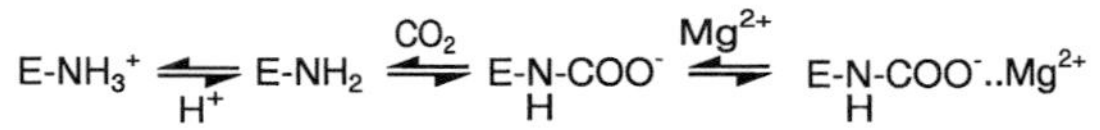

$$\text{E-NH}_3^+ \underset{\text{H}^+}{\rightleftharpoons} \text{E-NH}_2 \overset{\text{CO}_2}{\rightleftharpoons} \underset{\text{H}}{\text{E-N-COO}^-} \overset{\text{Mg}^{2+}}{\rightleftharpoons} \underset{\text{H}}{\text{E-N-COO}^-..\text{Mg}^{2+}}$$

Fig. 4.1 Activation of Rubisco. Carbamylation of K201 completes the metal-binding site.

Carboxylation

Once activated in this way, Rubisco is capable of catalysing the carboxylation of ribulose-P_2 by a molecule of CO_2 which is different from the one involved in the carbamylation reaction (Lorimer 1979). The sequence of intermediates through which the reaction proceeds at the active site is shown in Fig. 4.2. The central intermediate is the six-carbon, β-keto acid. It is shown in both its ketone and hydrated, *gem*-diol forms, although, if the carboxylation and oxygenation steps

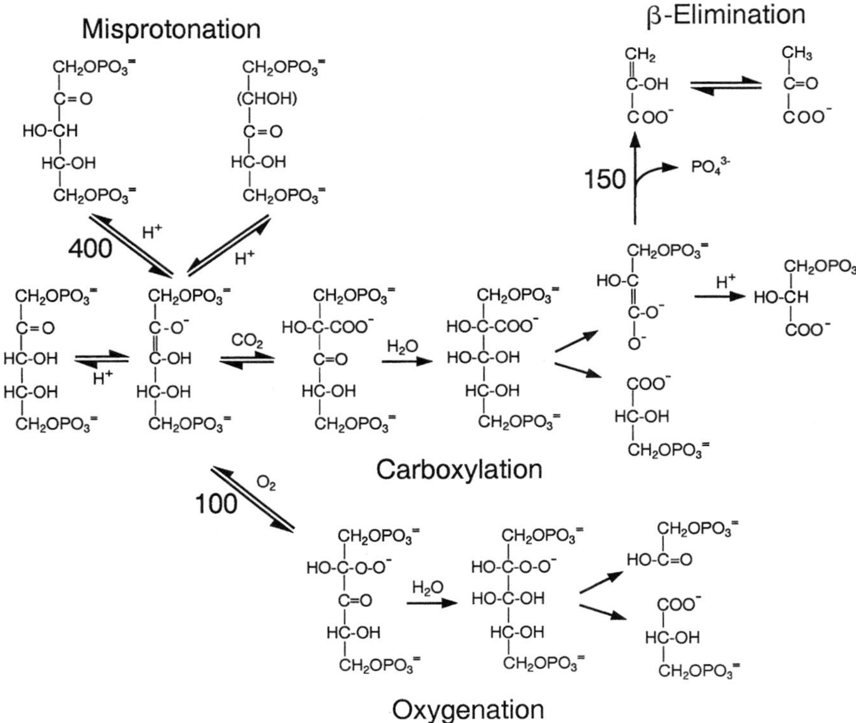

Fig. 4.2 The reactions catalysed by Rubisco. The relative specificity for each side reaction catalysed by the spinach enzyme (where known) is given beside the arrows.

were concerted (Cleland 1990), only the *gem*-diol would exist on the active site. In solution, this intermediate exists predominantly in the ketone form and it is relatively stable, decarboxylating at neutral pH and 25 °C with a half-life of approximately 1 h (Pierce *et al.* 1986). Flanking this central β-keto acid are two much more reactive nucleophilic intermediates. The first is the five-carbon 2,3-enediol which is produced by abstraction of a proton from ribulose-P_2. (We refer to it as an enediol, although it could also be an enediolate.) This is the species with which CO_2 reacts. Its presence was inferred from the exchange of the C3 proton of ribulose-P_2 with solvent which Rubisco catalyses (Saver and Knowles 1982) and confirmed by the detection of a very labile intermediate which eliminated inorganic phosphate when removed from the active site, but which could be stabilized by I_2 (Jaworowski *et al.* 1984). The three-carbon *aci*-carbanion intermediate produced by scission of the C2–C3 bond of the β-keto acid later in the reaction sequence would also be expected to have enediol character (Fig. 4.2). Its intermediacy was suspected from the pronounced isotope effect accompanying incorporation of solvent protons into the product, D-3-phosphoglycerate (P-glycerate) (Hurwitz *et al.* 1956; Saver and Knowles 1982). Such an isotope effect is consistent with competition between proton isotopes for an intermediate.

Oxygenation

The oxygenase reaction (Fig. 4.2) was the first abortive side reaction to be demonstrated with Rubisco and it was discovered many years ago (Bowes *et al.* 1971; Andrews *et al.* 1973). It occurs because the nucleophilic, 2,3-enediol intermediate, when bound in the active site, cannot distinguish between CO_2 and O_2 as electrophiles. Nor, apparently, can the active site avoid stabilizing the five-carbon hydroperoxy intermediate produced during the oxygenation reaction, presumably because of its resemblance to the six-carbon, β-keto acid intermediate of the carboxylase reaction (Fig. 4.2). At equal concentrations of CO_2 and O_2, the higher-plant enzyme catalyses one oxygenation of ribulose-P_2 for approximately every 100 carboxylations (Jordan and Ogren 1981; Andrews and Lorimer 1985). However, in the chloroplast stroma during active photosynthesis, dissolved oxygen is thirtyfold more abundant than dissolved CO_2. Therefore, oxygenation proceeds at approximately one-third of the rate of carboxylation, producing large quantities of its waste product, 2-phosphoglycolate. This encumbers all photosynthetic organisms with the necessity for the complicated metabolic sequence known as photorespiration. This process consumes energy in order to recycle three-quarters of the phosphoglycolate carbon back to photosynthetic metabolism. The remaining 25 per cent of the phosphoglycolate carbon is lost as photorespiratory CO_2 (for a review, see Artus *et al.* (1986)).

A slow and confused but improving catalyst

There are two major criteria for assessing the catalytic performance of an enzyme. These are the catalytic potency towards a particular substrate, which is reflected by the k_{cat}/K_m ratio (Fersht 1985), and the selectivity between similar substrates, which is the ratio between the k_{cat}/K_m ratios for the two substrates. On both criteria, Rubisco compares unfavourably with many other enzymes. For example, Rubisco's k_{cat}/K_m ratio for CO_2 is at least three orders of magnitude slower than the diffusion limit to catalysis which catalytic thoroughbreds like triosephosphate isomerase approach quite closely and its selectivity between CO_2 and O_2 seems feeble compared to the ability of tyrosyl-tRNA synthetase to distinguish between tyrosine and phenylalanine (Table 4.1).

It might be thought that Rubisco's troubles arise from the competing requirements of speed, on the one hand, and selectivity between CO_2 and O_2, on the other. However, when we look at a suite of natural Rubiscos drawn from a wide evolutionary divergence, plotting the carboxylation specificity versus the selectivity for CO_2, as opposed to O_2, we see no evidence of a negative correlation between these parameters which might reflect the necessity for a compromise between speed and selectivity (Fig. 4.3). Indeed, there seems to be some indication that the reverse is true, i.e. the most potent carboxylases are also the most selective. This is evidence for evolutionary progress in the right direction. And it must be remembered that only for the last 2 per cent of Rubisco's evolutionary history has the earth's atmospheric CO_2/O_2 ratio been low enough to make Rubiscos' selectivity between CO_2 and O_2 an important character. Before approximately 60 million years ago, the CO_2 partial pressure in the atmosphere was so high that oxygenation of ribulose-P_2 would have been an irrelevant curiosity (Ehleringer *et al.* 1991).

Table 4.1 Some catalytic parameters of Rubisco, tyrosyl-tRNA synthetase and triosephosphate isomerase

	Rubisco	Tyrosyl-tRNA synthetase*	Triosephosphate isomerase†
$k_{cat}(s^{-1})$	3–12	8	4300
$k_{cat}/K_m(M^{-1}s^{-1})$‡	$1–3 \times 10^5$	3.7×10^6	2.4×10^8§
	(CO_2)	(tyr)	(glycerald.-3P)
Relative specificity¶	15–100	1.5×10^5	–
	(CO_2/O_2)	(tyr/phe)	

* Data from Fersht *et al.* (1985).

† Data from Putman *et al.* (1972).

‡ Diffusion limit = 10^8–10^9 $M^{-1}s^{-1}$.

§ Corrected to exclude the contribution of the hydrated *gem*-diol form of glyceraldehyde-3-phosphate to the substrate concentration.

¶ $[(k_{cat}/K_m)_{substrate\ 1}]/[(k_{cat}/K_m)_{substrate\ 2}]$.

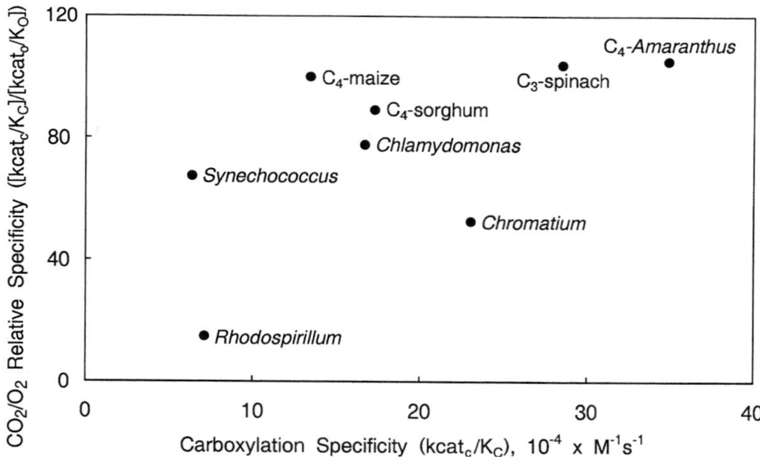

Fig. 4.3 The CO_2/O_2 relative specificities of Rubiscos from various sources plotted against their effectiveness as carboxylases (carboxylation specificity). Adapted from Badger and Andrews (1987) with permission.

Other side reactions originating from Rubisco's nucleophilic intermediates

Misprotonation of the five-carbon enediol

A second side reaction (or perhaps a group of side reactions) was revealed as a result of efforts to discover the cause of the anomalous behaviour of higher-plant Rubiscos during assay. As soon as ribulose-P_2 is added to fully activated enzyme to initiate catalysis, the activity declines progressively with a half-time of approximately 7 min until, eventually, a constant rate is achieved which is substantially less than the initial rate (Fig. 4.4) (Edmondson *et al.* 1990*a*). Under the conditions of the experiment described, the final rate was approximately half of the initial. However, the ratio between the final and initial rates is reduced as the pH or CO_2 concentration is lowered and it can be as low as 0.15 (Edmondson *et al.* 1990*a*).

We eventually showed that this kind of inhibition was caused by the accumulation of a tight-binding inhibitor at the active site during catalysis (Edmondson *et al.* 1990*b*). This inhibitor could be released from the enzyme with acid and used to inhibit fresh Rubisco. It bound relatively slowly (Fig. 4.5(a)), and apparently anticooperatively, as suggested by incomplete inhibition of activity at high inhibitor concentrations (Fig. 4.5(b)).

The inhibitor became labelled with tritium on reduction with borotritide and the label chromatographed as a bisphosphate (Edmondson *et al.* 1990*c*). After removal of the phosphate moieties, thin-layer chromatography revealed arabinitol and xylitol, the reduction products of xylulose, whereas similar treatment of ribulose-P_2 produced the expected mixture of arabinitol and ribitol (Edmondson *et al.* 1990*c*). Therefore, the inhibitor preparation must have contained D-xylulose-1,5-bisphosphate (xylulose-P_2). This compound would be formed at the active site of Rubisco by the attack of a proton on C3 of the enediol intermediate

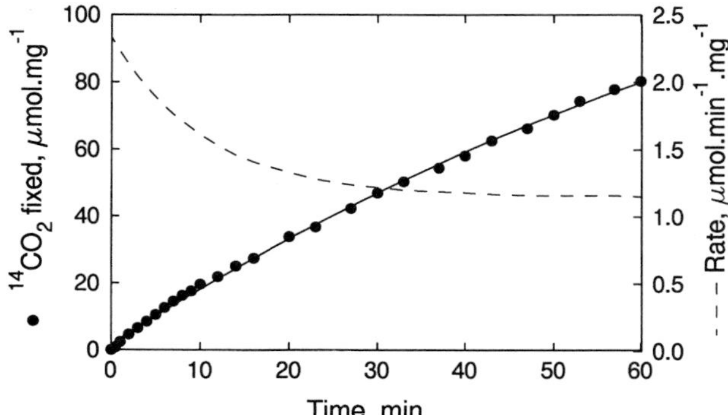

Fig. 4.4 Slow inactivation of Rubisco during carboxylation of ribulose-P_2. Spinach Rubisco (10 μg ml^{-1}) was preincubated for 10 min in a solution containing 100 mM Bicine–NaOH buffer, pH 8.3, 20 mM $MgCl_2$, 10 mM $NaH^{14}CO_3$, 0.1 mg ml^{-1} carbonic anhydrase and 1 mg ml^{-1} gelatin. At zero time, 1 mM ribulose-P_2 was added. The solid line represents the best fit of the data to an equation which models an exponential decline in activity to a final constant rate. The dotted line represents the first derivative of the solid line. From Edmondson *et al.* (1990*a*) with permission.

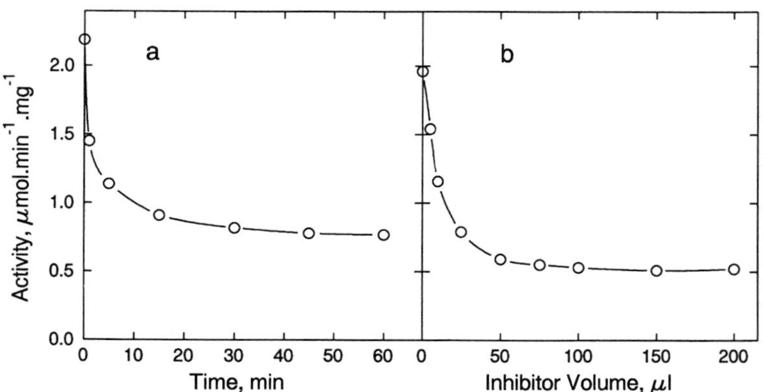

Fig. 4.5 Inhibition of fresh Rubisco with inhibitor obtained by perchloric acid precipitation of Rubisco after slow inactivation during catalysis. The inhibitor preparations were neutralized to pH 5 with KOH and the $KClO_4$ precipitate was removed. Aliquots of inhibitor preparations were mixed with preactivated Rubisco in 100 mM Bicine–NaOH buffer, pH 8.3, containing 20 mM $MgCl_2$, 10–15 mM $NaHCO_3$ and 1 mg ml^{-1} bovine serum albumin and incubated at 25 °C for the stated interval (a) or 60 min (b) before dilution and assay of the residual activity for 1 min. (a) Rate of inhibitor binding at constant inhibitor concentration. (b) Effect of varying inhibitor concentration. From Edmondson *et al.* (1990*b*) with permission.

from the *Si* face, i.e. the face opposite to the one from which attack of a proton would recreate the substrate, ribulose-P_2 (Fig. 4.2). Further experiments showed that not all of the inhibitor was xylulose-P_2 (Edmondson *et al.* 1990*c*). At least one other compound was also present which differed from xylulose-P_2 in being resistant to cleavage by aldolase. It was also relatively unstable and, in particular, much more labile to alkali than xylulose-P_2. This instability, after removal from the active site, has so far frustrated our attempts to identify this second inhibitor conclusively. However, we suspect that the second inhibitor could be one or both of the epimers at C2 of D-pent-3-ulose-1,5-bisphosphate. These compounds would also be formed by misprotonation of the enediol intermediate but at C2, rather than at C3 (Fig. 4.2). Reduction of the arabino epimer, D-3-keto-arabinitol-1,5-bisphosphate, followed by dephosphorylation, would yield only arabinitol because of the symmetry of this molecule. This could have contributed to the tritium-labelled arabinitol observed on reduction with borotritide. Recently, Zhu and Jensen (1991*a*,*b*) have reported further evidence for the production of this compound by Rubisco during catalysis.

One would expect that these misprotonations should be reversible, so these compounds should serve as substrates for the carboxylase as well as being products. However, the rate at which they are consumed must be less than the rate at which they are produced from ribulose-P_2, otherwise they would not accumulate during ribulose-P_2 consumption. This has also been confirmed recently, at least in the case of xylulose-P_2 (Yokota 1991).

These misprotonation reactions are not trivial quantitatively. Spinach Rubisco produces one molecule of xylulose-P_2 for approximately every 400 molecules of ribulose-P_2 consumed (Fig. 4.2) (Edmondson *et al.* 1990*c*). How does Rubisco cope with this kind of profligacy *in vivo*? Is it always strongly inhibited by these isomerization products? Work on Rubisco activase has shown that one of the roles of this mysterious protein is to relieve this kind of inhibition. Apparently, by some mechanism which remains to be determined, Rubisco activase facilitates the release of these inhibitors from the active site (Robinson and Portis 1989). Furthermore, in the case of xylulose-P_2, there appears to be a specific phosphatase to convert it to xylulose-5-phosphate for return to photosynthetic metabolism. (Portis, personal communication). As with the oxygenase reaction, Rubisco's ill-disciplined misprotonations also necessitate quite a complex mechanism, involving at least two other proteins, for disposing of the waste products.

β elimination of the five-carbon enediol?

It is possible that the enediol intermediate, when bound at the active site, might also abort by β elimination of the phosphate moiety attached to C1 to produce the diketone compound, 1-deoxy-D-glycero-2,3-pentodiulose-5-phosphate (Paech *et al.* 1978). Certainly, this fate swiftly befalls the enediol intermediate if it is removed from the active site (Jaworowski *et al.* 1984). At present there is no convincing evidence for such a reaction at the active site. Borotritide reduction of the inhibitor(s) after removal from the enzyme revealed little sign of a labelled monophosphate (Edmondson *et al.* 1990*c*). However, more direct experiments aimed at revealing such a β elimination reaction of this intermediate at the active site are clearly warranted.

β elimination of the three-carbon aci-*carbanion*

We recently discovered that Rubisco produced another ^{14}C-labelled product, in addition to P-glycerate, from $^{14}CO_2$ (Andrews and Kane 1991). When the products were resolved on a column which separates organic acids, a small peak of [^{14}C]pyruvic acid was observed (Fig. 4.6(b)). Its identity was confirmed by show-

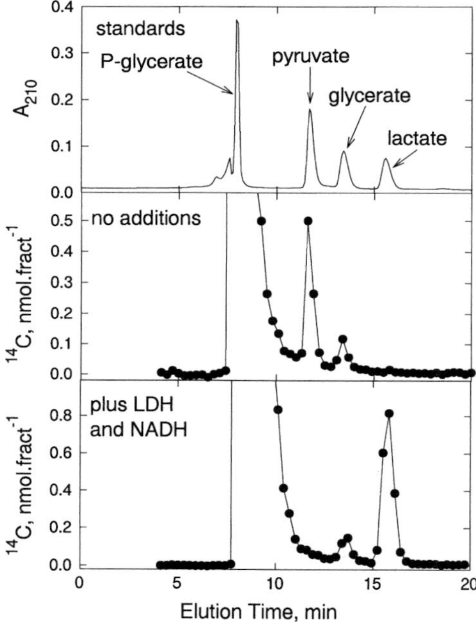

Fig. 4.6 Chromatographic separation on a 0.78 × 30 cm Bio-Rad Aminex HPX-87H column of the ^{14}C-labelled products of the ribulose-P_2 carboxylase reaction carried out in the presence of $^{14}CO_2$. The mobile phase was 0.013 N H_2SO_4 and the flowrate was 0.6 ml min^{-1}. Spinach Rubisco (0.16 mg) was preactivated for 10 min at 25 °C in a 1 ml solution containing 60 mM Tris-HCl buffer, pH 8.0, 15 mM $MgCl_2$, 10 mM $NaH^{14}CO_3$ and, where indicated, 0.15 mM NADH and 3 units of rabbit muscle lactate dehydrogenase. Ribulose-P_2 was then added to 2.2 mM and, after a further 60 min during which all of the ribulose-P_2 was consumed, cations were removed with cation exchange resin (H^+ form) and the solution was dried, dissolved in the mobile phase and applied to the column. From Andrews and Kane (1991) with permission.

ing that it was converted to lactic acid by lactate dehydrogenase and NADH- (Fig. 4.6(c)). A very small peak which cochromatographed with glyceric acid was also seen. It presumably resulted from the major product, P-glycerate, by virtue of a trace of contaminating phosphatase activity. Continuous spectrophotometric measurement of pyruvate production showed that it continued until the ribulose-P_2 was exhausted and that it was completely and quickly inhibited by the analogue of the β-keto acid intermediate, 2′-carboxyarabinitol-1,5-bisphosphate (Fig. 4.7). Furthermore, no pyruvate was produced from the major product, P-glycerate, which excluded the possibility that it arose as a result of contaminating enzymes. A side reaction of the Rubisco carboxylation mechanism must be responsible.

The fraction of carboxylation product partitioned towards pyruvate was found to be extraordinarily constant, remaining at approximately 0.7 per cent in a wide variety of conditions (Table 4.2). It was not affected by varying the pH

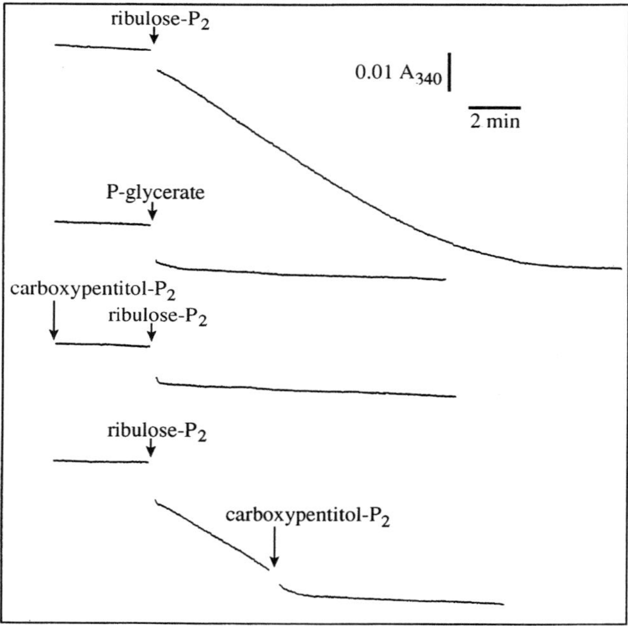

Fig. 4.7 Spectrophotometric measurement of pyruvate produced during ribulose-P_2 carboxylation. Absorbance was monitored at 340 nm. Spinach Rubisco (67 μg ml^{-1}) was preactivated in a solution containing 80 mM Tris-HCl buffer, pH 8.0, 17 mM $MgCl_2$, 10 mM $NaHCO_3$, 50 μM NADH, and 3 unit ml^{-1} of rabbit muscle lactate dehydrogenase. Ribulose-P_2 (1.1 mM), P-glycerate (2 mM) and 2′-carboxypentitol-1,5-bisphosphate (a mixture of 2′-carboxyarabinitol-1,5-bisphosphate and 2′-carboxyribitol-1,5-bisphosphate, 40 μM) were added where indicated. From Andrews and Kane (1991) with permission.

(between 6.4 and 9.1) or the CO_2 concentration or by substituting manganese ions for magnesium ions. It was the same for Rubiscos from a bacterium, a cyanobacterium, and a higher plant. However, substituting ^2H_2O as the solvent stimulated the ratio approximately threefold (Table 4.2). The ratio also varied with temperature, increasing approximately twofold between 10 and 40 °C. A plot of the logarithm of the ratio *versus* reciprocal absolute temperate was linear and its slope indicated that the activation energy for pyruvate production was 4.8 kcal mol^{-1} greater than for P-glycerate production.

Since the pyruvate product was labelled with ^{14}C, it must originate from an intermediate occurring after the carboxylation step and the obvious candidate is the three-carbon *aci*-carbanion. β elimination of the phosphate moiety would produce *enol*-pyruvate and, ultimately, pyruvate (Fig. 4.2). This provides a ready explanation for the stimulatory effect of solvent ^2H_2O on the partitioning towards pyruvate. The reduced activity of the deuteron, compared to the proton, would slow the final protonation step that produces P-glycerate, thus providing the *aci*-carbanion with more opportunity to β eliminate.

Table 4.2 Ratio of pyruvate produced to ribulose-P_2 consumed by various Rubiscos under various conditions

Enzyme source	Solvent	Metal	pH	$[HCO_3^-]$ (mM)	Pyruvate formed (per cent of ribulose-P_2 consumed)
Spinach	H_2O	Mg^{2+}	6.4	10	0.62
Spinach	H_2O	Mg^{2+}	6.9	10	0.66
Spinach	H_2O	Mg^{2+}	7.3	5	0.72
Spinach	H_2O	Mg^{2+}	8.0	10	0.68
Spinach	H_2O	Mg^{2+}	8.3	11	0.67
Spinach	H_2O	Mg^{2+}	8.6	50	0.66
Spinach	H_2O	Mg^{2+}	9.1	50	0.76
Spinach	H_2O	Mg^{2+}	8.3	10	0.64
Spinach	H_2O	Mg^{2+}	8.3	1.0	0.64
Spinach	H_2O	Mg^{2+}	8.3	0.5	0.63
Spinach	H_2O	Mg^{2+}	7.8	10	0.70
Spinach	H_2O	Mn^{2+}	7.8	10	0.76
Spinach	98.5%^2H_2O	Mg^{2+}	8.3*	9	2.2
Rhodospirillum	H_2O	Mg^{2+}	7.8	46	0.65
Synechococcus	H_2O	Mg^{2+}	7.8	69	0.71

The extent of pyruvate formation accompanying complete consumption of the supplied ribulose-P_2 was measured spectrophotometrically as shown in Fig. 4.7.

*Uncorrected pH meter reading.

From Andrews and Kane (1991) with permission.

A stereo-electronic difficulty

All of these side reactions stem from Rubisco's inability to restrain its two nucleophilic intermediates entirely to the productive catalytic pathway. It would be instructive to compare the magnitude of the abortive reactions of these intermediates with abortive reactions of intermediates of other enzymatic reactions. Such data are scarce but it is known that triosephosphate isomerase employs an enediol intermediate that is susceptible to a β-elimination reaction which produces inorganic phosphate and methyl glyoxal (Campbell *et al.* 1979; Iyengar and Rose 1981*a*,*b*). However, this β elimination occurs very infrequently at the active site of the isomerase. Calculations based on the data of Iyengar and Rose (1981*a*) indicate that it occurs less than once in every million turnovers of the enzyme at pH 7.5. Therefore, the isomerase is at least four orders of magnitude tidier, in this respect, than Rubisco.

Although pyruvate is not toxic and its disposal does not present a metabolic problem, its production by Rubisco nevertheless represents a finite wastage of energy. The energetic equivalent of one ATP molecule is wasted for every pyruvate molecule produced by β elimination. Therefore, one suspects that mechanistic difficulties may have retarded Rubisco's evolutionary progress in suppressing this reaction. A stereochemical consideration of the carboxylation reaction provides a clue about one such difficulty. Structural information about the active site of spinach Rubisco complexed to the analogue of the β-keto acid intermediate (Knight *et al.* 1990) has been useful in this context. The analogue is bound with the phosphate ester at C1 (P1) in the extended configuration shown in Fig. 4.8(c). When we use that configuration as a basis for inferring the configuration of the preceding five-carbon enediol intermediate, by assuming minimum movement of the atoms involved in the chemistry, we arrive at a structure (shown in Fig. 4.8(b)) which has the bond between C1 and the bridge O atom of the P1 phosphate ester in the same plane as the unsaturated bond joining C3 and C2. This is the configuration where β elimination would be most strongly disfavoured, because it minimizes orbital overlap between the C1–0 bond and the π system (Rose 1981). But when we attempt to predict the configuration of the *aci*-carbanion intermediate later in the catalytic sequence using similar assumptions about minimum movement, we arrive at an unacceptable result. If the C1–bridge O bond was to retain something like its angle in the β-keto acid, it would find itself at right angles to the unsaturation which is now between C2 and the carboxylate carbon. This would be the worst possible angle because it would facilitate β elimination by maximizing orbital overlap with the π system. So, to avoid β elimination, the C1–bridge O bond must move through an angle of 90° to adopt the position shown in Fig. 4.8(d). Either the orientation of P1 must change, or C1 must move, or both. This could be achieved either by movement of the P1 phosphate group, as shown in Fig. 4.8, or by rotation of C2 and its attached hydroxyl and carboxyl groups. Either way, considerable movement

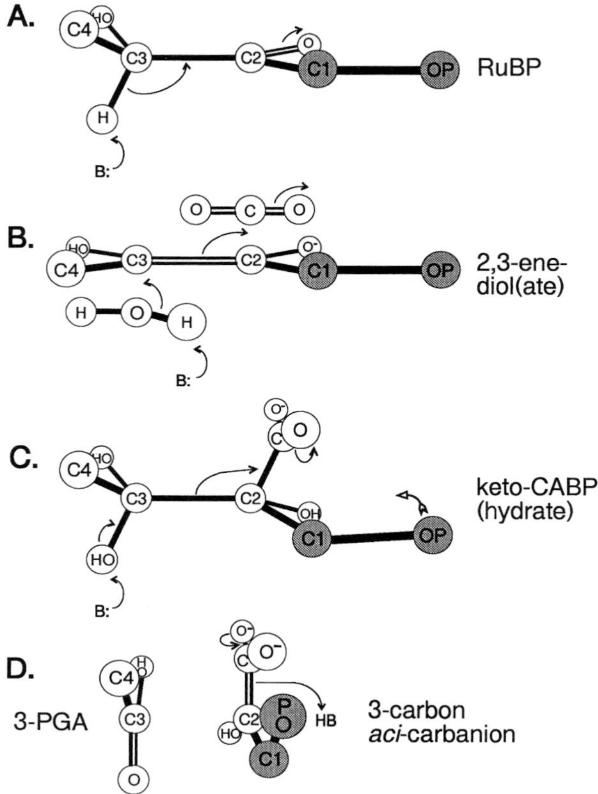

Fig. 4.8 A possible stereochemical course for Rubisco's carboxylation reaction. A concerted attack of CO_2 and H_2O on the five-carbon enediol is shown. However, sequential attack of CO_2 followed by H_2O, as shown in Fig. 4.2, is also possible. RuBP, ribulose-P_2; keto-CABP, 2′-carboxy-3-keto-arabinitol-1,5-bisphosphate; 3-PGA, P-glycerate.

must occur within the active site during each and every catalytic cycle. Perhaps this movement simply cannot occur fast enough to suppress β elimination completely. Or perhaps the structural features within the active site which are required to stabilize the two different nucleophilic intermediates differ so much that only an uneasy compromise between them has been found at the present stage of evolution of the enzyme.

Directed mutagenesis within a phosphate-binding site

We consider that these concepts could be tested using site-directed mutagenesis. For instance, changes which impair the ability of the active site to control the movement of the P1 phosphate group ought to influence the relative rates of these side reactions. In the case of triosephosphate isomerase, deletion of the phosphate-binding helix greatly increased the tendency towards β elimination (Pompliano *et al.* 1990). In the Rubisco large subunit, a small helix in loop 8 of the α/β barrel domain is very similar in sequence and structure to the triosephosphate isomerase phosphate-binding helix and it provides part of the binding site of the P1 phosphate of the analogue of the β-keto acid intermediate (Knight *et al.* 1990). However, the hydroxyl oxygen atom of a threonine residue, T65, in an absolutely conserved loop of the aminoterminal domain which folds over the active site on the α/β barrel domain of its companion large subunit in the large-subunit dimer, also makes particularly obvious hydrogen-bonded contacts with two of the oxygen atoms of the P1 phosphate (Fig. 4.9). It also has a similar contact with the ϵ-amino nitrogen atom of the catalytically essential lysine, K334.

To determine whether the interactions with T65 are important in binding and orientating the P1 phosphate, we mutated this residue to valine in the Rubisco from the cyanobacterium, *Synechococcus* PCC 6301, expressed in *E. coli*. This removed the hydrogen-bonding capacity of the side chain of residue 65 with

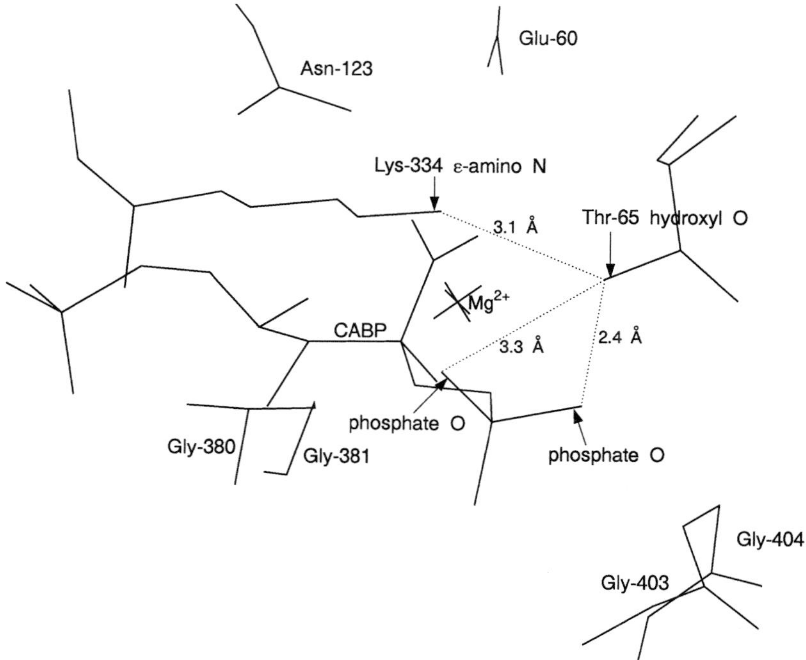

Fig. 4.9 A view of the active site of spinach Rubisco with 2′-carboxyarabinitol-1, 5-bisphosphate (CABP), an analogue of the β-keto acid intermediate, bound. The hydroxyl oxygen atom of threonine-65 and the distances to neighbouring atoms in hydrogen-bonding distance are indicated.

minimal alteration to its bulk. The T65V mutant was expressed abundantly by *E. coli* and was readily purified. It bound the reaction-intermediate analogue quite tightly and the quaternary enzyme-CO_2–Mg^{2+}-analogue complex could be isolated by gel filtration without difficulty. The mutant enzyme had a full complement of small subunits and additional small subunits did not increase its activity. However, the k_{cat} of the mutant enzyme was only approximately 1 per cent of that of the wild type and the substrate affinities were weakened, particularly for ribulose-P_2 (Table 4.3). This is, perhaps, the expected consequence of interference with the binding of the phosphate group. However, the partitioning of product towards pyruvate was not increased. Rather it was reduced approximately five fold. That is, contrary to expectations, the resistance of the *aci*-carbanion to β elimination was substantially improved. So our speculations about the role of the T65 hydrogen bonds in binding and orientating the P1 phosphate group are clearly wrong. Furthermore, the CO_2/O_2 relative specificity was reduced about 2.5 fold, compared to the wild-type enzyme, indicating an entirely different kind of impact of the mutation of the reaction mechanism (Table 4.3).

Perhaps the hydrogen bonds from T65 to the phosphate oxygen atoms are not necessary to correctly position the phosphate moiety. Rather, the data appear to indicate that the reverse may be true, i.e. the phosphate moiety is firmly held in position by the phosphate-binding helix in loop 8 (G403 and G404 in Fig. 4.9) and other interactions with loop 7 (G380 and G381 in Fig. 4.9) and the hydrogen bonds to T65 serve to position the hydroxyl oxygen atom of T65 correctly. The T65 hydroxyl oxygen atom may act as a bridge which fixes, through hydrogen-bonded interactions, the correct position of the ϵ-amino nitrogen atom of K334 with respect to the P1 phosphate group and the carboxyl group of the intermediate. Alteration of the orientation of K334 could explain the large reduction in k_{cat} as well as the reduction in CO_2/O_2 specificity. The effects on k_{cat} and CO_2/O_2 specificity of mutations in loop 6 of the α/β barrel domain have previously been interpreted in terms of alterations in the orientation of K334 (Chen *et al.* 1991).

Table 4.3 Properties of variants of *Synechococcus* Rubisco

	L_8S_8 (wild type)	L_8S_8 (T65V)	L_8 (wild type)
$k_{cat}(s^{-1})$	11.6	0.091	0.12*
$K_m(CO_2)$ (μM)	284	633	480*
K_m(ribulose-P_2) (μM)	21	174	200*
Pyruvate/ribulose-P_2†(%)	0.71	0.14	0.75
CO_2/O_2 specificity	43	18.5	46‡

* Data from Andrews (1988).

† Pyruvate produced as a percentage of ribulose-P_2 consumed.

‡ Data from Gutteridge (1991) recalculated to compensate for differences in assumptions about the pK' of the CO_2/HCO_3^- equilibrium by normalizing the data for the wild-type L_8S_8 enzyme.

The hydrogen bonding of T65 might also be involved in holding the whole aminoterminal domain in correct juxtaposition with the rest of the active site.

Effect of removal of the small subunits

There are some similarities between the properties of the T65V mutant of the *Synechococcus* holoenzyme and the wild-type L_8 core, devoid of small subunits. Both are approximately 1 per cent active and both have weakened substrate affinities compared to the wild-type holoenzyme (Table 4.3). Schneider *et al.* (1990) suggested that interactions of the small subunit with helix 8 of the α/β barrel domain of the large subunit might influence the orientation of the P1 phosphate-binding site in loop 8. If so, we might expect differences in the *aci*-carbanion's tendency to β eliminate when the small subunits are absent. However, measurements of the pyruvate-producing side reaction catalysed by the L_8 core of *Synechococcus* Rubisco (which had to surmount difficulties associated with the diminished activity and limited solubility of the L_8 core) revealed that the partitioning towards pyruvate was indifferent to the presence or absence of the small subunits (Table 4.3). Apparently, even in the absence of small subunits, the active site's ability to control the movements of the P1 phosphate group during catalysis is not impaired. However, when we look at the CO_2/O_2 relative specificity, the correlation between T65V and L_8 breaks down. Gutteridge (1991) measured the relative specificity of L_8 (also a technically demanding task) and found it unchanged compared to the holoenzyme (Table 4.3). Therefore, unlike the situation with the residue 65 mutation, removal of the small subunits appears not to alter the orientation of K334.

Conclusion

It is unlikely that the complete list of Rubisco's abortive side reactions is yet known and some of the reactions which have been discovered so far are yet to be fully characterized. Further studies of such reactions will undoubtedly have more to contribute to our understanding of the carboxylation mechanism.

Acknowledgement

The authors are grateful to C.-I. Brändén and I. Andersson for supplying them with the structural coordinates for spinach Rubisco.

References

Andrews, T.J. (1988). *J. Biol. Chem.*, **263**, 12 213–20.
Andrews, T.J. and Kane, H.J. (1991). *J. Biol. Chem.*, **266**, 9447–52.

Andrews, T.J. and Lorimer, G.H. (1985). *J. Biol. Chem.*, **260**, 4632–6.

Andrews, T.J. and Lorimer, G.H. (1987). In *The biochemistry of plants: a comprehensive treatise*. Vol. 10. *Photosynthesis* (ed. M.D. Hatch and N.K. Boardman), pp. 131–218. Academic Press, New York.

Andrews, T.J., Lorimer, G.H., and Tolbert, N.E. (1973). *Biochemistry*, **12**, 11–18.

Artus, N.N., Somerville, S.C., and Somerville, C.R. (1986). *CRC Crit. Rev. Plant Sci.*, **4**, 121–47.

Badger, M.R. and Andrews, T.J. (1987). In *Progress in photosynthesis research*. Vol. III (ed. J. Biggins), pp. 601–9. Martinus Nijhoff, Dordrecht.

Bowes, G., Ogren, W.L., and Hageman, R.H. (1971). *Biochem. Biophys. Res. Commun.*, **45**, 716–22.

Campbell, I.D., Jones, R.B., Kiener, P.A., and Waley, S.G. (1979). *Biochem. J.*, **179**, 607–21.

Chen, Z., Yu, W., Lee, J.-H., Diao, R., and Spreitzer, R.J. (1991). *Biochemistry*, **30**, 8846–50.

Cleland, W.W. (1990). *Biochemistry*, **29**, 3194–7.

Edmondson, D.L., Badger, M.R., and Andrews, T.J. (1990*a*). *Plant Physiol.*, **93**, 1376–82.

Edmondson, D.L., Badger, M.R., and Andrews, T.J. (1990*b*). *Plant Physiol.*, **93**, 1390–7.

Edmondson, D.L., Kane, H.J., and Andrews, T.J. (1990*c*). *FEBS Lett.*, **260**, 62–6.

Ehleringer, J.R., Sage, R.F., Flanagan, L.B., and Pearcy, R.W. (1991). *Trends Ecol. Evol.*, **6**, 95–9.

Fersht, A.R. (1985). *Enzyme structure and mechanism* (2nd edn). W.H. Freeman, New York.

Fersht, A.R., Shi, J.-P., Knill-Jones, J., Lowe, D.M., Wilkinson, A.J., Blow, D.M., *et al.* (1985). *Nature*, **314**, 235–8.

Gutteridge, S. (1991). *J. Biol. Chem.*, **266**, 7359–62.

Hurwitz, J., Jakoby, W.B., and Horecker, B.L. (1956). *Biochim. Biophys. Acta*, **22**, 194–5.

Iyengar, R. and Rose, I.A. (1981a). *Biochemistry,* **20**, 1223–9.

Iyengar, R. and Rose, I.A. (1981b). *Biochemistry*, **20**, 1229–35.

Jaworowski, A., Hartman, F.C., and Rose. I.A. (1984). *J. Biol. Chem.*, **259**, 6783–9.

Jordan, D.B. and Ogren, W.L. (1981). *Nature*, **291**, 513–15.

Knight, S., Andersson, I., and Brändén, C.-I. (1990). *J. Mol. Biol.*, **215**, 113–60.

Lorimer, G.H. (1979). *J. Biol. Chem.*, **254**, 5599–601.

Paech, C., Pierce, J., McCurry, S.D., and Tolbert, N.E. (1978). *Biochem. Biophys. Res. Comm.*, **83**, 1084–92.

Pierce, J., Andrews, T.J., and Lorimer, G.H. (1986). *J. Biol. Chem.*, **261**, 10 248–56.

Pompliano, D.L., Peyman, A., and Knowles, J.R. (1990). *Biochemistry*, **29**, 3186–94.

Putman, S.J., Coulson, A.F.W., Farley, I.T.R., Riddleston, B., and Knowles, J.R. (1972). *Biochem. J.*, **129**, 301–10.

Robinson, S.P. and Portis, A.R., Jr. (1989). *Plant Physiol.*, **90**, 968–71.

Rose, I.A. (1981). *Phil. Trans. R. Soc. Lond. Series B*, **293**, 131–43.

Saver, B.G. and Knowles, J.R. (1982). *Biochemistry*, **21**, 5398–403.

Schneider, G., Knight, S., Andersson, I., Brändén, C.-I., Lindqvist, Y., and Lundqvist, T. (1990). *EMBO J.*, **9**, 2045–50.

Yokota, A. (1991). *Plant Cell Physiol.*, **32**, 755–62.

Zhu, G. and Jensen, R.G. (1991*a*). *Plant Physiol.*, **97**, 1348–53.

Zhu, G. and Jensen, R.G. (1991*b*). *Plant Physiol.*, **97**, 1354–8.

5

Tests and models of mechanisms for biotin-dependent carbon dioxide fixation

Ronald Kluger and Belinda Tsao

Biotin and carbon dioxide utilization

Carbon dioxide is the primary by-product of biological oxidative metabolism. This product is part of the biosphere and is dealt with by living systems in a number of ways, all of which provide useful materials for the organisms involved. Plants provide a general bank for carbon dioxide acquisition and a reductive photosynthetic apparatus to convert carbon dioxide into carbohydrates with oxygen as a by-product. There also exist nonreductive biological processes in which carbon dioxide is formally extracted from bicarbonate, at the expense of ATP hydrolysis, and transferred to a substrate where it replaces a proton. Since carbon dioxide is easily transferred between species in the air or in aqueous solution (as bicarbonate), there is ample opportunity for carbon dioxide which is given off by one species to be utilized by another. Enzymes which promote this class of reaction require special chemical assistance. This assistance is provided by the coenzyme biotin and the mechanism by which this assistance is administered is a fascinating problem in enzymic reaction mechanisms. While many of the characteristics of enzymic reactions involving biotin are understood, the detailed pathway of the reaction is inferred rather than established and a number of ambiguities remain to be resolved. This chapter will review the status of some research on the mechanism by which carbon dioxide from dissolved bicarbonate is transferred to biotin.

Biotin fulfils two roles in the utilization of carbon dioxide. The first is in the removal of bicarbonate from solution and the safe storage of the activated carbon dioxide equivalent. The second is the transfer of the carbon dioxide equivalent to an acceptor substrate in place of one of its protons.

$$O{=}C{=}O \longleftrightarrow O{=}C^{+}{-}O^{-}$$

$$^{-}O{-}C({=}O){-}OH \longrightarrow O{=}C^{+}{-}O^{-} \longleftrightarrow O{=}C{=}O$$

$$OH^{-} + ATP \longrightarrow ADP + P_i$$

We can visualize the reaction in terms of the removal of an equivalent of hydroxide from bicarbonate and its transfer to ATP accompanied by hydrolysis (Scheme 5.1). The formal mechanism must have an equivalent in reality and it is a challenge to unravel how this sequence can occur and then to develop efficient chemical systems which can accomplish the same task.

Mechanisms for carboxylation of biotin

Labelling and exchange studies have limited choices of the possible mechanisms by which the steps can occur. An oxygen atom from bicarbonate is transferred to the inorganic phosphate produced from ATP (Kaziro *et al.* 1962). Enzymes that promote biotin-dependent carboxylations do not catalyse partial exchange reactions. Thus, a mechanism in which ATP, biotin, and bicarbonate combine in a concerted process is in accord with these properties but the transition state for such a process is chemically unreasonable (Wood 1976; Dugas and Penney 1981; Knowles 1989).

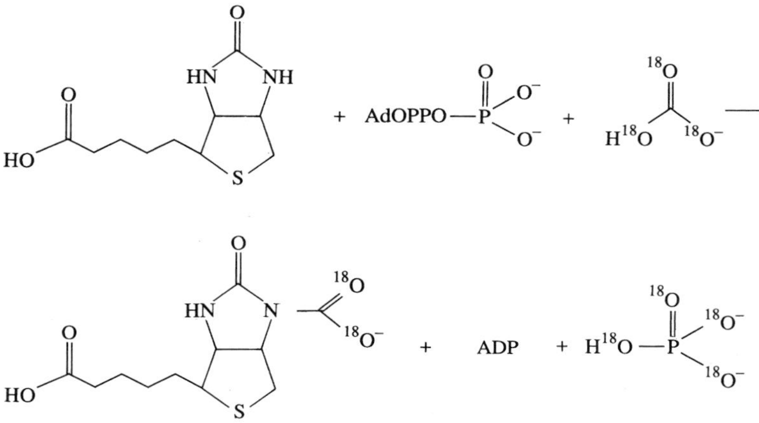

Scheme 5.1 The transfer of carbon dioxide from dissolved bicarbonate requires concomitant hydrolysis of ATP. Labelling indicates that the oxygen removed by dehydration of bicarbonate is transferred to phosphate.

Pairwise mechanisms

Alternatively, it can be proposed that intermediates form but are not accessible for exchange with external species. A widely accepted mechanism involves an initial reaction between ATP and bicarbonate to form carboxyphosphate, followed by transfer of the carboxyl group to biotin (Wood 1976; Knowles 1989). This is consistent with available data from stereochemical studies in which it was shown that substitution at the γ phosphorus of ATP occurs with inversion

(Hansen and Knowles 1985), extrapolations from related chemical reactions (Sauers *et al.* 1975), and enzymatic reactions with alternative substrates (Polakis *et al.* 1972; Guchait *et al.* 1974; Ogita and Knowles 1988). However, other pairwise combinations of the three reactants could produce carboxybiotin by what may well be reasonable mechanisms that are also consistent with the data (Kluger *et al.* 1979; Blonski *et al.* 1984; Kluger 1989). We have done a number of studies of chemical reactions related to two of the less popular pairwise combination mechanisms and those results will be reviewed in this section (Kluger and Taylor 1991; Tsao 1990).

Mechanisms involving phosphorylated biotin

The interaction of ATP and biotin to produce an activated derivative which might more readily react with bicarbonate was first proposed by Calvin and Pon (1959), although no specific mechanism was put forward. Later, the idea of a phosphorylated biotin intermediate and a concerted cyclic mechanism was proposed (Knappe 1970). However, the phosphorylation of biotin was an unknown process and no chemical reaction that parallels such a process was available for analysis. To test the reasonableness of such a route involving a phosphorylated biotin intermediate, we must be able to suggest how biotin is likely to become phosphorylated by ATP and how the phosphorylated product would be a productive intermediate on the pathway of formation of carboxybiotin. What advantages and disadvantages would such a route have in terms of efficiency of catalysis? Would such a route accommodate all known experimental observations?

Enzymes can promote reactions by binding substrates in proximity and thus reducing the large entropic barrier to reaction (Page and Jencks 1971). Since reactions in solution normally have such an entropic barrier, a bimolecular process may often appear to be unlikely while on an enzyme, with reactants in proximity, reaction will occur without a significant barrier. Thus, in order to test the validity of a proposed mechanism of reaction on an enzyme, we use an intramolecular reaction of the groups of interest since this also has a greatly reduced entropic barrier. Since the oxygen of the urea group of biotin is its most nucleophilic site (Bruice and Hegarty 1970), would phosphorylation of this site lead to carboxybiotin by a reasonable route? Is a urea reactive as a nucleophile toward an activated phosphate such as ATP? Thus, we synthesized a diethyl phosphonate adduct of *N,N′*-dimethyl urea. The hydrolytic reaction of the diethyl ester is accelerated in acidic solutions and the reaction was shown to involve nucleophilic reaction of the urea oxygen (Kluger *et al.* 1979). Since the model substance has relatively poor leaving groups, acid catalysis was required for the reaction. This prevented the true magnitude of the participation reaction to be observed since the reaction was limited by cleavage of the leaving groups. In the enzymic reaction, where ADP would be the leaving group, no such assistance by acid would be needed (a metal ion would fulfil that function in neutral solutions).

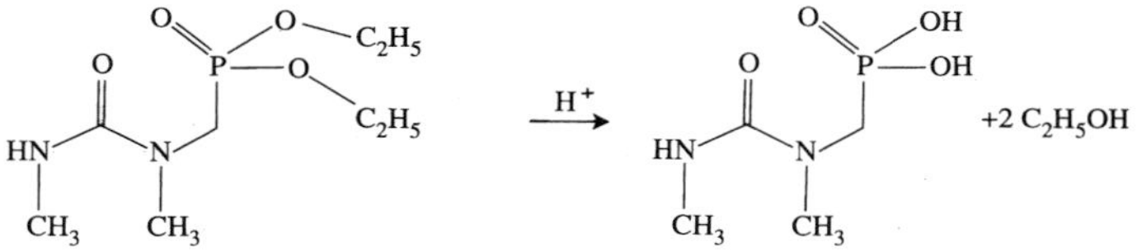

A more relevant model would not require catalysis by strong acids and would be assisted by buffers in neutral solution.

Model for N-phosphorylation in neutral solution

An improved model was prepared in which better leaving groups were esterified to phosphorus (Tsao 1990). Further, the imidazolidinone ring of biotin was modelled by imidazolidinone itself, rather than the acyclic species. Since a planar conformation of the urea group is an inherent part of biotin's structure, the ring assures that this aspect of the structure is closely followed. The better leaving group simulates the interaction of biotin with an activated phosphate derivative, such as ATP. We have found in this system, the reaction in which participation occurs does not require acid catalysis. The reactant, diphenyl (imidazolidinone) methylphosphonate, **1**, undergoes rapid hydrolysis in neutral and basic solutions, indicating that the reaction involves participation of the urea group in the hydrolysis. A mechanism is shown in Scheme 5.2. This process is similar to that proposed for activation of biotin by ATP.

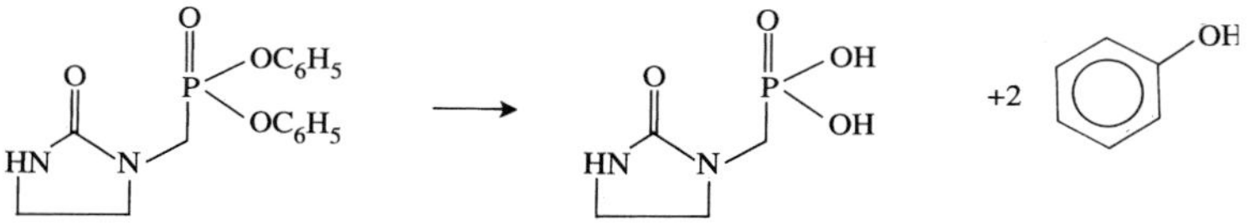

Scheme 5.2 The hydrolysis occurs rapidly due to participation of the ureido group.

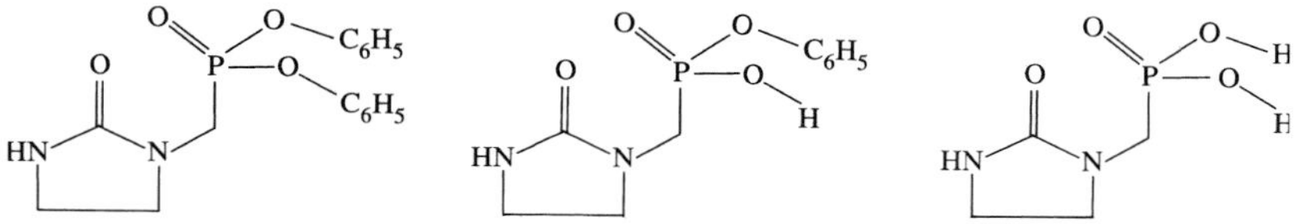

The hydrolysis of **1** was readily followed at 265 nm where the reactant (1), monoester (2), and diacid (3) have different extinction coefficients.

Acid and base catalysis in a model for N-phosphorylation

The observed rate constants increase linearly with buffer concentration, permitting a pH rate profile to be obtained for the reaction at zero buffer and for buffer catalysis rate constants to be determined. The rate decreases below pH 3, is

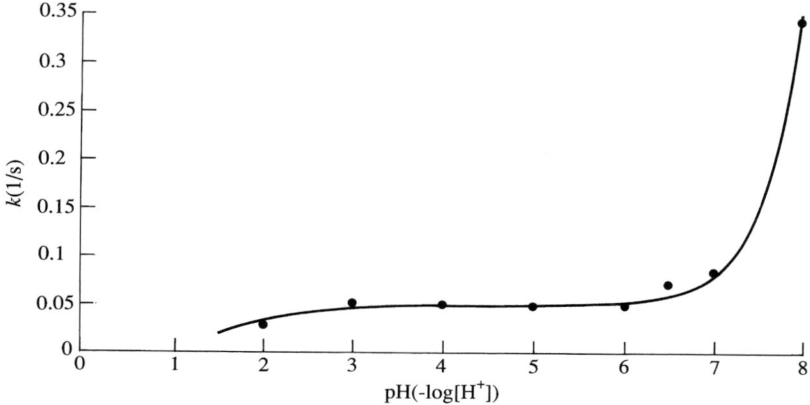

Fig. 5.1. Observed first-order rate constant for hydrolysis of **1** at 40 °C.

independent of pH between pH 3 and pH 6 and increases above pH 6. The data are summarized in Fig. 5.1 and plotted according to the equation

$$k_{obs} = k_o\ ([A]/[A] + [HA^+]) + k_B\ [OH^-]$$

where $k_o = 0.47\ s^{-1}$, $k_B = 3.0 \times 10^5\ M^{-1}\ s^{-1}$, $K_A = 1.7$ M.

The reaction is general-base-catalysed with a Brönsted β slope of 0.5. A mechanism involving general-base catalysis is shown in Scheme 5.3. The uncatalysed reaction would have water in place of 'B' and the specific-base-catalysed reaction would replace 'B' with 'OH^-'.

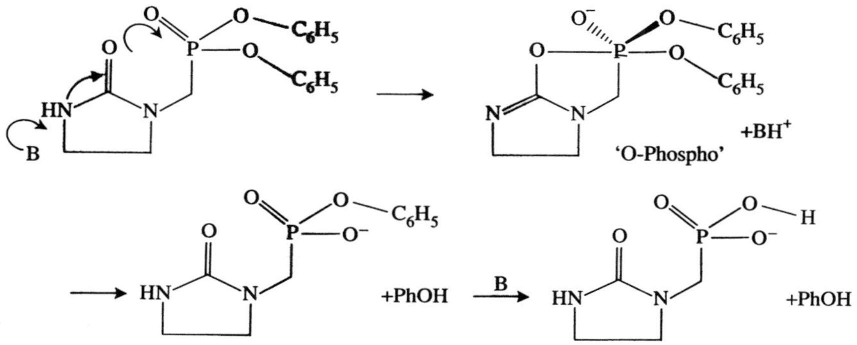

Scheme 5.3 The rapid general-base-catalysed hydrolysis of **1** is due to nucleophilic catalysis. The implicates the covalent intermediate '*O*-Phospho'.

The rate of hydrolysis of **1** from pH 2 to pH 6 is 10^4 times that of diphenyl methylphosphonate which hydrolyses without neighbouring group participation ($k = 3 \times 10^{-5}\ s^{-1}$). Such a large rate enhancement requires nucleophilic catalysis (we have determined that the inductive effect of the urea group is very small using linear free energy relationships).

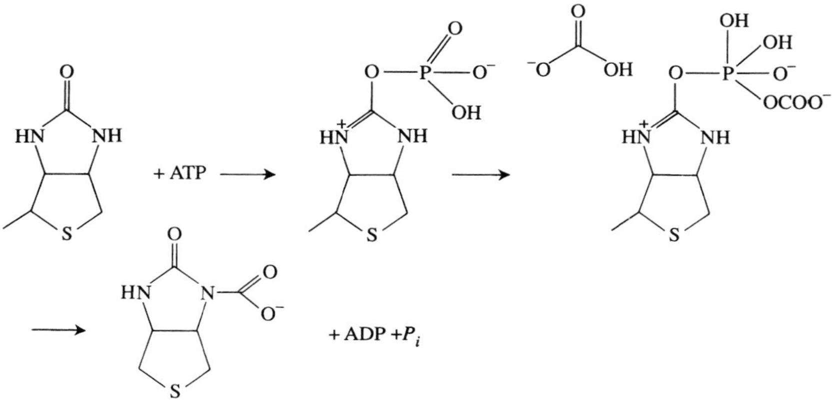

Scheme 5.4 Mechanism for carboxylation of biotin with hydrolysis of ATP and reaction of bicarbonate via *O*-phosphorylated intermediate.

Consequences of biotin phosphorylation mechanisms

A reaction in which biotin first reacts with ATP to become phosphorylated on oxygen followed by reaction with bicarbonate is shown in Scheme 5.4. The initial attack of biotin upon ATP proceeds with inversion at phosphorus (Thatcher and Kluger 1989) to form O-phosphobiotin. Attack of bicarbonate must place the carboxyl group adjacent to the urea nitrogen of biotin which becomes carboxylated (*N*1′). In order for this stereochemistry to function, the substitution at phosphorus must occur with retention of configuration since attack by bicarbonate is on the face of the phosphate group adjacent to the urea of biotin. This requires pseudorotation of the five-coordinate phosphorus intermediate which forms on addition. After addition, the bond to the urea can break and give an activated carbonate group which can readily transfer carbon dioxide to the urea nitrogen. The overall stereochemistry at phosphorus is inversion, which is what was observed by Hansen and Knowles (1985).

A route through O-phosphobiotin was considered as a possibility by Hansen and Knowles in their stereochemical study (1985). Although their results were consistent with the mechanism as it was presented here and in our earlier paper (Kluger *et al.* 1979), these authors and Knowles in his recent review (1989) dismiss mechanisms involving phosphorylation of biotin. They write the reaction as the concerted process proposed by Knappe (1970). It is ruled out because they consider a concerted process unlikely. However, there is no reason why the reaction should not be stepwise, involving one or more intermediates. Therefore the grounds for dismissing O-phosphorylation mechanisms in general are not solid. It would be a simple matter to write the stepwise mechanism and remove the objection. We have shown that other evidence cited by Knowles to distinguish between this and a mechanism involving reaction of bicarbonate with ATP prior to reaction with biotin is not conclusive (Kluger 1990).

A mechanism involving initial reaction of biotin with bicarbonate and ATP

The direct reaction of biotin with bicarbonate can also lead to carboxylation, with ATP serving as a dehydrating agent. The electrophilic reactivity of bicarbonate should not be significantly different from that of carboxyphosphate. The advantage of a reaction occurring via carboxyphosphate is that this molecule possesses a good leaving group. However, phosphorylation of the adduct of biotin with bicarbonate can be concerted with its formation or proceed in a step after initial addition. Scheme 5.5 shows the stepwise variant of this mechanism.

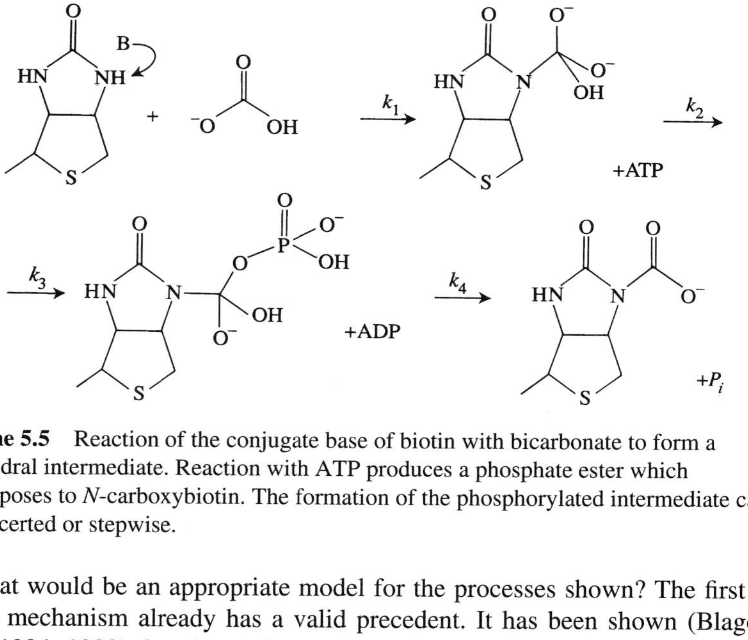

Scheme 5.5 Reaction of the conjugate base of biotin with bicarbonate to form a tetrahedral intermediate. Reaction with ATP produces a phosphate ester which decomposes to *N*-carboxybiotin. The formation of the phosphorylated intermediate can be concerted or stepwise.

What would be an appropriate model for the processes shown? The first step in the mechanism already has a valid precedent. It has been shown (Blagoeva *et al.* 1984, 1989) that the conjugate base of a urea will add to an adjacent carboxylate, a reaction which parallels the addition of the conjugate base of biotin to bicarbonate. Studies of proton exchange reactions show that the conjugate base of biotin can be expected to be a kinetically feasible intermediate in an enzymic reaction (Perrin and Dwyer 1987). Therefore, the initial pairwise reaction of biotin and bicarbonate to form a covalent adduct is chemically reasonable. The formation of carboxybiotin by this mechanism requires the adduct to react with ATP to form a phosphorylated tetrahedral intermediate and decompose by expulsion of the phosphate to regenerate the carbonyl group of the carboxy substituent. A model must then answer the question: will an intermediate generated from the addition of biotin to bicarbonate react with a phosphate derivative to form a carboxylated product by subsequent C–O cleavage to form the carbonyl group and eliminate phosphate?

The tetrahedral adduct formed from the nucleophilic addition of bicarbonate to a urea derivative, such as biotin, would be an unstable species which would only form in small steady-state amounts since the trigonal state is much lower in energy than the four-coordinate state. Thus, a study of such a species would be impractical. We chose to study the reaction of a carbonyl hydrate with a neighbouring phosphate. Such an analogue would form in larger quantity and still would be able to regenerate a carbonyl group by elimination of a phosphate group from a phosphorylated adduct. The early studies of Ramirez showed that acetoin dimethyl phosphate hydrolyses in mild basic solution 10^6 times faster than does trimethyl phosphate (Ramirez *et al.* 1962). One mechanism that Ramirez proposed to account for the rapid hydrolysis rate involved attack of the hydrate form of the carbonyl group on the adjacent phosphate ester. However, no direct evidence for the mechanism was provided and later workers proposed alternative mechanisms which would not be appropriate models for carboxylation of biotin (Cox and Ramsay 1964). These involved enolization of the substrate and direct reaction at phosphorus.

Models for reaction of bicarbonate with biotin followed by reaction with ATP

We conducted the hydrolysis of diethyl methylacetoin phosphate in water labelled with ^{18}O. This material cannot enolize and it undergoes hydrolysis rapidly (Kluger and Taylor 1991). The addition of solvent water to the carbonyl group forms a covalent hydrate, serving as a model for the tetrahedral adduct that would result from the addition of biotin to bicarbonate. A combination of rate factor relationships in the model permits the adduct formed from the carbonyl hydrate and the neighbouring phosphate to react in a manner that exactly parallels that which is proposed in the mechanism for carboxylation of biotin.

The amount of isotope incorporated into the phosphate product is less than that which is in the solvent but more than what is originally present in the reactant. Therefore, exchange in the substrate must be competitive with the hydrolysis reaction. The phosphate ester cannot undergo exchange under the reaction conditions but the carbonyl group can via formation and partial equilibration of a covalent hydrate with the solvent. We have shown through labelling and kinetic studies on a series of model compounds that the Ramirez mechanism is correct. Criticism of the Ramirez mechanism failed to note the actual course of such a reaction. Thus, the reaction involves initial attack of the carbonyl hydrate upon the adjacent phosphate ester to give a phosphorylated carbonyl hydrate. This breaks down by expulsion of the phosphate (Scheme 5.6).

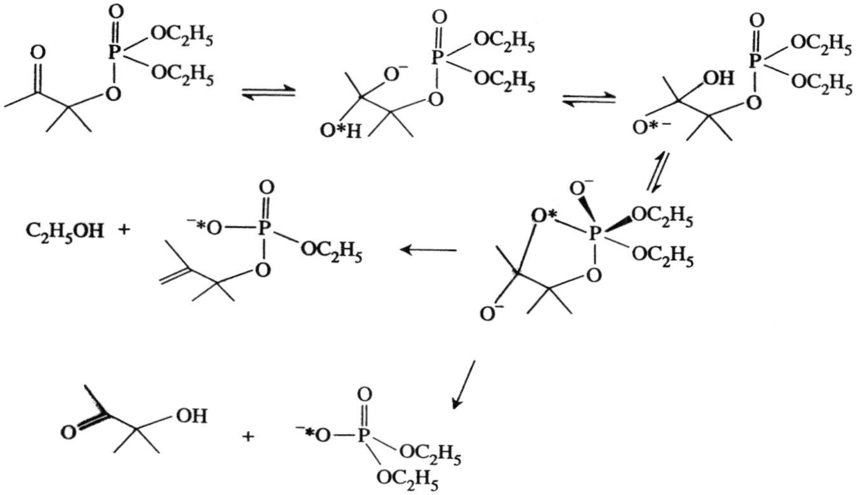

Scheme 5.6 Detailed mechanism for participation of a carbonyl hydrate in phosphate ester hydrolysis.

By having the reaction with bicarbonate occur first, no cleavage of ATP is necessary until after reaction with bicarbonate has occurred. Thus the system can be efficient in providing a direct route to utilize the most abundant species.

Carboxyl transfer from carboxy biotin

The transfer of carbon dioxide from carboxybiotin to a carbanionic acceptor is the final step in the extraction of carbon dioxide from dissolved bicarbonate into formation of an acyl derivative of a species containing a carbon–carbon bond. After the initial reaction which utilizes ATP to form carboxybiotin the transfer to a substrate occurs without further energy inputs.

The transfer reaction was studied in terms of stereochemical relationships in the substrate. The carboxyl group which is added replaces a proton and the relative stereochemistry of the proton and the carboxyl which replaces it is the same: the reaction proceeds with retention of relative configuration at the site of substitution (Rétey and Lynen 1965). The mechanistic conclusions from such an observation should be straightforward. The process must be complex: electrophilic substitution with front side attack and departure must involve an intermediate because the site will be too crowded for everything to react at once. Cram and Haberfield (1961) have studied the relationship of stereochemistry and mechanism in similar anionic substitution reactions. They found that the stereochemical outcome is a function of the nature of the reactants and solvent. In all cases, the reaction involves at least one intermediate. Stubbe *et al.* (1980) have advocated a mechanism in which a base on the enzyme removes a proton from the α position of a keto-acid. The carbanion adds to the carboxyl group of carboxybiotin. In a variant of this mechanism, in analogy to Jencks' proposal (Sauers *et al.* 1975) for the carboxylation of biotin, carboxybiotin first loses its carboxyl group as carbon dioxide and this is trapped by the substrate carbanion. The details of the process remain uncertain.

Triggering the release of carbon dioxide from carboxybiotin

A more general question arises from the observation reported by Goodall and coworkers: the reactivity of carboxybiotin is 'triggered' by the binding of a substrate or by an analogue of a substrate (Goodall *et al.* 1983). Carboxybiotin must hold the carboxy group in a stable situation until the material which is to receive the carbon dioxide equivalent it available. Carboxybiotin must change from providing stable storage for carbon dioxide to being a rapid supplier of carbon dioxide. The model is analogous to that of the 'just-in-time' parts supplier in the automobile industry. The material needed is the responsibility of the specialist who stores it or makes it for ready availability and is prepared to activate the transfer process immediately upon notification of a need. The parts supplier can be triggered by a communicated message; how does the message get to carboxybiotin and how is the response made?

Mechanism of triggering

We proposed that a conformation control system could readily account for such a process (Thatcher *et al.* 1986). If the carboxyl group of carboxybiotin is held in the same plane as the urea group of biotin there are strong stabilizing forces which can be summarized as a resonance structure in which there is a double bond between the carboxyl group and the urea. Rotation of the carboxyl group out of the plane of the urea promotes the breaking of the C–N bond and the transfer of the carbon dioxide equivalent (Scheme 5.7).

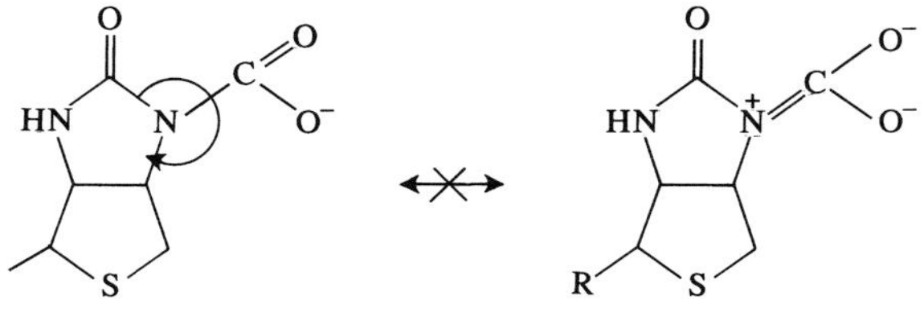

Scheme 5.7 Rotation of the carboxyl group out of the plane of the urea removes a resonance interaction which stabilizes the C–N bond.

Examination of reported structures of biotin and analogues of biotin suggests that the fused ring containing sulphur holds the imidazolidinone ring planar. X-ray studies of analogues in which sulphur is replaced by oxygen or carbon result in structures in which the imidazolidinone ring is distorted (DeTitta *et al.* 1976). It is likely that the long carbon–sulphur bond serves to produce the correct bridge length while shorter bonds cause distortion.

Models for carboxyl transfer and triggering

If rotation of the carboxyl of carboxybiotin out of the plane of the urea group will enhance reactivity towards decarboxylation, we propose that host–guest chemistry can be used to produce a catalyst that will promote decarboxylation of *N*-carboxyimidazolidinone. We have prepared the macrocyclic compound (MAC) shown below.

MAC

The molecule binds imidazolidinone tightly (K_{ass} = 57 000 M^{-1}) while *N*-carboxymethyl imidazolidinone binds weakly (K = 100). *N,N′*-dimethyl imidazolidinone does not bind at all. (Fig 5.2).

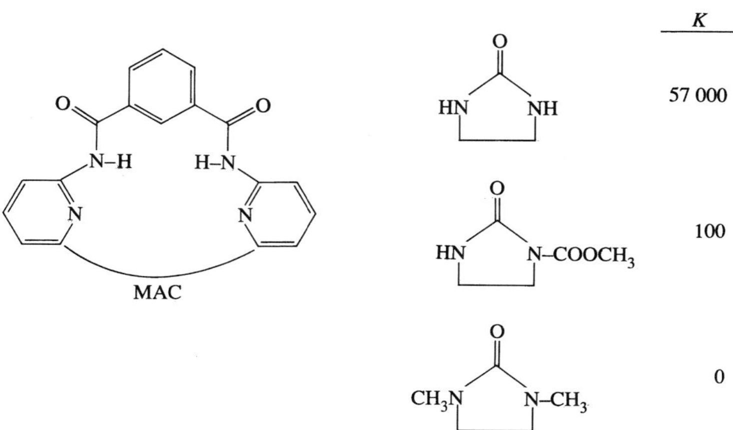

Fig. 5.2 Binding of imidazolidinones to MAC.

The selectivity of MAC in strongly favouring binding of the derivative lacking a carboxy group indicates that it will promote the decarboxylation of *N*-carboxyimidazolidinone. Attractive interactions will develop in the transition state for decarboxylation as the N–C bond is broken. Such a change follows the model in which we proposed that destabilization of the ground state by forcing the carboxyl group of carboxybiotin out of the plane of the urea will promote decarboxylation upon binding of the substrate. We are in the process of testing this hypothesis.

Conclusion

The efficient utilization of bicarbonate as a source of carbon dioxide for biosynthetic purposes is widespread in nature and commonly involves reaction with ATP to provide a carboxylated biotin intermediate with transfer of water to ATP in a hydrolysis product. Although some mechanisms have gained general acceptance, the basis for the acceptance has not been critical testing but rather the accumulation of circumstantial evidence. Other mechanisms which have been considered here deserve further consideration. Structural studies on a biotin-dependent enzyme will be particularly useful in elucidating the mechanism if the location of specific binding sites can be determined.

Acknowledgements

The authors thank the Natural Sciences and Engineering Research Council of Canada for support through an operating grant, and the Swedish Academy of Sciences for inviting their participation.

References

Blagoeva, I.B., Pojarlieff, I., and Kirby, A.J. (1984). *J. Chem. Soc. Perkin Trans. 2,* 745–751.

Blagoeva, I.B., Pozharliev, I., Tashev, D., and Kirby, A.J. (1989). *J. Chem. Soc. Perkin Trans. 2*, 347–53.

Blonski, C., Gasc, M.B., Hegarty, A.F., Klaebe, A., and Perié, J. (1984). *J. Am. Chem. Soc.*, **106**, 7523–7529.

Calvin, M. and N.G. Pon, (1959). *J. Cell. Comp. Physiol.*, **54** (Suppl. 1), 51–54.

Cox, J.R. and Ramsay, B. (1964). *Chem. Rev.*, **64**, 317–352.

Cram. D.J. and Haberfield, P.J. (1961). *Am. Chem. Soc.*, **83**, 2354–2362.

DeTitta, G.T., Edmonds, J.W., Stallings, W.C., and Donohue, J. (1976). *J. Am. Chem. Soc.*, **98**, 1920–1926.

Dugas, H. and Penney, C. (1981). *Bioorganic chemistry*. Springer, New York.

Goodall, G.J., Prager, R., Wallace, J.C., and Keech, D.B. (1983). *FEBS Lett.*, **163**, 6–9.

Guchait, R.B., Polakis, S.E., Hollis, D., Fenselau, C., and Lane, M.D. (1974). *J. Biol. Chem.*, **249**, 6646–6656.

Hansen, D.E. and Knowles, J.R. (1985). *J. Am. Chem. Soc.*, **107**, 8304–8305.

A.F. Hegarty, and Bruice, T.C. (1970). *J. Am. Chem. Soc.*, **92**, 6561–6567.

Karizo, Y., Hase, L.F., Boyer, P.D., and Ochoa, S. (1962). *J. Biol. Chem.*, **237**, 1460–68.

Kluger, R. (1989). *Bioorg. Chem.*, **17**, 287–93.

Kluger, R. (1990). *Chem. Rev.*, **90**, 1151–69.

Kluger, R. and Tsao, B.F. (1993). *J. Am. Chem. Soc.*, **115**, 2089–90.

Kluger, R. and Taylor, S.D. (1991). *J. Am. Chem. Soc.*, **113**, 996–1001.

Kluger, R., Davis, P.P., and Adawadkar, P.D. (1979). *J. Am. Chem. Soc.*, **101**, 5995–6000.

Knappe, J. (1970). *Ann. Rev. Biochem.*, **39**, 757–76.

Knowles, J.R. (1989). *Ann. Rev. Biochem.*, **58**, 195–221.

Ogita, T. and Knowles, J.R. (1988). *Biochemistry*, **27**, 8028–33.

Page, M.I. and Jencks, W.P. (1971). *Proc. Nat. Acad. Sci. USA*, **68**, 1678–83.

Perrin, C.A. and Dwyer, T.J. (1987). *J. Am. Chem. Soc.*, **109**, 5163–67.

Polakis, S.E., Guchait, R.B., and Lane, M.D. (1972). *J. Biol. Chem.*, **254**, 1335–37.

Ramirez, F., Hansen, B., and Desai, N.B. (1962). *J. Am. Chem. Soc.*, **84**, 4588.

Rétey, J. and Lynen, F. (1965). *Biochem. Z.*, **342**, 256–71.

Sauers, C.K., Jencks, W.P., and Groh, S. (1975). *J. Am. Chem. Soc.*, **97**, 5546–53.

Stubbe, J.A., Fish, S., and Abeles, R.H. (1980). *J. Biol. Chem.*, **255**, 236–42.

Thatcher, G.R.J. and Kluger, R. (1989). *Adv. Phys. Org. Chem.*, **25**, 99–266.

Thatcher, G.R.J., Poirier, R. and Kluger, R. (1986). *J. Am. Chem. Soc.*, **108**, 2699–2704.

Tsao, B.F. (1990). M.Sc. thesis, University of Toronto.

Wood, H.G. (1976). *Trends Biochem. Sci.*, **1**, 4–6.

6

Chemistry of B_{12} derivatives related to their roles in bacterial carbon dioxide fixation

Bernhard Kräutler

Introduction

Vitamin B_{12} derivatives play several central roles in the metabolism of micro-organisms (Friedrich 1988; Dolphin 1982; Golding and Rao 1987), which are also believed to possess the capacity unique in nature to build up the B_{12} structure. Corrinoids related to vitamin B_{12} appear to have a particularly essential function in some methanogenic, acetogenic, sulphur- and sulphate-reducing bacteria, where they have been assigned a role in the fixation of carbon dioxide via the recently discovered acetyl coenzyme A pathway (Fuchs 1986; Wood *et al.* 1986; Thauer 1988; Ragsdale *et al.* 1990).

In the course of this pathway of autotrophic fixation of carbon dioxide in bacteria, enzyme-bound methyl corrinoids related to methylcobalamin are hypothetical intermediates. These are formed by abstraction of the methyl group of N^5-methyltetrahydrofolate (or of related N^5-methyltetrahydropterines), which in turn is provided in several steps from carbon dioxide by a sequence of enzymatic reduction steps (Ljungdahl and Wood 1982); in addition to the reduction of carbon dioxide by the enzyme carbon monoxide dehydrogenase a 'bound' carbon monoxide is also formed (Thauer 1988). The two reduced one-carbon units (the cobalt-bound methyl group and carbon monoxide) are finally combined to the acetyl group of acetyl coenzyme A in an as yet uncharacterized step, presumably a carbonylation reaction at a metal centre. This latter reaction apparently takes place in the Ni–Fe–S enzyme carbon monoxide dehydrogenase (Gorst and Ragsdale 1991). The assembly of acetyl coenzyme A occurs with overall retention of the configuration of a chiral methyl group from N^5-H,D,T-methyltetrahydrofolate in the acetogen *Clostridium thermoaceticum* (Lebertz *et al.* 1987; Raybuck *et al.* 1987). An alternative path with formation of the acetyl group by carbon monoxide insertion on the enzyme-bound methyl corrinoid correspondingly appears inoperative in the bacterial systems (Pezacka and Wood 1988; Ragsdale *et al.* 1990) (in homogenous aqueous solution and at ambient temperatures CO-insertion into the Co–CH_3-bond of methylcobalamin has been studied earlier (Kräutler 1984)).

Apparently, the function of the enzyme-bound corrinoid cobalt complexes in the course of the assembly of acetyl coenzyme A from carbon dioxide merely concerns the catalysis of a methyl group transfer from an N^5-methyltetrahydropterine to the as yet unidentified acceptor (but possibly a thiol group) on the carbon monoxide dehydrogenase complex (Lu *et al.* 1990).

A related role of enzyme-bound corrinoids in the catalysis of a methyl group transfer lately has been well studied for the methylation of homocysteine to methionine, catalysed by enzymes from bacterial and mammalian sources (Matthews 1984; Banerjee and Matthews 1990): there, an enzyme-bound Co(l)-corrin accepts the methyl group of the N^5-methyltetrahydropterine, and the methyl corrinoid formed this way then transfers the cobalt bound methyl group to the thiol homocystein (Banerjee *et al.* 1990). In this way, formally as a methyl cation, a methyl group is transferred with overall retention of configuration at the methyl carbon (but presumably via two substitution steps, each occurring with stereochemical inversion (Zydowsky *et al.* 1986)).

Structure and reactivity of natural cobalt corrins

The structure of vitamin B_{12} (see Fig. 6.1) is made up of:

1 the corrin ligand, structurally and biosynthetically related to the ligand of natural porphyrins (Eschenmoser 1988; Leeper 1989);
2 the nucleotide function, that coordinates to the corrin-bound metal centre in a 'nucleotide loop', unique to 'complete corrinoids' among the metal-containing cofactors known today;
3 the ligand-bound cobalt ion; and
4 the cobalt-bound cyanide ion (a nonphysiological artefact) (Fig. 6.1)

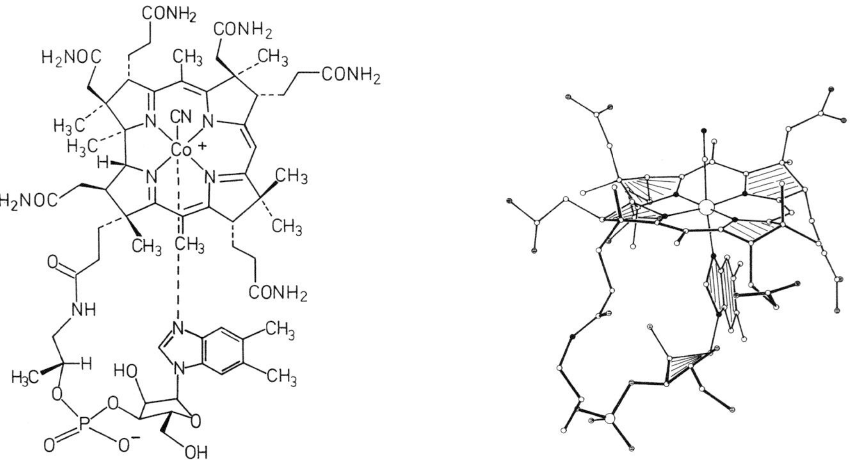

Fig. 6.1 Structural formula of vitamin B_{12} (**1**) and three-dimensional structure of **1** according to X-ray analysis by Hodgkin *et al.* (1955).

Most metabolic functions of the corrinoids are associated with their capacity for organometallic chemistry (under physiological conditions), discovered some 30 years ago, first with the analysis of the structure of coenzyme B_{12} (**2**, deoxyadenosyl-cobalamin) (Lenhert and Hodgkin 1961; Lenhert 1968). In the coenzyme **2** an organometallically coordinated deoxyadenosyl ligand, and in methylcobalamin (**3**), a methyl group, take up the upper axial position at the Co(III) centre.

The biologically relevant reactivity of the coenzyme **2** is associated with the ease of homolysis of its Co–C bond (homolytic Co–C bond-dissociation energy of *ca.* 30 kcal mol^{-1}; Halpern 1985; Hay and Finke 1986). For methylcobalamin (**3**) in turn, the heterolysis of the Co–methyl bond (formal loss of a CH_3^+ ion) should be considered most relevant for the cofactor role of the corrinoids in enzymatic methyl group transfer reactions (Kräutler 1988*a*): formation of a methylcorrin by the methylation of an enzyme-bound Co(I)-corrin and its subsequent demethylation by thiolates to a Co(I)-corrin are believed to be the crucial two steps of the catalytic cycle in cobalamin-dependent methylation of homocysteine to methionine (Banerjee and Matthews 1990).

Structural characteristics of naturally occurring Co(III)-corrins

In bacterial metabolism (according to the presently available evidence) corrinoids function as enzymatic cofactor at the oxidation levels Co(I), Co(II), and Co(III) of the bound cobalt centre. Due to the ease of aerial oxidation of the reduced Co(II)- and Co(I)-corrins, the early structural investigations concerned

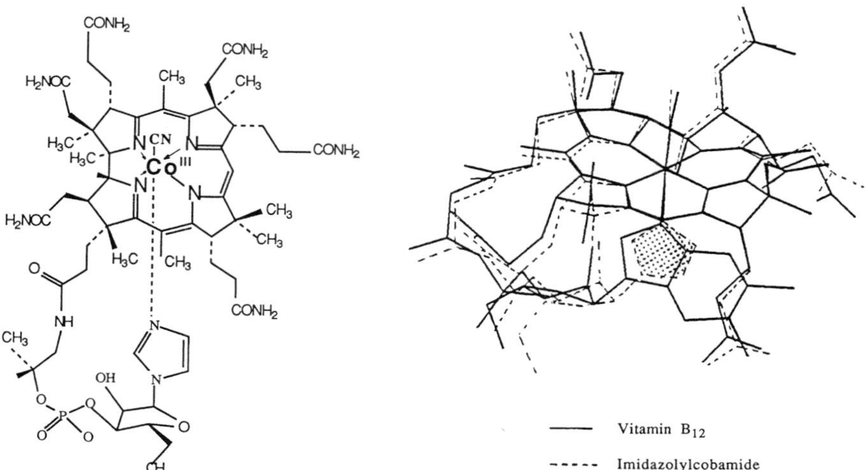

Fig. 6.2 Structural formula of Coα-imidazolyl-Coβ-cyanocobamide and comparison of its three-dimensional structure (dotted lines) with that of vitamin B_{12} (solid lines).

the (diamagnetic) Co(III) corrins (Hodgkin *et al.* 1955; Glusker 1982). Throughout, the corrin ligand structure was found to vary little and the corrin-bound metal centre to carry two axial ligands (to be hexacoordinate) in the Co(III) corrins. Structural variability is minimal (Pett *et al.* 1987) and the most variable factor concerns the lengths of the bonds from the cobalt centre to the axial ligands (Glusker 1982).

In anaerobic bacteria, the natural 'complete corrins' (i.e. those carrying an intact nucleotide loop) feature an intriguing variability of the constitution of the nucleotide base (Kräutler 1988*b*). This appears to be a consequence of the specific biosynthetic origin of the nucleotide base rather than of a species-specific selection of the cofactor reactivity (Kräutler *et al.* 1988). With the structure of Coα-imidazolyl-Coβ-cyano-cobamide, first reported by Eberhard *et al.* (1988) and now obtained from guided biosynthesis in bacteria, a significant effect of the structure of the metal-coordinated nucleotide base on the corrin ligand was uncovered by X-ray analysis (Kräutler *et al.*, unpublished). In the imidazolyl-cobamide, the 'folding' of the corrin ligand is considerably reduced (folding angle: 11.7°) compared to the situation in vitamin B_{12} (folding angle *ca.* 18°, see Fig. 6.2). Apparently the steric bulk of the 5,6-dimethylbenzimidazole base of the cobalamins (such as vitamin B_{12}) contributes to the intriguing 'folding' of the corrin ligand there (Fig. 6.2).

Structure and reactivity of the Co(II)-corrins

For the paramagnetic Co(II)-corrins (that typically are sensitive to air oxidation to Co(III)-corrins), evidence has been available for some time (from electron spin resonance and electrochemical experiments) for a pentacoordinated Co(II) centre (reviewed by Pratt (1975)). We have more recently been able to crystallize several Co(II)-corrins. The structures of a Co(II)-heptamethyl cobyrinate and Co(II)-cobalamin (**4**), the Co(II)-analogue of vitamin B_{12}, were examined by X-ray analysis in Kratky's group in Graz (Kräutler *et al.* 1987; Kräutler *et al.* 1989). For these compounds, which are persistent radicaloids and excellent radical traps, a pentacoordinated Co(II) centre was indeed found. In addition, Co(II)-cobalamin proved to differ little in the structure of the corrin ligand and of the nucleotide loop from the corresponding alkyl-Co(III)-cobalamins **2** and **3** (see Fig. 6.3).

Accordingly, the homolytic cleavage of the Co(III)-alkyl bond of coenzyme B_{12} and of methylcobalamin, as well as the very rapid reaction of Co(II)-corrins with free alkyl radicals (Endicott and Netzel 1979) occur with little change in the structure of the cobalt-corrin part. Correspondingly, also for an enzyme-bound alkyl-Co(III)-corrin, a relevant contribution to the enigmatic protein induced activation of **2** for Co–C-bond homolysis cannot be based on a conformational deformation of the corrin ligand (Kräutler *et al.* 1989). A complex of **4** with molecular oxygen, shown to form in preparations of corrinoid enzymes

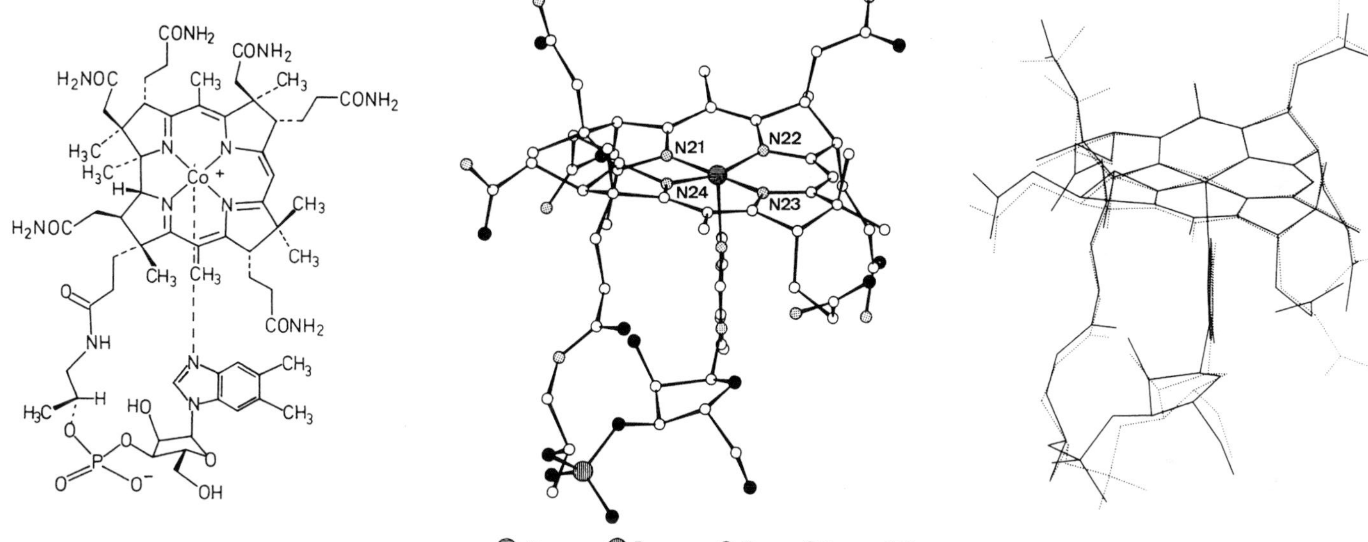

Fig. 6.3 Structural formula of cob(II)alamin (**4,** *left*), three dimensional structure of **4** (*middle*), and comparison of its structure (**4**, solid lines, *right*) with that of the corrin part of coenzyme B_{12} (**2**, dotted lines).

upon exposure to air (Frasca *et al.* 1988), can also be formed reversibly in the crystal and was studied by X-ray analysis (Hohenester *et al.* 1991).

Structure and reactivity of Co(I)-corrins

So far, of the highly oxygen sensitive Co(I) forms of corrinoid cobalt complexes (such as Co(I)-cobalamin), information from X-ray analysis on their three-dimensional structure is not available. Analysis by nuclear magnetic resonance of the structure of the diamagnetic Co(I)-heptamethyl cobyrinate (Puchberger and Kräutler, unpublished) did not reveal any unexpected coordination patterns at the Co(I) ion: in the strongly nucleophilic (Schrauzer *et al.* 1968), highly reduced (Lexa and Saveant 1983) Co(I) forms of corrinoid cobalt complexes, the diamagnetic Co(I) centre (a d^8-metal ion, isoelectronic to a Ni(II) ion) presumably is bound by the corrin ligand only, in a square planar tetracoordinated fashion (Pratt 1982). With high diastereoselectivity, alkylation via a nucleophilic substitution reaction (at carbon) occurs on one side, at the ('upper') β face, apparently even without the (kinetic) assistance of the nucleotide function on the ('lower') α face (Kräutler and Caderas 1984). The Co(I)-corrins can also be alkylated in more complex reactions believed to proceed via (one) electron transfer induced radical processes (Breslow and Khanna 1976; Kräutler and Caderas 1984).

Structure and reactivity of methyl-Co(III)-corrins

Methylcobalamin (**3**), the simplest organometallic representative of the cobalamins, has been examined by X-ray analysis and nuclear magnetic resonance spectroscopy (Rossi *et al.* 1985). In aqueous solution it can be demethylated by thiolates (Hogenkamp *et al.* 1985), supporting in a model reaction the hypothetical role of methyl corrinoids as methyl group donors to homocysteine and similar thiols in enzymatic methyl group transfer (Banerjee and Matthews 1990). A methyl group transfer between methyl-Co(III)-cobinamide (cobinamides lack a nucleotide function) and Co(I)-cobalamin rapidly occurs in aqueous solution to result in an equilibrium, where methylcobalamin (**3**) and Co(I)-cobinamide prevail, indicating the nucleotide coordination to stabilize **3** (by *ca.* 4 kcal mol^{-1}) against abstraction by nucleophiles of the cobalt-bound methyl group (Kräutler 1987).

According to our present understanding, the protein-bound corrinoid species involved in enzymatic methyl group transfer are Co(I)-corrins and methyl-Co(III)-corrins, the latter with an intact nucleotide coordination (Matthews *et al.* 1990). From knowledge of solution and crystal data, a major structural factor will involve the arrangement of the nucleotide loop with respect to the metal centre (coordination intact in the methyl-Co(III)-corrin, coordination absent in

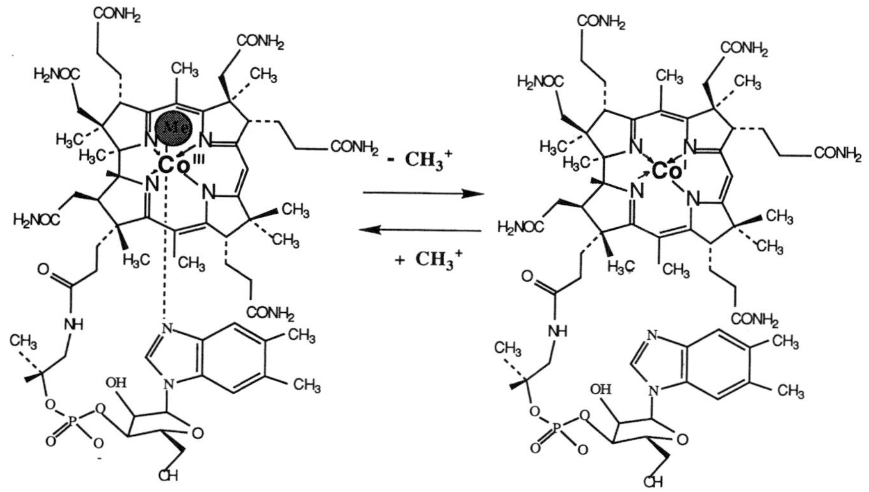

Fig. 6.4 Transition between methylcob(III)alamin and cob(I)alamin by transfer of a methyl group (formally a methyl cation).

the Co(I)-corrin, see (Fig. 6.4). A major structural change involving the nucleotide loop will also have to accompany the enzymatic methylation/demethylation steps, therefore (potentially) subject to enzymatic control. A weakened nucleotide coordination will activate the more oxidized states of the protein-bound corrinoid factor (methyl-Co(III)- and Co(II)-corrin) and correspondingly would ease the access to the Co(I) state, both by single electron reduction of a Co(II)-corrin precursor (see for example Lu *et al.* 1990) and by demethylating the methyl-Co(III)-corrin due to abstraction of its cobalt-bound methyl group by a nucleophile (Fig. 6.5).

Summary

The vitamin B_{12} derivatives correspondingly possess a series of important reactivities in organometallic transformations, to which they owe their function as methyl-group transferring and activating cofactors in the bacterial fixation of carbon dioxide via the acetyl coenzyme A path (see Fig. 6.5). As the highly oxygen sensitive Co(I)-corrins they can be provided by biological reducing agents and are both very powerful nucleophiles and good leaving groups, crucial for the catalysis of enzymatic methyl-group transfer reactions. As Co(II)-corrins, furthermore, they provide persistent radicaloids and powerful radical traps, while their alkyl-Co(III)-forms, in reverse, can represent latent alkyl radicals. More specifically, as methyl-Co(III)-corrins they are a group of versatile methylating agents, that can transfer their cobalt-bound methyl group to nucleophilic, radicaloid, and electrophilic alkyl group acceptors.

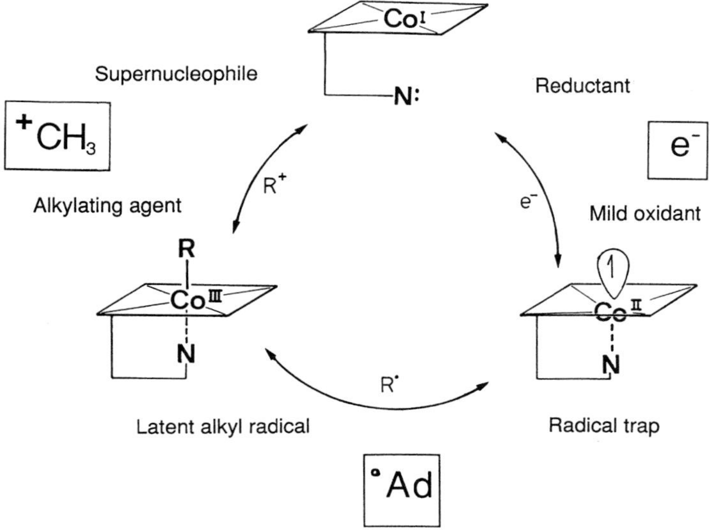

Fig. 6.5 Reactivity patterns of vitamin B_{12} derivatives in organometallic transformations.

Acknowledgements

The authors would like to thank Prof. Dr R.K. Thauer (University of Marburg), Dr E. Stupperich, and Prof. Dr G. Fuchs (University of Ulm) for their fruitful cooperations and helpful discussions, and M. Puchberger (ETH Zürich) for his excellent experimental contributions. They are also particularly grateful to Doz. Dr C. Kratky and his group at the University of Graz for their important X-ray crystallographic cooperation, so crucial to their research on vitamin B_{12} derivatives. This work was mostly carried out at the Laboratory of Organic Chemistry of the ETH in Zürich and was financially supported by the Swiss National Science Foundation.

References

Banerjee, R.V. and Matthews, R.G. (1990). *FASEB J.*, **4**, 1450–9.

Banerjee, R.V., Frasca, V., Ballou D.P., and Matthews, R.G. (1990). *Biochemistry*, **29**, 11101–9.

Breslow, R. and Khanna, P.L. (1976). *J. Am. Chem. Soc.*, **98**, 1297–9.

Dolphin, D. (ed.) (1982). *B_{12}*, Vols 1 and 2. Wiley, New York.

Eberhard, G., Schlayer, H., Joseph, H., Fridrich, E., Urz, B., and Müller, O. (1988). *Biol. Chem. Hoppe–Seyler*, **369**, 1091–8.

Eschenmoser, A. (1988). *Angew. Chem.*, **100**, 5–40.

Endicott, J.F. and Netzel, T.L. (1979). *J. Am. Chem. Soc.*, **101**, 4000–2.

Frasca, V., Banerjee, R.V., Dunham, W.R., Sands, R.H., and Matthews, R.G. (1988). *Biochemistry*, **27**, 8458–65.

Friedrich, W. (1988). Vitamin B_{12}. In *Handbook of vitamins*, pp. 839–928. Walter de Gruyter, New York.

Fuchs, G. (1986). *FEMS Microbiol. Rev.*, **39**, 181–213.

Glusker, J.P. (1982). X-ray crystallography of B_{12} and cobaloximes. In B_{12} (ed. D. Dolphin), Vol. 1, pp. 24–106. Wiley, New York.

Golding, B.T. and Rao, D.N.R. (1987). Adenosylcobalamin-dependent enzymic reactions. In *Enzyme mechanisms* (ed. M.I. Page and A. Williams). pp. 404–28. Royal Soc. Chem., London.

Gorst, C.M. and Ragsdale, S.W. (1991). *J. Biol. Chem.*, **266**, 20 687–93.

Halpern, J. (1985). *Science*, **227**, 869–75.

Hay, B.P. and Finke, R.G. (1986). *J. Am. Chem. Soc.*, **108**, 4820–9.

Hodgkin, D.C., Johnson, A.W., and Todd, A.R. (1955). The structure of vitamin B_{12}. In *Recent work on naturally occurring nitrogen hetero-cyclic compounds. Spec. publ. No. 3* (ed. K. Schoffield), pp. 109–23. The Chemical Society, London.

Hogenkamp, H.P.C., Bratt, G.T., and Sun, S. (1985). *Biochemistry*, **24**, 6428–32.

Hohenester, E., Kratky, C., and Kräutler, B. (1991). *J. Am. Chem. Soc.*, **113**, 4523–30.

Kräutler, B. (1984). *Helv. Chim. Acta*, **67**, 1053–9.

Kräutler, B. (1987). *Helv. Chim. Acta*, **70**, 1268–78.

Kräutler, B. (1988*a*). Structural effects on cobalt-methylation and demethylation of vitamin B_{12}-derivatives. In *The biological alkylation of heavy elements* (ed. P.J. Craig and F. Glockling), pp. 31–45. Royal Soc. Chem., London.

Kräutler, B. (1988*b*). *Chimia*, **42**, 91–4.

Kräutler, B. and Caderas, C. (1984). *Helv. Chim. Acta*, **67**, 1891–6.

Kräutler, B., Keller, W., Hughes, M., Caderas, C., and Kratky, C. (1987). *J. Chem. Soc., Chem. Commun.*, **672**, 1678–80.

Kräutler, B., Kohler, H.P., and Stupperich, E. (1988). *Eur. J. Biochem.*, **176**, 461–9.

Kräutler, B., Keller, W., and Kratky, C. (1989). *J. Am. Chem. Soc.*, **111**, 8936–8.

Lebertz, H., Simon, H., Courtney, L.F., Benkovic, S.J., Zydowsky, L.D., Lee, K. *et al.* (1987). *J. Am. Chem. Soc.*, **109**, 3173–4.

Lenhert, P.G. (1968). *Proc. R. Soc. London*, **303**, 45–84.

Lenhert, P.G. and Hodgkin, D.C. (1961). *Nature*, **192**, 937–8.

Leeper, F.J. (1989). *Nat. Prod. Rep.*, **6**, 171–203.

Lexa, D. and Saveant, J.-M. (1983). *Acc. Chem. Res.*, **16**, 235–43.

Ljungdahl, L.G. and Wood, H.G. (1982). Acetate biosynthesis. In B_{12} (ed. D. Dolphin), Vol. 2, pp. 166–202. Wiley, New York.

Lu, W.-P., Harder, S.R., and Ragsdale, S.W. (1990). *J. Biol. Chem.*, **265**, 3124–33.

Matthews, R.G. (1984). Methionine biosynthesis. In *Folates and pterins* (ed. R.L. Blakley and S.J. Benkovic), Vol. 1, pp. 497–54. Wiley, New York.

Matthews, R.G., Banerjee, R.V. and Ragsdale, S.W. (1990). *BioFactors*, **2**, 147–52.

Pett, V.B., Liebmann, M.N., Murray-Rust, P., Prasad, K., and Glusker, J.P. (1987). *J. Am. Chem. Soc.*, **109**, 3207–15.

Pezacka, E. and Wood, H.G. (1988). *J. Biol. Chem.*, **263**, 16 000–6.

Pratt, J.M. (1982). Coordination chemistry of the B_{12} dependent isomerase reactions. In B_{12} (ed. D. Dolphin), Vol. 1, pp. 326–92.

Ragsdale, S.W., Baur, J.R., Gorst, C.M., Harder, S.R., Lu, W.-P., and Roberts, D.L. *et al.* (1990). *FEMS Microbiol. Rev.*, **87**, 397–402.

Raybuck, S.A., Bastian, N.R., Zydowsky, L.D., Kobayashi, K., Floss, H.G., and Orme-Johnson, W.H. *et al.* (1987). *J. Am. Chem. Soc.*, **109**, 3171–3.

Rossi, M., Glusker, J.P., Randaccio, L., Summers, M.F., Toscano, P.J., and Marzilli, L.G. (1985). *J. Am. Chem. Soc.*, **107**, 1729–38.

Schrauzer, G.N., Deutsch, E., and Windgassen, R.J. (1968). *J. Am. Chem. Soc.*, **90**, 2441–2.

Thauer, R.K. (1988). *Nachr. Chemie Techn. Lab.*, **36**, 993–7.

Wood, H.G., Ragsdale, S.W., and Pezacka, E. (1986). *FEMS Microbiol. Rev.*, **39**, 345–62.

Zydowsky, T.M., Courtney, L.F., Frasca, V., Kobayashi, K., Shimizu, H., and Yuen, L. *et al.* (1986). *J. Am. Chem. Soc.*, **108**, 3152–3.

7

Stoichiometric and catalytic activation of carbon dioxide on transition-metal complexes

Arno Behr

A very promising approach to the activation of carbon dioxide is offered by its coordination to transition-metal complexes. In this chapter a brief insight is given into both stoichiometric and catalytic activation of CO_2. Both parts are linked together very closely, because catalytic reactions can only occur if stoichiometric activation steps first start the reaction.

In the first part some information is given about the coordination modes of carbon dioxide to transition metals. How can CO_2 add to the metal and of what kind are the bonding interactions?

In the second part some stoichiometric reactions of carbon dioxide with other molecules under the influence of the transition metal will be presented. A special focus will be on carbon–carbon-linkage reactions, because C–C bond formation is a key step in the synthesis of larger molecules. Both oxidative coupling reactions and insertion reactions of CO_2 into the metal–alkyl and metal–allyl bonds will be considered in some more detail.

If carbon dioxide can be activated by the metal and if a CC-linkage reaction takes place, then the next question will be whether this reaction also run in a catalytic way. Therefore, in the third part of the review some examples of the catalytic CC linkage of CO_2 with alkynes and dienes will be given (Behr 1988).

Figure 7.1 shows a general scheme with different possible routes of interaction between carbon dioxide and a substrate S on the transition metal M. In cycle A the substrate S adds first on the metal and yields an activated intermediate M–S. Then CO_2 coordinates to the intermediate and the linkage product S–CO_2 can be formed. In cycle B carbon dioxide first coordinates to the metal and then reacts with the substrate S in a second step. If the final product is a metal complex, then the reaction is only stoichiometric. If the linkage product S–CO_2 can be set free and if the metal can enter again into one of the cycles, then the reaction is catalytic.

The first step in cycle B is the coordination of CO_2 to the metal. There exist different types of metal/CO_2 complexes because carbon dioxide has several potential coordination modes. The carbon atom of CO_2 can be described as a Lewis acid centre and a metal–carbon bond can be formed. Structure 1 shows such a metallacarboxylate complex. On the other hand, the CO_2 molecule has two CO π bonds which can interact with the metal, and hence also a side-on coordination can happen yielding a π complex (2) or a three-membered metallacycle (3).

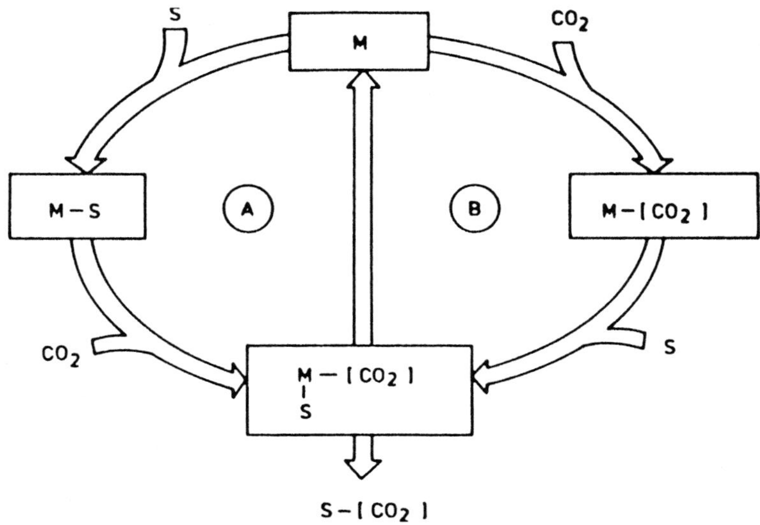

Fig. 7.1 General scheme of possible interactions of carbon dioxide with a substrate S (M = transition metal).

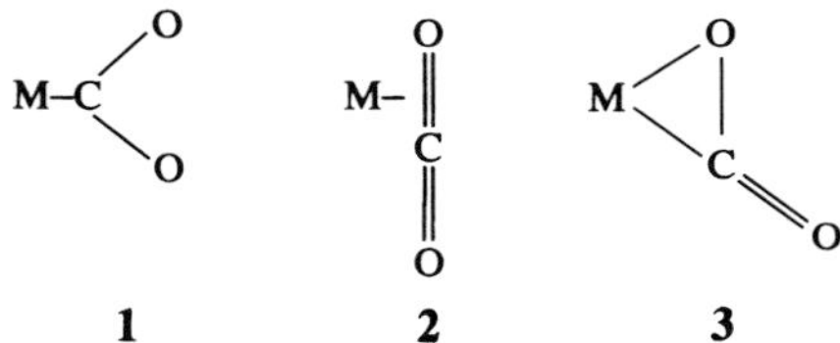

Figure 7.2 shows that indeed a broad number of different CO_2 complexes exist, and this survey gives only a few examples of the numerous complexes now identified by X-ray analysis. Figure 7.2 contains both a typical metallacarboxylate complex (Calabrese *et al.* 1983) and an example of a complex with side-on coordination (Aresta *et al.* 1975; Aresta and Nobile 1977; Aresta *et al.* 1990). In both cases, the linear CO_2 molecule loses its high symmetry and changes into an activated bent molecule. The other examples in Fig. 7.2 show that CO_2 can also coordinate to several metals, and that two carbon dioxide molecules can add together to one transition metal (Audett *et al.* 1982; Jolly *et al.* 1971; Döhring *et al.* 1985). Obviously, the structures of isolated metal/CO_2 complexes must not be identical with intermediates in catalytic reactions, however, these structures give us some idea about the general coordination behaviour of CO_2.

Important stoichiometric reactions of CO_2 on transition metals are the so-called oxidative coupling reactions.

Oxidative coupling means the reaction of carbon dioxide and an unsaturated compound with a transition metal yielding a metallaheterocycle (Fig. 7.3). If at

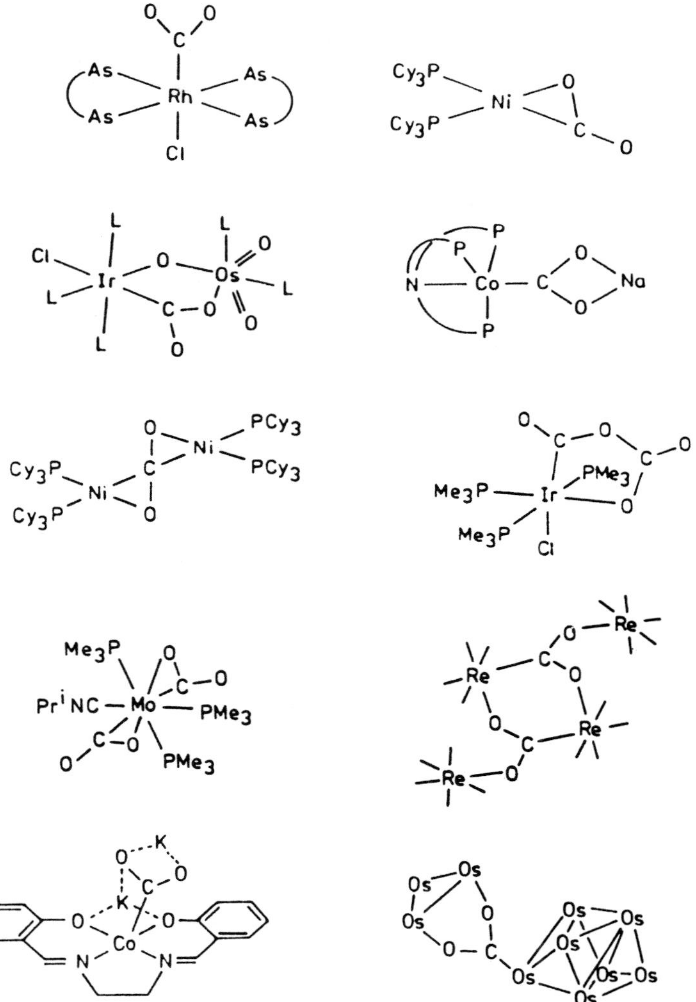

Fig. 7.2 Examples of complexes of carbon dioxide.

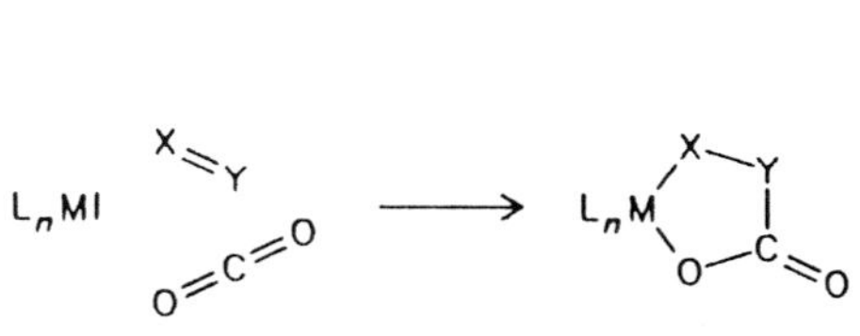

L_nMl X=Y O=C=O ⟶ L_nM(X–Y–C(=O)–O)

Fig. 7.3 Oxidative coupling of carbon dioxide with an unsaturated compound X = Y and a transition metal complex.

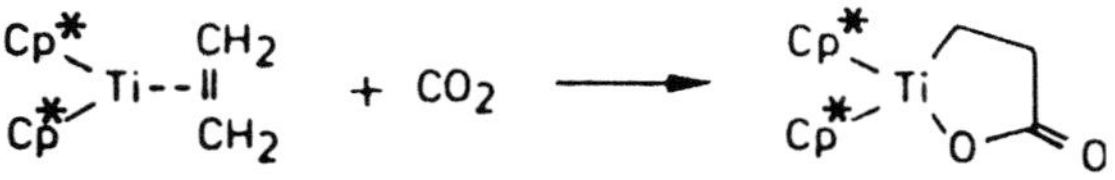

Fig. 7.4 Synthesis of a titanalactone.

least one atom of the unsaturated compound is carbon, a new carbon–carbon bond is formed. This reaction, which can be interpreted as a (2+2+2) cycloaddition, can be performed, for instance, with alkynes, alkenes, cycloalkanes or 1,3-dienes.

An example is the titanium ethene complex with pentamethylcyclopentadienyl ligands which reacts with CO_2 yielding a titanalactone (Fig. 7.4). This is a nice example of what was called cycle A in the general reaction scheme. First the substrate, ethene, is bound to titanium, and then CO_2 causes the oxidative coupling yielding the five-membered ring (Cohen and Bercaw 1985).

Another example (Fig. 7.5) starts with an ironbis (ethene) complex. With CO_2 the iron five-membered ring system is formed. Reaction with a further CO_2 molecule yields ironbis (carboxylate) complexes which can be hydrolysed by hydrochloric acid in methanol with formation of a succinate and a methylmalonate. This example demonstrates clearly the influence of complex ligands on the CC-linkage reaction with CO_2. Addition of the chelate ligand bis (dicyclohexylphosphino) ethane (dcpe) leads preferentially to succinate; if, however, trimethylphosphine (PMe_3) is used as ligand, methylmalonate is formed exclus-

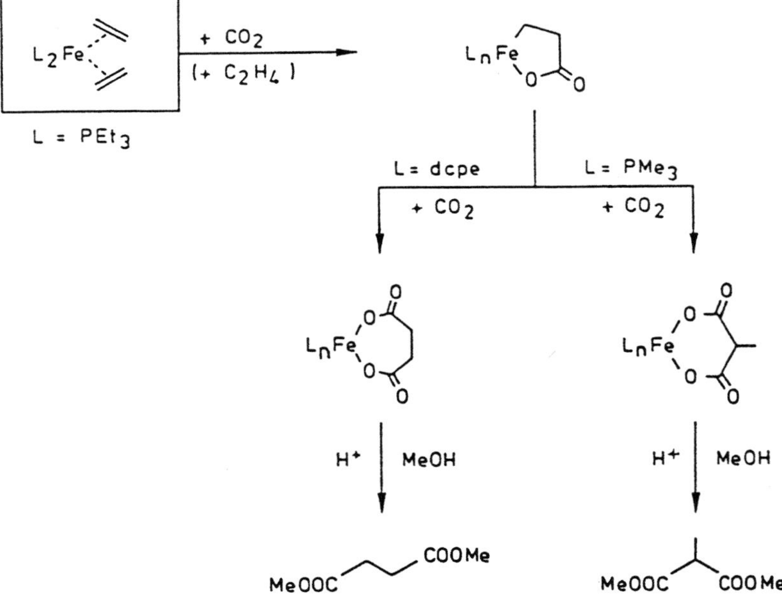

Fig. 7.5 Reaction of carbon dioxide with an ironbis(ethene) complex.

ively (Hoberg *et al.* 1987). Also in catalytic CC-linkage reactions the ligand has a significant influence on the product distribution.

Most oxidative coupling reactions have been studied with nickel complexes. As shown in Fig. 7.6, linear and cyclic mono- and polyens could be coupled with carbon dioxide yielding – after hydrolysis – a great number of carboxylic acids. Thus oxidative coupling with carbon dioxide proved to be a generally applicable reaction concerning both the metal and the unsaturated compound (Behr and Thelen 1984; Behr and Kanne 1986*a*; Walther *et al.* 1984).

In the next section the insertion reactions of CO_2 will be presented (Arafa *et al.* 1988; Vivanco *et al.* 1990). Once again reactions with new CC-bond formation and especially insertions of CO_2 into metal–alkyl bonds will be considered. Generally the insertion of carbon dioxide into the M–C bond yields metal carboxylates. If this stoichiometric reaction can also be carried out catalytically an interesting access to acids, esters or lactones is given. Once again nickel chemistry proved to be very promising. Figure 7.7 shows the reaction of a nickel–phenyl complex with ethene and CO_2, and particular attention should be drawn to the pathway where both molecules are inserted together into the nickel–carbon bond (Behr *et al.* 1983). First, ethene is inserted yielding a 2-phenylethyl nickel complex. Then CO_2 is inserted and forms a carboxylate complex that releases hydrocinnamic ester after treatment with methanol and borontrifluoride. Consequently, in one reaction two new CC bonds are formed successively.

Another interesting example is shown in Fig. 7.8. Quadricyclane and a nickel complex form *in situ* a nickelacylobutane with two nickel–carbon bonds. At room temperature carbon dioxide is inserted into one bond and yields a

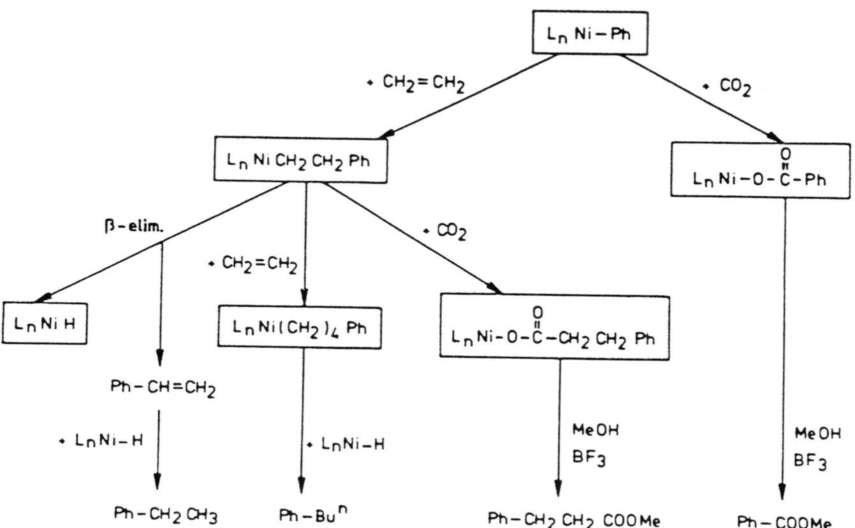

Fig. 7.7 Insertion reactions into a nickel phenyl complex

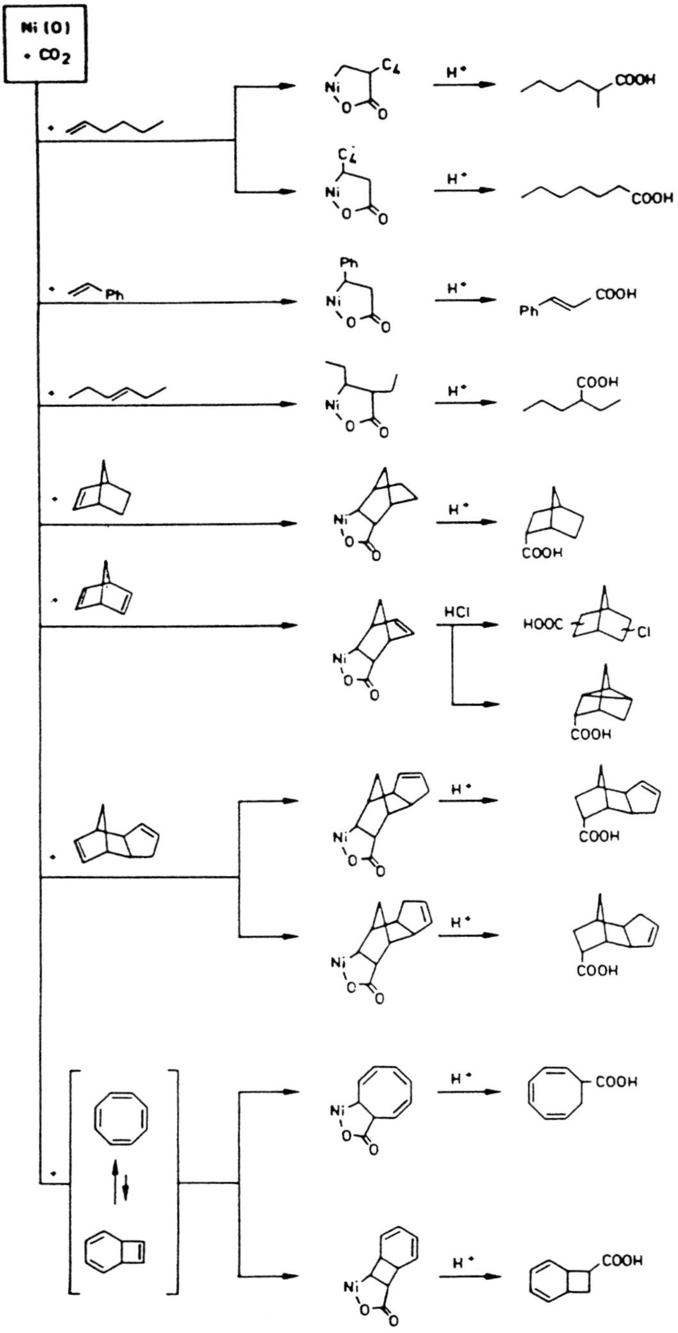

Fig. 7.6 Oxidative coupling reactions with carbon dioxide and nickel complexes.

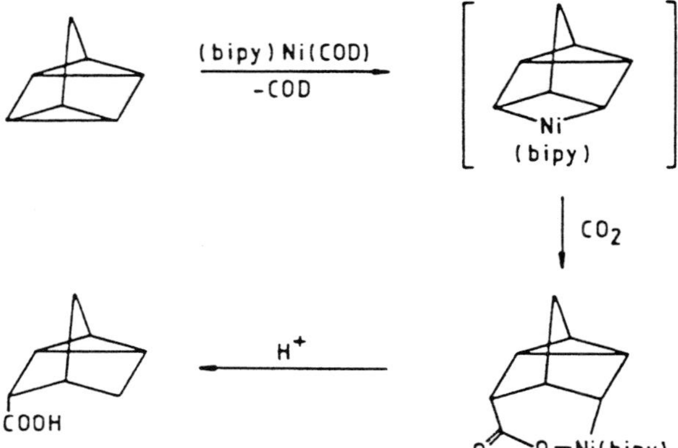

Fig. 7.8 Insertion of carbon dioxide into a nickelacyclobutane complex.

carboxylate complex. Decomposition of the complex releases the corresponding carboxylic acid. This CO_2 insertion occurs at very mild conditions with high yields, and it can be presumed that the high strain energy of the starting cycloalkane is responsible for the fast and complete reaction (Behr and Thelen 1984).

In the next section, some insertion reactions of carbon dioxide into the allylic transition-metal–carbon bond will be presented. This stoichiometric reaction is of high importance as a model reaction for the catalytic reaction of carbon dioxide with 1,3-dienes, because the dienes often reacts via allylic intermediates.

An example is the titanium–crotyl complex (Fig. 7.9), which is formed from titanocenedichloride and butadiene. Reaction with carbon dioxide yields the

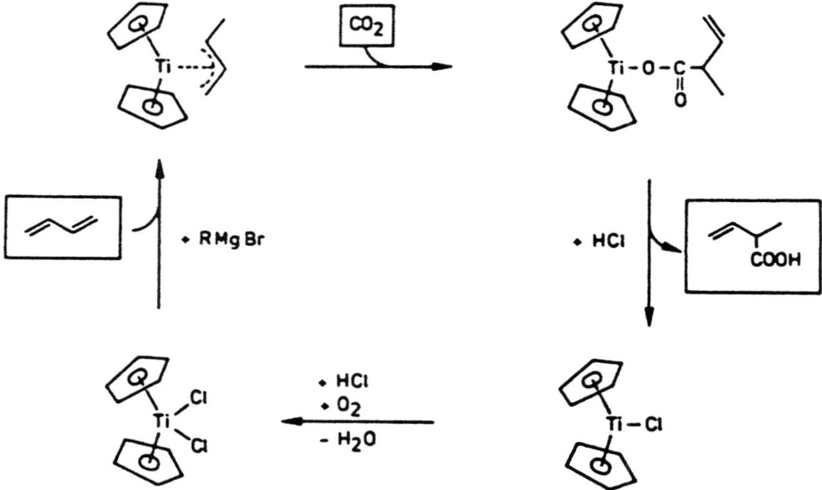

Fig. 7.9 Insertion of carbon dioxide into a titanium crotyl complex.

crotonatotitanocene complex which can be hydrolysed to methylbutenoic acid. When the titanium complex is transformed back into the starting complex, this stoichiometric reaction sequence mimics, as a whole, the catalytic synthesis of methylbutenoic acid from butadiene and CO_2 (Sato *et al.* 1981).

In the catalytic chemistry of 1,3-dienes, very often palladium complexes are of great importance. In Fig. 7.10 a palladium–allyl complex is shown which contains further ligands X and Y. By addition of a basic phosphine one coordination place in the complex is occupied by the phosphine, thus shifting the η^3-allyl system into a η^1-allyl system. Now, CO_2 can be inserted into the Pd–carbon bond, and a carboxylate complex is formed which is decomposed into unsaturated acids and esters or into lactones. It is obvious that the phosphine ligand has once again a great influence on the reactivity of the complex and on the ability of CO_2 to insert into the allyl complex. A special example for the ligand influence is given in Fig. 7.11. The starting model compound is a binuclear palladium complex in which the two metal atoms are linked together by a bisallylic octadienyl chain. If triisopropylphosphine is added to the complex one η^3-allyl system changes into the η^1-allyl-system. It is noteworthy that by this allylic rearrangement the octadienyl chain is bound by its terminal carbon on the palladium. Thus the subsequent insertion of carbon dioxide yields a complex with a linear carboxylate ligand, and the work up leads to a straight-chain product, the pelargonic acid (Behr and Ilsemann 1984; Behr *et al.* 1986*a*). This example demonstrates again the decisive control of CC-linkage reactions by the ligand field of transition metal complexes.

After these considerations of stoichiometric model reactions with CO_2 some catalytic CC-linkage reactions of carbon dioxide will be presented (Behr 1988; Braunstein *et al.* 1988*a*; Darensbourg 1990; Walther 1987). Various unsaturated

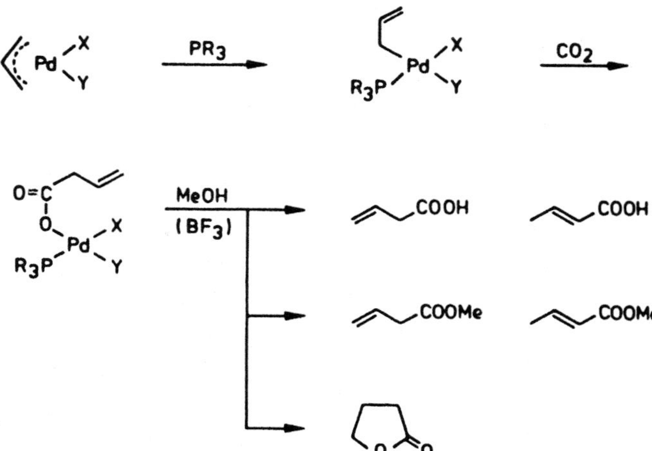

Fig. 7.10 Insertion of carbon dioxide into palladium allyl complexes.

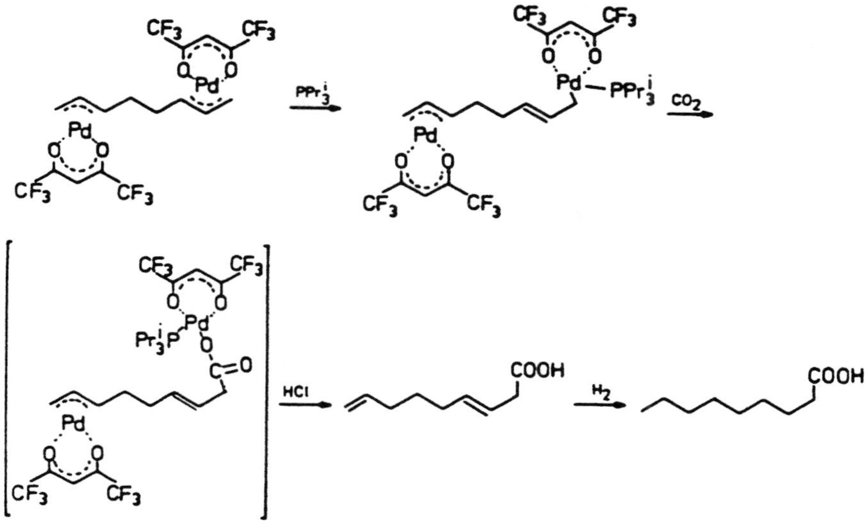

Fig. 7.11 Insertion of carbon dioxide into a binuclear palladium bisallyl complex.

hydrocarbons are able to react with CO_2, for instance alkynes, alkenes, allenes, 1,3-dienes or methylene cyclopropanes. The products formed are lactones, acids or esters, and the best transition-metal catalysts proved to be rhodium and nickel, and especially palladium complexes. Using hexynes (Fig. 7.12) substituted pyrones are produced in yields up to 50 per cent. Nickel complexes stabilized by chelate phosphine ligands are very effective catalysts. Also cationic rhodium complexes catalyse this pyrone formation; however, only in minor amounts (Inoue *et al.* 1980; Albano and Aresta 1980; Walther *et al.* 1987; Tsuda *et al.* 1989; Pillai *et al.* 1990).

The mechanism of the nickel-catalysed reaction has been discussed in detail and Fig. 7.13 shows the two proposals. In path A on the left, the first step is the oxidative coupling of two alkynes thus forming a nickelacyclopentadiene species. Insertion of CO_2 into one nickel–carbon bond would then form a cyclic carboxylate complex, which would release the pyrone by reductive elimination. However, all stoichiometric model studies with CO_2 and nickelacyclopentadiene complexes have so far failed. Path B on the right involves the reaction of one alkyne and one CO_2 molecule to form an oxanickelacyclopentene complex.

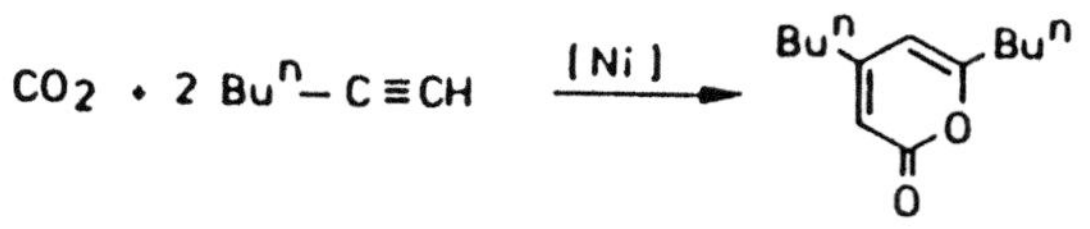

$$CO_2 + 2\,Bu^n\text{–}C\equiv CH \xrightarrow{[Ni]}$$

Fig. 7.12 Reaction of carbon dioxide with hexyne.

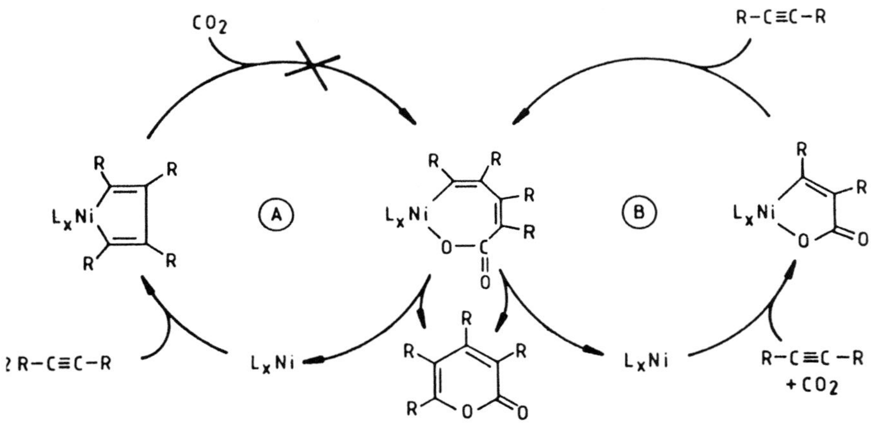

Fig. 7.13 Mechanism of the nickel-catalysed reaction of carbon dioxide with alkynes.

Further alkyne insertion yields a seven-membered ring complex which could be isolated and identified when a chelate ligand was used to stabilize this intermediate complex. This is reliable evidence that path B is the actual mechanism in nickel catalysis (Hoberg *et al.* 1984).

In the last section, catalytic CC-linkage reactions of CO_2 with 1,3-dienes will be presented (Ionue *et al.* 1978; Musco *et al.* 1978; Behr and Juszak 1983; Behr and He 1984; Behr 1985; Behr *et al.* 1986*b*; Behr 1988; Braunstein *et al.* 1988*b*). The chemistry of butadiene will be considered in more detail because of the high activity of this model substance. Figure 7.14 shows that the reaction of butadiene with CO_2 yields – depending on the transition-metal catalyst – different products, in particular γ lactones, one δ lactone, and some octadienyl esters. This reaction was investigated with the aim of a high yield of the δ lactone and palladium catalysis especially, was studied in detail. Different palladium compounds can be used as starting complexes which have ligands with

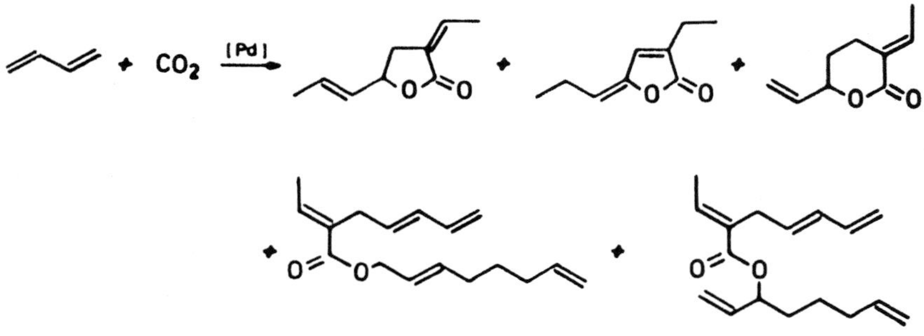

Fig. 7.14 Reaction of carbon dioxide with butadiene: Reaction products.

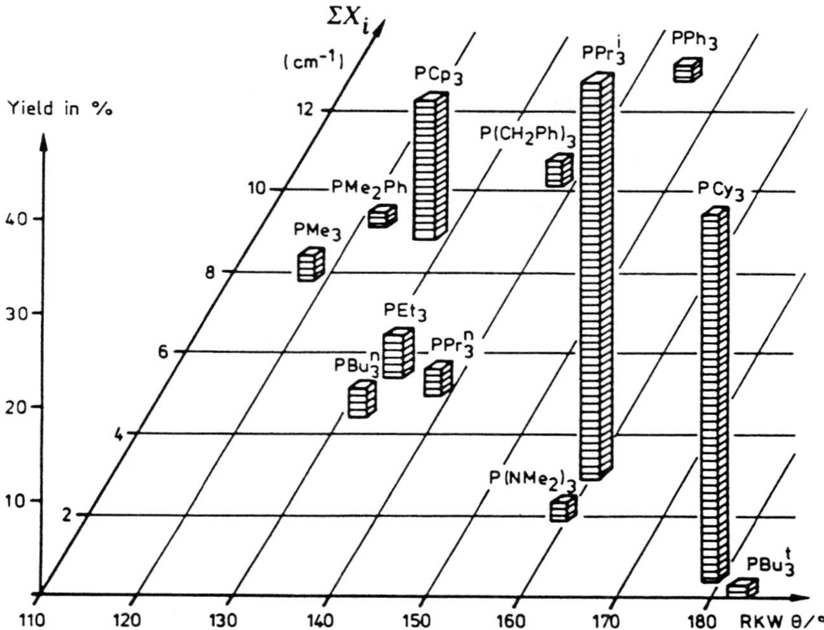

Fig. 7.15 Reaction of carbon dioxide with butadiene: Influence of the phosphine ligands on the yield of the δ-lactone.

low coordination capability such as acetylacetonate. To reach a high yield of the δ lactone, further phosphine ligands with a stronger coordination ability must be added. Figure 7.15 demonstrates that only special phosphine ligands are able to steer the reaction into the right direction. In Fig. 7.15 the electronic and steric abilities of the phosphine ligands are plotted against the yield of the δ lactone, and it is obvious that only very few phosphines, especially tricyclohexylphosphine and triisopropylphosphine, enable high yields of the desired product. Also the amount of the added ligand (Fig. 7.16) is of great importance. The figure shows the ratio of triisopropylphosphine to palladiumbis(acetylacetonate). At a ratio of 3:1 a maximum of yield and selectivity of the δ lactone can be noticed whereas an excess of phosphine blocks the free coordination sites of the catalytic intermediate. In Fig. 7.17 a simplified mechanistic proposal for the reaction of carbon dioxide and butadiene is given. The mechanistic cycle starts with the palladium phosphine complex PdL_n, in which the number of phosphine ligands cannot be determined exactly. Two molecules of butadiene coordinate and dimerize yielding a palladium bis- η^3-allyl complex with still one phosphine ligand L. This ligand changes one η^3-allyl bond into a η^1-allyl bond and thus enables the coordination of carbon dioxide. CO_2 inserts into the η^1-allyl bond and forms an allyl–carboxylate complex from which the products, lactones or esters are released. By this elimination step the starting intermediate PdL_n is

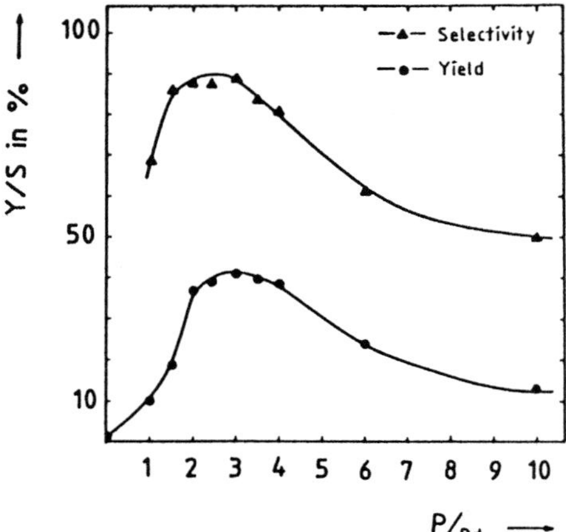

Fig. 7.16 Reaction of carbon dioxide with butadiene: Influence of the phosphine/-palladium-ratio on yield (Y) and selectivity (S) of the δ-lactone.

regenerated and the catalytic cycle is closed. The central molecule in this catalytic cycle is the palladium carboxylate complex, and therefore it was interesting to try to isolate such a compound in a stoichiometric model experiment. In the reaction of the starting complex palladiumbis (acetylacetonate), the ligand triisopropylphosphine, and the δ lactone a palladium–phosphine–carboxylate complex was indeed formed (Fig. 7.18). The X-ray structure in Fig. 7.19 shows a palladium atom in the centre of the complex with a square planar ligand coordination field. The carboxylate groups are bound to the palladium by one oxygen atom, whereas the diene chains of the carboxylate ligands do not undergo any interaction with the palladium. Obviously, this isolated complex is not identical with the catalytic intermediate discussed in the mechanistic proposal; however, its easy formation supports strongly the proposed catalytic cycle.

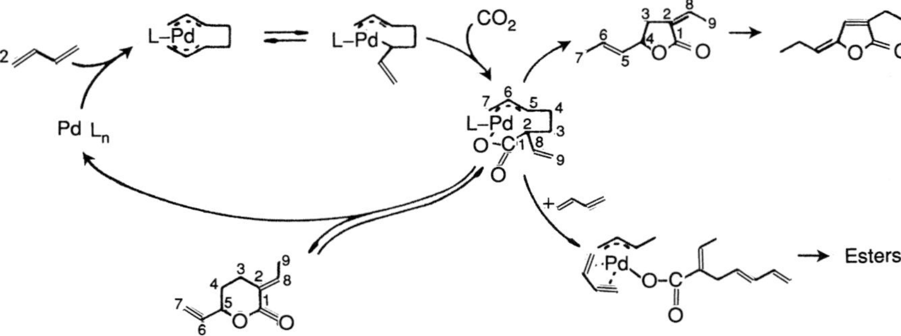

Fig. 7.17 Reaction of carbon dioxide with butadiene: Mechanistic proposal for the palladium-catalysed reaction.

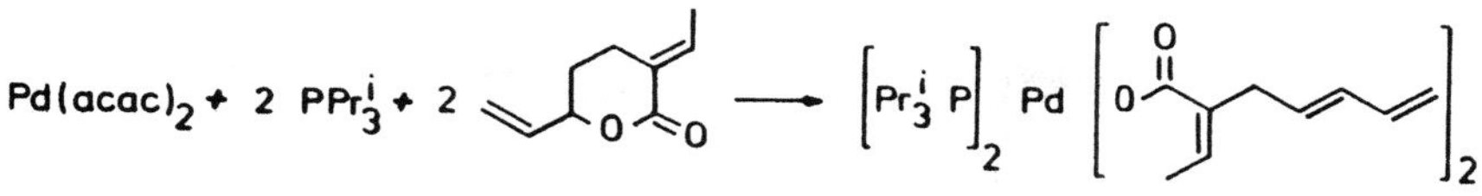

Fig. 7.18 Synthesis of the palladium phosphine carboxylate complex.

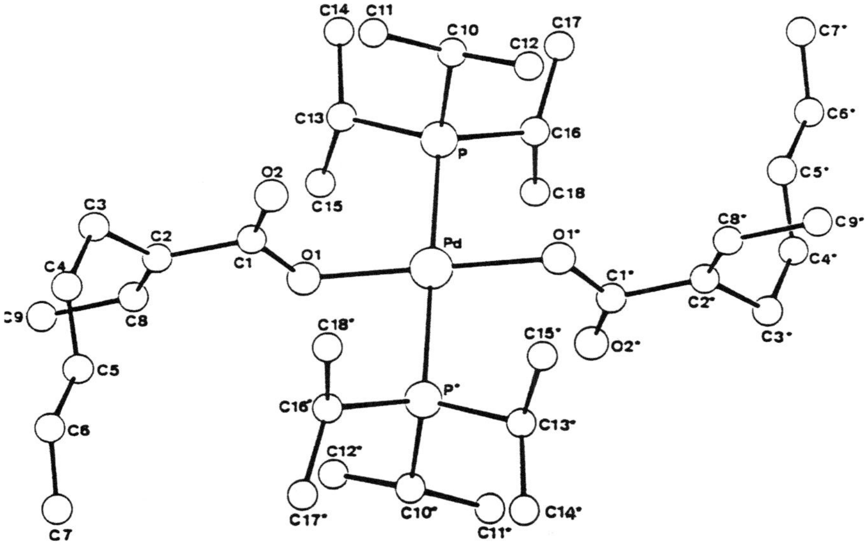

Fig. 7.19 X-ray structure of the palladium phosphine carboxylate complex.

Not only palladium, but also other transition-metals enable CC-linkage reactions between CO_2 and butadiene. Figure 7.20 gives a summary: the starting complex ML_x is located in the middle, and on the left side the already described palladium cycle A is shown with its different products. On the upper right cycle B describes the catalysis with rhodium compounds. Using rhodium catalysts three molecules of butadiene coordinate to the metal. Further insertion of CO_2 yields a γ lactone as the only observed product. Cycle C on the right describes a variant of the rhodium catalysis: the δ lactone is able to coordinate to rhodium via opening of the lactone ring. By binding of a further molecule of butadiene and by a further CC-bond formation the same γ lactone can be formed as via cycle B.

Figure 7.21 shows that dienes other than butadiene can be included in the reaction with CO_2. Isoprene and piperylene, which contain an additional methyl group in the diene molecule, can be linked with butadiene and carbon dioxide and methyl-substituted δ lactones are the isolated catalytic products.

Another interesting extension of these CO_2 reactions is a 'three component reaction' with the substrates carbon dioxide, butadiene, and epoxides. Figure

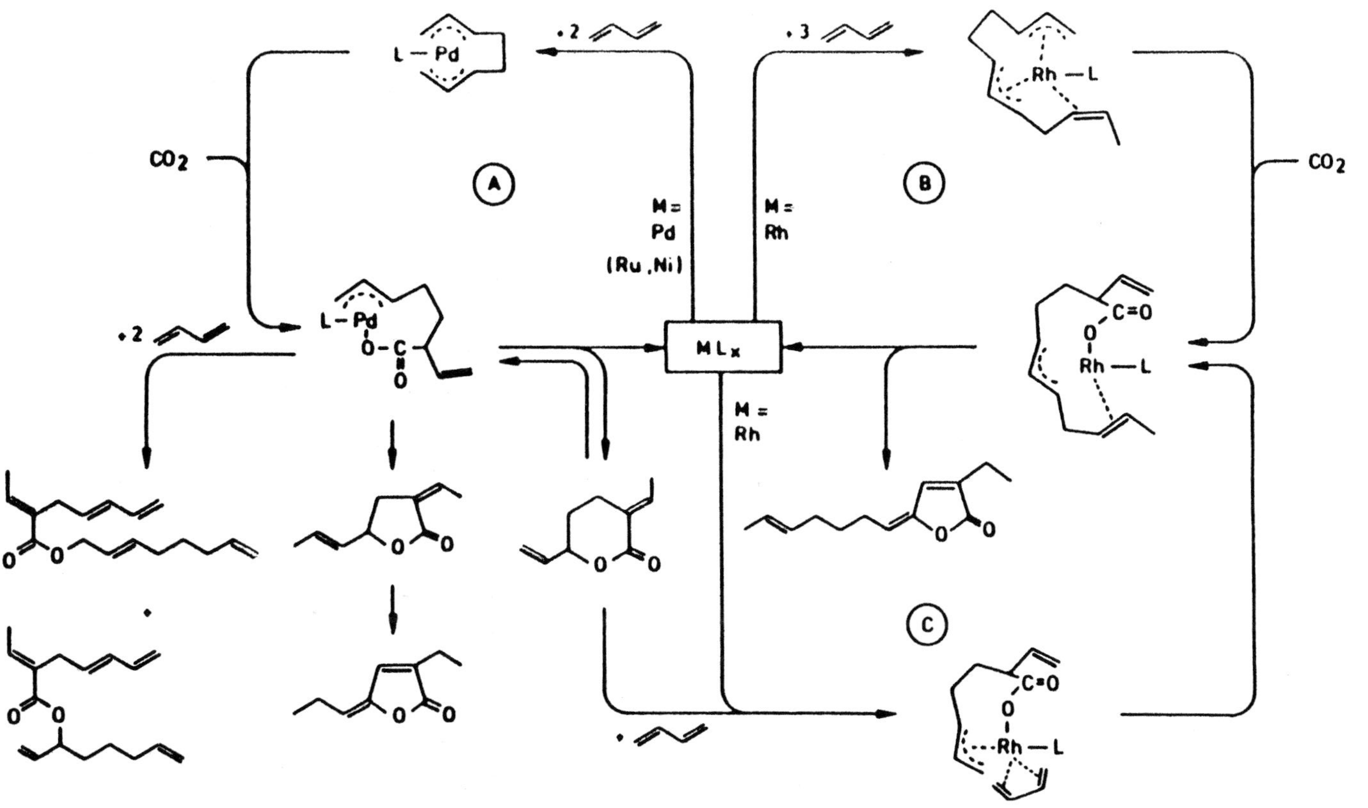

Fig. 7.20 Reaction of carbon dioxide with butadiene: General mechanistic scheme.

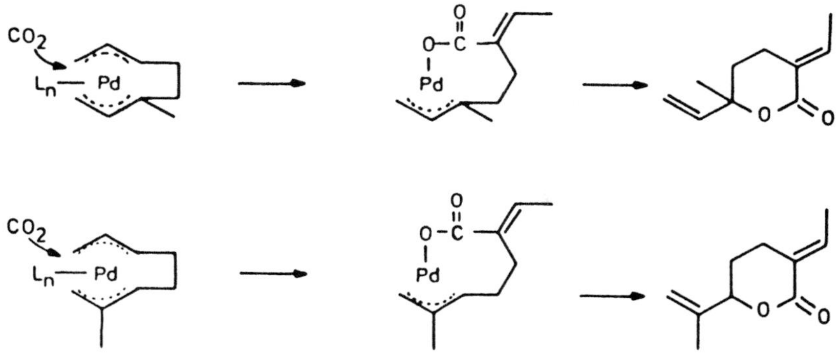

Fig. 7.21 Reactions of carbon dioxide with isoprene and piperylene.

7.22 demonstrates that these three reactants link together in a 'one step reaction' yielding glycol esters of a C_9-carboxylic acid. Both ethylene oxide and propylene oxide are possible reaction partners. Figure 7.23 shows the mechanistic proposal for the reaction with ethylene oxide. The first step is addition of butadiene and CO_2 to the palladium catalyst yielding the already well discussed carboxylate complex. Ethylene oxide can coordinate to this carboxylate complex, the oxirane ring opens and inserts into the metal–oxygen bond. An alkoxy–allyl complex is formed which then releases the glycol ester with reformation of the starting catalytic complex (Behr and Kanne 1986*b*).

This short review set out to demonstrate the narrow relationships between stoichiometric and catalytic reactions of carbon dioxide on transition-metal complexes. Stoichiometric model reactions of CO_2 with metal–carbon complexes proved to be a good method of understanding the intermediates in the catalytic reactions of unsaturated hydrocarbons and carbon dioxide. However, more work still needs to be done if a better understanding of this interdisciplinary field of research is to be achieved, and many further discoveries are yet to be made in this fascinating area of CO_2 chemistry.

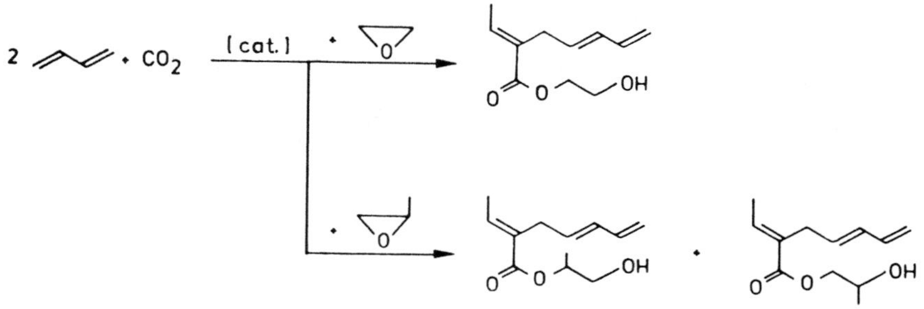

Fig. 7.22 Reactions of carbon dioxide, butadiene, and epoxides.

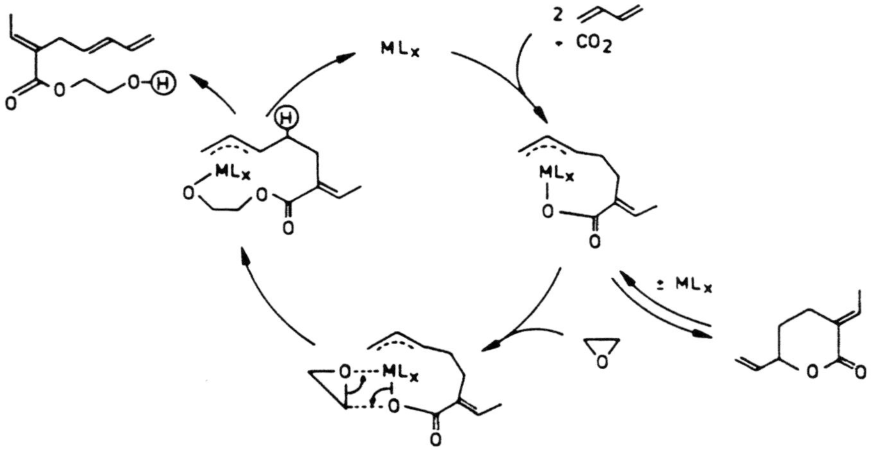

Fig. 7.23 Reaction of carbon dioxide, butadiene and ethylene oxide: Mechanistic proposal.

References

Albano, P. and Aresta, M. (1980). *J. Organomet. Chem.*, **190**, 243–46.

Arafa, I.M., Shin, K., and Gof, H.M. (1988). *J. Am. Chem. Soc.*, **110**, 5228–9.

Aresta, M. and Nobile, C.F. (1977). *J. Chem. Soc., Chem. Commun.*, 708.

Aresta, M., Nobile, C.F., Albano, V.G., Forni, E., and Manassero, M. (1975). *J. Chem. Soc., Chem. Commun.*, 636–7.

Aresta, M., Quaranta, E., and Tommasi, T. (1990). In *Enzymatic and model carboxylation and reduction reactions for carbon dioxide utilization* (ed. M. Aresta and J.V. Schloss), pp. 429–30. Kluwer, Dordrecht, The Netherlands.

Audett, J.D., Collins, T.J., Santarsiero, B.D., and Spies, G.H. (1982). *J. Am. Chem. Soc.*, **104**, 7352–3.

Behr, A. (1985). *Bull. Soc. Chim. Belg.*, **94**, 671–83.

Behr, A. (1988). *Carbon dioxide activation by metal complexes*. VCH, Weinheim.

Behr, A. and He, R. (1984). *J. Organomet. Chem.*, **276**, C69–C72.

Behr, A. and v. Ilsemann, G. (1984). *J. Organomet. Chem.*, **276**, C77–C79.

Behr, A. and Juszak, K.-D. (1983). *J. Organomet. Chem.*, **255**, 263–8.

Behr, A. and Kanne, U. (1986*a*). *J. Organomet. Chem.*, **317**, C41–C44.

Behr, A. and Kanne, U. (1986*b*). *J. Organomet. Chem.*, **309**, C15–C23.

Behr, A. and Thelen, G. (1984). *C_1-Mol. Chem.*, **1**, 137–53.

Behr, A., Keim, W., and Thelen, G. (1983). *J. Organomet. Chem.*, **249**, C38–C40.

Behr, A., v. Ilsemann, G., Keim, W., Krüger, C., and Tsay, Y.-H. (1986*a*). *Organometallics*, **5**, 514–18.

Behr, A., He, R., Juszak, K.-D., Krüger, C. and Tsay, Y.-H. (1986*b*). *Chem. Ber.*, **119**, 991–1015.

Braunstein, P., Matt, D., and Nobel, D. (1988*a*). *Chem. Rev.*, **88**, 747–64.

Braunstein, P., Matt, D., and Nobel, D. (1988*b*). *J. Am. Chem. Soc.*, **110**, 3207–12.

Calabrese, J.C., Herskovitz, T., and Kinney, J.B. (1983). *J. Am. Chem. Soc.*, **105**, 5914–15.

Cohen, S.A. and Bercaw, J.E. (1985). *Organometallics*, **4**, 1006–14.

Darensbourg, D.J. (1990). In *Enzymatic and model carboxylation and reduction reactions for carbon dioxide utilization* (ed. M. Aresta and J.V. Schloss), pp.43–64, Kluwer. Dordrecht, The Netherlands.

Döhring, A., Jolly, P.W., Krüger, C., and Romao, M.J. (1985). *Z. Naturforsch.*, **40b**, 484–8.

Hoberg, H., Schäfer, D., Burkhart, G., Krüger, C., and Romao, M.J. (1984). *J. Organomet. Chem.*, **266**, 203–24.

Hoberg, H., Jenni, K., Angermund, K., and Krüger, C. (1987). *Angew. Chem.*, **99**, 141–2.

Inoue, Y., Sasaki, Y., and Hashimoto, H. (1978). *Bull. Chem. Soc. Jpn.*, **51**, 2375–8.

Inoue, Y., Itoh, Y., Kazama, H., and Hashimoto, H. (1980). *Bull. Chem. Soc. Jpn.*, **53**, 3329–33.

Jolly, P.W., Jonas, K., Krüger, C., and Tsay, Y.-H. (1971). *J. Organomet. Chem.*, **33**, 109–22.

Musco, A., Perego, C., and Tartiari, V. (1978). *Inorg. Chim. Acta*, **28**, L147–L148.

Pillai, S.M., Ohnishi, R., and Ichikawa, M. (1990). *J. Chem. Soc., Chem. Commun.*, 246–7.

Sato, F., Iijima, S., and Sato, M. (1981). *J. Chem. Soc., Chem. Commun.*, 180–1.

Tsuda, T., Morikawa, S., and Saegusa, T. (1989). *J. Chem. Soc., Chem. Commun.*, 9–10.

Vivanco, M., Ruiz, J., Floriani, C., Chiesi-Villa, A., and Guastini, C. (1990). *Organometallics*, **9**, 2185–7.

Walther, D. (1987). *Coordination Chem. Rev.*, **79**, 135–74.

Walther, D., Dinjus, E., Sieler, J., Andersen, L., and Lindqvist, O. (1984). *J. Organomet. Chem.*, **276**, 99–107.

Walther, D., Schönberg, H., Dinjus, E., and Sieler, J. (1987). *J. Organomet. Chem.*, **334**, 377–88.

8

Organometallic chemistry of carbon dioxide pertinent to catalysis

Donald J. Darensbourg

Introduction

Transition-metal mediated transformations of carbon dioxide have attracted increased attention from the organometallic chemist. The search for alternate uses of carbon dioxide is motivated by two principal concerns: the increase in carbon dioxide concentration levels over the past four centuries and the depletion of fossil fuels. Indeed the two problems represent both the cause and effect, where the increased burning of fossil fuels has been implicated as one of the primary causes of the rise of CO_2 concentration levels (Idso 1988; Abrahamson 1989). The controversy over global climate changes are subsiding, with the consensus of opinion being that increasing concentrations of greenhouse gases in the atmosphere will sooner or later lead to global warming (~0.3 °C per decade). Of the greenhouse gases carbon dioxide contributes most to global warming, i.e. approximately 55 per cent. An issue which is now open to debate is how to reduce emissions of greenhouse gases in an economically feasible manner. Some of the easier measures to take would be to control CO_2 emissions in manufacturing and electricity generation processes, and to recover part or all of the cost of doing this with production of value added chemicals.

The exploitation of carbon dioxide as an alternative source of chemical carbon is intimately associated with a better understanding of the challenging chemistry accompanying this ubiquitous molecule. Central reactions in this chemistry are the insertion of CO_2 into M–H, M–C, and M–O bonds. The authors are presently investigating the mechanistic aspects of these reaction processes and will herein describe their current level of understanding of these insertion and deinsertion reactions. It is worth noting that these reactions are of importance to both chemical and biochemical processes. As indicated in eqns (1)–(3), reactions (1) and (3) are reversible, whereas reaction (2) is, in general, irreversible. In all three of these processes there is no evidence for coordination of CO_2 at the *metal centre* prior to insertion. Hence, CO_2 is best thought of as an external reagent which couples to a metal coordinated substrate, concomitantly entering the metal's coordination sphere as a carboxylated ligand.

$$[M]H + CO_2 \quad [M]O_2CH \qquad (1)$$
$$[M]R + CO_2 \rightarrow [M]O_2CR \qquad (2)$$
$$[M]OR + CO_2 \quad [M]O_2COR \qquad (3)$$

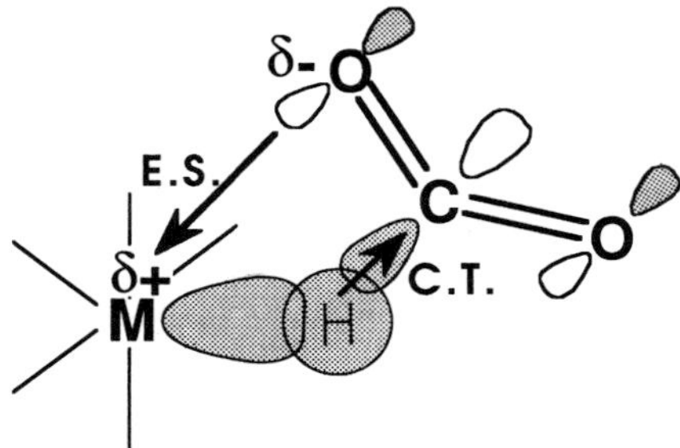

Fig. 8.1 Strong charge-transfer interaction between metal hydride and carbon dioxide.

Insertion of carbon dioxide into the metal-hydride bond

A well studied theme in our laboratories has been the mechanistic aspects of the insertion of carbon dioxide into anionic metal hydrides of chromium and tungsten to afford metalloformates. Reactions of these particular metal derivatives occur readily at ambient temperature and reduced pressures of carbon dioxide (Darensbourg *et al.* 1981; Darensbourg and Rokicki 1982*a*; Slater *et al.* 1982), and are thought to be prototypic of carboxylation reactions involving low valent, saturated metal centres. There are no documented cases of CO_2 insertion into the metal-hydride bond to provide the alternative, metallocarboxylic acid isomer, [M]COOH. Recent theoretical studies ascribe this preference to a strong charge-transfer interaction between the M–H σ orbital and the π^* orbital of CO_2, supplemented by an attractive electrostatic interaction between H——CO_2 and O——M (Sakaki and Ohkuba 1989). This interaction is illustrated in Fig. 8.1. Nevertheless, metallocarboxylic acid complexes are well known in organometallic chemistry resulting from the direct nucleophilic addition of hydroxide ion at a metal–bound carbon monoxide ligand (eqn. (4)) (Ford and Rokicki 1988). Reaction (4) is reversible as shown by oxygen atom exchange processes when $^{18}OH^-$ is employed, where the metal derivative becomes enriched in oxygen-18 labelled CO (Darensbourg *et al.* 1980)

$$M–CO + OH^- \rightleftharpoons M–COOH^- \tag{4}$$

The reaction profile for CO_2 insertion/deinsertion processes involving the group 6-metal carbonyl hydride anions is depicted in Scheme 8.1. The activation parameters for decarboxylation were determined by way of the $^{13}CO_2$ exchange reactions defined in eqn (5) (Darensbourg *et al.* 1990*a*). This process was monitored by means of ^1H NMR, where the formate resonance at $\delta 8.28$ ppm is split by $^{13}C(J_{C\text{-}H} = 186$ Hz) upon $^{13}CO_2$ exchange. The reaction was observed to be first order in $HCO_2Cr(CO)_5^-$, zero order in CO_2, and retarded by addition of carbon monoxide. As indicated in Scheme 8.1 the O, H-dihapto structure represents the transition state of the rate determining step and the O,O-dihapto structure a reac-

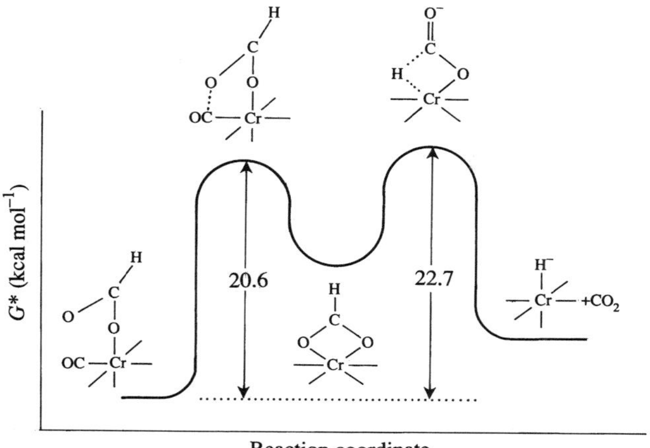

Scheme 8.1

tion intermediate. SCF calculations of the relative energies of these various species are totally consistent with these experimental observations (Bo and Dedieu 1989).

$$Cr(CO)_5O_2CH^- + (\text{excess})\ ^{13}CO_2 \rightleftharpoons Cr(CO)_5O_2{}^{13}CH^- + CO_2 \quad (5)$$

The barrier for *cis*-CO loss from $Cr(CO)_5O_2CH^-$ was obtained by measuring the kinetic parameters for reaction (6) involving the closely related derivative, $Cr(CO)_5O_2CCH_3^-$, which does not undergo decarboxylation. For a series of $W(CO)_5O_2CR^-$ analogues the rate of *cis*-CO loss was shown to be highly dependent on the nature of the R substituent, with electron-releasing R groups resulting in enhanced rates of CO dissociation (Darensbourg *et al.* 1991*a*). The *cis*-CO labilizing ability of the monodentate-bound carboxylate ligand is attributed to an intramolecular nucleophilic displacement of a CO ligand by the distal oxygen atom of the carboxylate ligand as depicted in Fig. 8.2, with attendant formation of an η^2-O_2CR intermediate (Scheme 8.2).

$$Cr(CO)_5O_2CCH_3^- + P(OCH_3)_3 \rightarrow cis\text{-}Cr(CO)_4(P(OCH_3)_3)O_2CCH_3^- \quad (6)$$

It should be pointed out here that at higher CO pressures (>600 psi) the decarboxylation reaction (eqn (5)) is not completely quenched. This behaviour is indicative of an alternative, slightly higher energy pathway for decarboxylation which does not involve CO dissociation. Such a transition state is illustrated in Fig. 8.3 which involves an interaction of an unoccupied metal orbital and the occupied formate fragment (consisting of the π orbitals of CO_2 and the a_1 orbital of the hydride). Indeed, Sullivan and Meyer (1984) have noted such a mechanism involving an osmium formate derivative where the metal-ancillary ligand bonds are strong. By way of contrast the ruthenium analogue, which possesses weaker metal-ligand bonds, proceeds by the dissociative route.

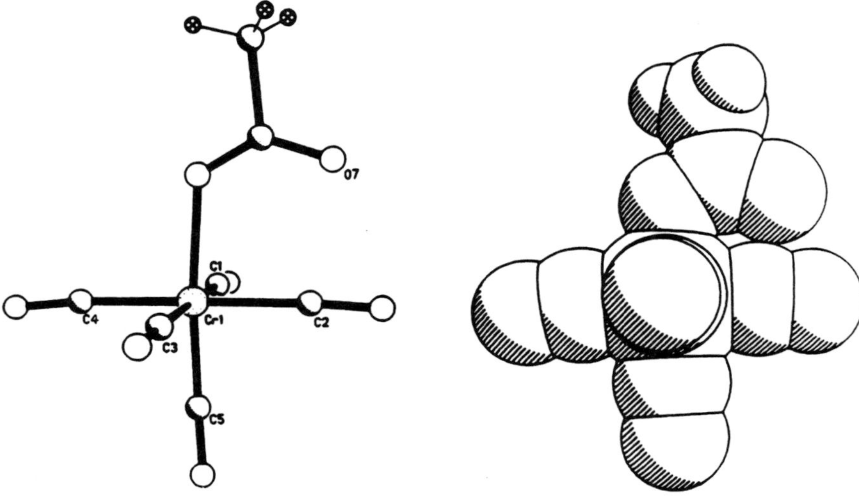

Fig. 8.2 Model of $[Cr(CO)_5O_2CCH_3]^-$ with bond rotation through the chromium–oxygen bond to minimize the O7 equatorial-C distance, as well as the corresponding space-filling model.

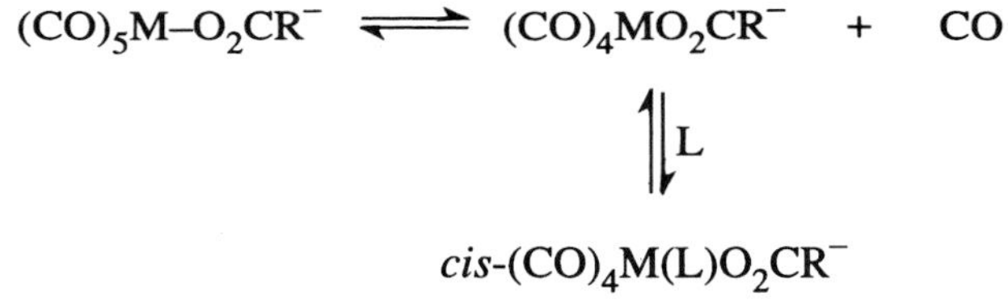

$$(CO)_5M\text{–}O_2CR^- \rightleftharpoons (CO)_4MO_2CR^- + CO$$

$$(CO)_4MO_2CR^- \underset{}{\overset{L}{\rightleftharpoons}} cis\text{-}(CO)_4M(L)O_2CR^-$$

Scheme 8.2

Fig. 8.3 Orbital interaction in the non-dissociative transition state for decarboxylation of the formate ligand in an octahedral complex.

The paucity of ancillary-ligand dissociation prior to formate decarboxylation is most apparent in reactions involving unsaturated metal centres, eqn (7). The rate of CO_2 exchange was demonstrated to be first order in nickel complex and zero order in CO_2, consistent with a rate determining step involving CO_2 extru-

Fig. 8.4 Orbital interaction between low lying empty b_2 orbital on $[Ni(H)(PCy_3)_2]$ fragment which interacts with the doubly occupied orbital of the formate ion.

sion to afford a metal dihydride intermediate which rapidly reinserts CO_2 (Darensbourg *et al.* 1990*b*). The reaction was found to be independent of added tricyclohexylphosphine concentrations. This is not unanticipated since in these d^8 metal species there is a low-lying empty p orbital for interacting in a bonding manner with the doubly occupied orbital of HCO_2^- (Fig. 8.4).

$$HNi(O_2CH)(PCy_3)_2 + (\text{excess})\ {}^{13}CO_2 \rightleftharpoons HNi(O_2{}^{13}CH)(PCy_3)_2 + CO_2 \quad (7)$$

Similarly the activation parameters ($\Delta H^* = 22$ kcal mol^{-1} and $\Delta S^* = -4.8$ cal mol-deg^{-1}) for reaction (8) were observed to be nearly identical with those determined for reaction (7), indicating a common reaction intermediate in the two processes. The latter intramolecular C–H/Ni–D exchange process exhibits an equilibrium isotope effect, with $K_{eq} = 1.47$.

$$DNi(O_2CH)(PCy_3)_2 \rightleftharpoons HNi(O_2CD)(PCy_3)_2 \quad (8)$$

Insertion of carbon dioxide into metal–carbon bonds

Anionic alkyl and aryl complexes of group 6 metal carbonyls generally undergo irreversible CO_2 insertion forming metallocarboxylates as indicated in eqn (9) for $RW(CO)_5^-$ derivatives (Darensbourg and Rokicki 1982*b*, Darensbourg and Kudaroski 1984). The rate of reaction (9) was dependent on the nature of the X

$$XCH_2^-W(CO)_5^- + CO_2 \rightarrow XCH_2CO_2W(CO)_5^- \quad (9)$$

substituent, decreasing along the series X = CH_3>H>Ph>>CN. The reactions were shown to obey second order kinetics, first order in metal substrate and first order in carbon dioxide, with activation parameters indicative of an associative interchange (I_a) mechanism (Darensbourg *et al.* 1985). For example, when X = H, $\Delta H^* = 10.2$ kcal mol^{-1} and $\Delta S^* = -43.3$ e.u. Compatible with an I_a pathway the stereochemistry about the α-carbon atom is maintained during the carbon dioxide insertion reaction (Darensbourg and Grotsch 1985). That is, *threo*-$W(CO)_5CHDCHDPh^-$ underwent CO_2 insertion to afford *threo*-$W(CO)_5O_2CCHDCHDPh^-$ (eqn (10)).

Furthermore, the second-order rate constant for CO_2 insertion into the metal-carbon bond was found to be independent of added CO pressure, indicative of a transition state containing a saturated metal centre. The rate of CO_2 insertion

Ph, H, D, D, H, OTs — *erthro* $\xrightarrow[-NaOTs]{Na_2W(CO)_5}$ Ph, D, H, D, H, $W(CO)_5$ — *threo* $\xrightarrow{CO_2}$ Ph, D, H, D, H, $O{=}C{-}O{-}W(CO)_5$ — retention (10)

$^3J_{HCCH}$=3.6 Hz $^3J_{HCCH}$=4.3 Hz

was greatly accelerated by enhancing the nucleophilic character of the metal centre by replacing CO ligands with more electron donating, sterically nonencumbering phosphine ligands. For example, the rate of CO_2 insertion into *cis*-$CH_3W(CO)_4PMe_3^-$ is 243 times greater than that in $CH_3W(CO)_5^-$. At the same time the W–CH_3 bond distance observed in *cis*-$CH_3W(CO)_4PMe_3^-$ of 2.18(3) Å is somewhat shorter (and presumably stronger) than the corresponding bond length in the $CH_3W(CO)_5^-$ anion (Darensbourg *et al.* 1987).

In summary, the proposed transition state for the carbon–carbon bond-forming reaction involving carbon dioxide, where there is simultaneous interaction of CO_2 with the nucleophilic metal centre and the alkyl or aryl ligand, is illustrated in **A**. The nature of the counter ion (M^+) associated with the anionic metal alkyl or aryl derivative has an effect on the rate of CO_2 insertion. To cite an instance, the carboxylation reaction of $CH_3W(CO)_5^-$ was found to occur ten times faster for the Na^+ salt as compared with the kryptofix-2.2.1 encapsulated Na^+ salt (Darensbourg and Pala 1985). This rate enhancement is attributed to the cation (M^+) stabilizing the negative charge build up on the distal oxygen atom of the incipient carboxylate ligand. That alkali metal cations are capable of strong interactions with the distal oxygen atom of a monodentate carboxylate ligand

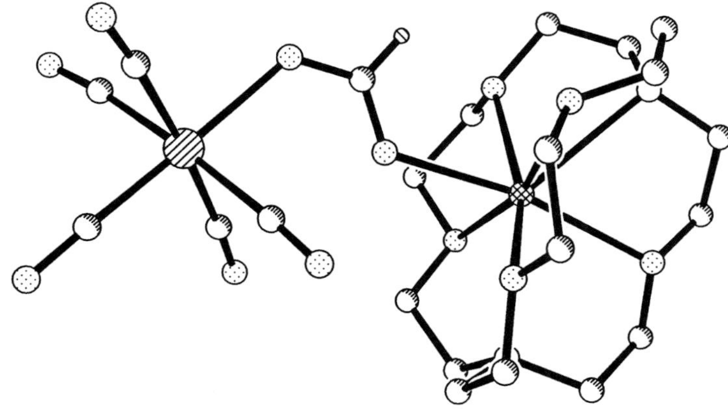

Fig. 8.5 X-ray structure of [Na-kryptofix-2.2.1][$W(CO)_5O_2CH$].

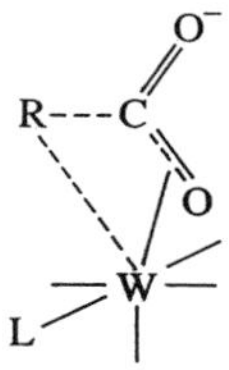

A

is dramatically seen in the solid-state structure of [Na–kryptofix-2.2.1][$HCO_2W(CO)_5$], where the shortest Na···O interaction involves the distal oxygen atom of the formate ligand (Fig. 8.5).

Dedieu *et al.* (1992) have rationalized the mechanistic differences observed for the C–H versus C–C bond forming reactions of carbon dioxide on the basis of orbital interactions. In the C–H bond forming process there is a stabilizing interaction between the empty d_π orbital of the $M(CO)_4$ fragment and the doubly occupied valence orbital of the incipient HCO_2^- ligand (**B**), whereas, in the C–C bond-forming reaction there is unfavourable overlap between the orbital of RCO_2^- and d_π(**C**). In this instance the upper lobe of the d_π orbital points towards the nodal plane of the p component on the carbon atom, hence there is no advantage in proceeding via a dissociative mechanism. Consequently, the rates of these latter reactions (eqn (2)) are considerably slower than the corresponding processes (eqn (1)) involving the analogous metal hydrides (see earlier). In addition we have compared the rates of CO_2 insertion into M–H versus M–R bonds directly, i.e. employing the complex $Ni(H)(Ph)(PCy_3)_2$. As expected carbon dioxide insertion occurred quantitatively and exclusively at the Ni–H bond to yield the $(HCO_2)Ni(Ph)(PCy_3)_2$ derivative. However, it has not been established whether the formate complex is the thermodynamic product.

As mentioned in the introduction the carbon–carbon bond-forming reaction (eqn (2)) is generally irreversible; however, in specific instances metal–carboxylates undergo decarboxylation with formation of the corresponding metal–carbon bonded complex (Mehrota and Bohra 1983; Deacon *et al.* 1986). A notable

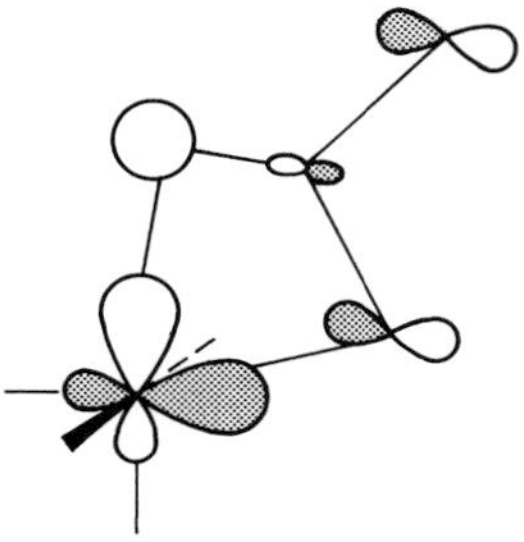

B 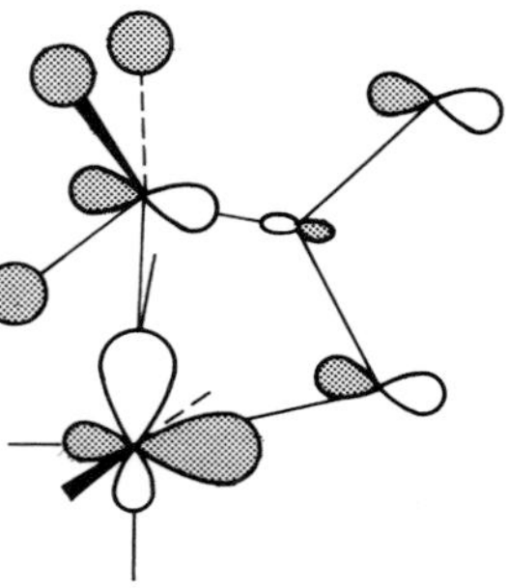 **C**

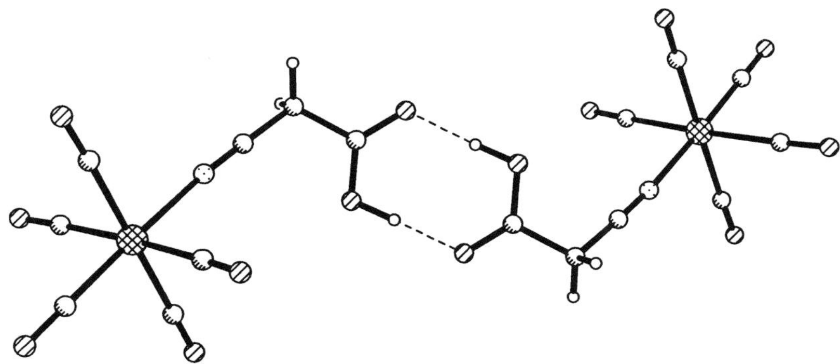

Fig. 8.6 X-ray structure of $W(CO)_5NCCH_2COOH$ illustrating strong intermolecular hydrogen bonding.

example is copper (I) cyanoacetate in the presence of tri-*n*-butylphosphine, eqn (11) (Tsuda *et al.* 1974, 1976*a*,*b*, 1978*a*,*b*, 1980). Although we have an ongoing investigation of the mechanistic aspects of this process involving copper (I) derivatives (Darensbourg *et al.* 1991*b*), perhaps of more relevance to the current discussion are decarboxylation reactions involving Group VI metal carboxylates.

$$NCCH_2CO_2Cu(P\text{-}n\text{-}Bu_3)_x \quad NCCH_2Cu(P\text{-}n\text{-}Bu_3)_x + CO_2 \tag{11}$$

Despite the fact that the $W(CO)_5O_2CCH_2CN^-$ derivative does not sustain decarboxylation under any reasonable set of conditions in either solutions or the solid state, it is a very effective catalyst for the homogeneous decarboxylation of cyanoacetic acid (eqn (12))(Darensbourg *et al.* 1991*c*). The cyanoacetic acid substrate inhibits catalysis via the formation of the N-bound $W(CO)_5NCCH_2COOH$ species, which exhibits strong intermolecular hydrogen-bonding in the cystalline state (Fig. 8.6). That the $W(CO)_5O_2CCH_2CN^-$ complex is the active species is manifested in the observations that the rate of decarboxylation decreases as the initial acid concentration increases and increases as the acid is consumed, with concomitant decrease and increase in the concentration of $W(CO)_5O_2CCH_2CN^-$, respectively. Furthermore, the activation parameters of CO dissociation in $W(CO)_5O_2CCH_2CN^-$ are nearly identical to those determined for the decarboxylation process.

$$HOOCH_2CN \xrightarrow[\text{DME (65°C)}]{W(CO)_5O_2CCH_2CN^-} CH_3CN + CO_2 \tag{12}$$

Scheme 8.3 depicts our current interpretation of these experimental findings, where the rate-determining step is loss of CO from $W(CO)_5O_2CCH_2CN^-$ with subsequent binding of the acid via the nitrile functionality. Figure 8.7 illustrates this intermediate by means of a molecular model derived from known structural

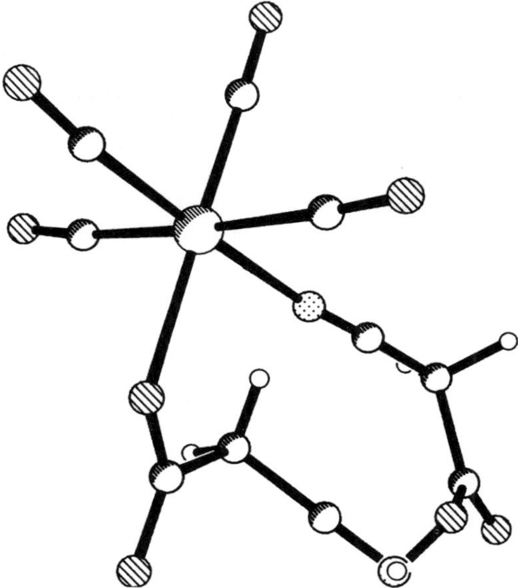

Fig. 8.7 Molecular model of the proposed intermediate in the decarboxylation reaction of cyanoacetic acid, *cis*-$W(CO)_4(O_2CCH_2CN)(NCCH_2COOH)^-$.

parameters for the $W(CO)_5O_2CCH_2CN^-$ and the $W(CO)_5NCCH_2COOH$ derivatives. It is apparent from this modelling study that the unbound nitrile group in the intermediate is poised to accept the proton from the N-bound acid.

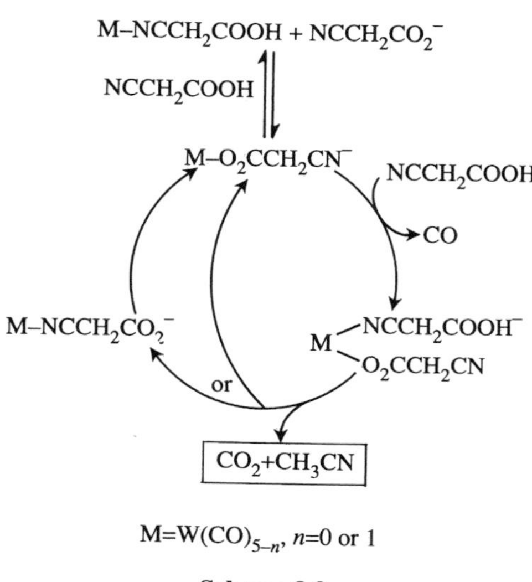

$M=W(CO)_{5-n}$, n=0 or 1

Scheme 8.3

Following proton transfer, CO_2 is lost and reprotonation of the N-bound carbanion moiety occurs with formation of the acetonitrile complex. Consistent with this analysis addition of acetonitrile inhibits the decarboxylation reaction. Further studies are underway to examine the role of other metal-anchored bases which can behave as proton acceptors from the N-bound cyanoacetic acid, for example $W(CO)_4(\eta^2\text{–}PPh_2CH_2CO_2)^-$. Hopefully, mechanistic knowledge of the decarboxylation process will aid in the development of catalysts for the carboxylation of select C–H bonds to provide carboxylic acids.

Insertion of carbon dioxide into metal–oxygen bonds

Mechanistic investigations of carbon dioxide insertion reactions into metal–oxygen bonds (eqn (3)) involving anionic octahedral complexes analogous to the metal-hydride and metal-alkyl complexes previously discussed are hampered by the reactivity patterns of monomolecular metal alkoxides. Namely, monomolecular metal alkoxides are unstable with respect to formation of the metal hydride and corresponding aldehyde or ketone, eqn (13) (Tooley *et al.* 1986). In addition, these complexes exhibit a strong tendency to aggregate to dimers or higher nuclearity clusters (McNeese *et al.* 1984, 1985; Darensbourg *et al.* 1990*c*). Scheme 8.4 summarizes the behaviour of these derivatives towards aggregation. Both of these chemical properties are a direct consequence of these $M(CO)_5OR^-$ complexes having very labile CO ligands, as demonstrated by reaction (14) which occurs on a time scale of minutes at ambient temperature

$$[M]OCHR_2^- \rightleftharpoons [M]H^- + R_2CO \qquad (13)$$

$$M(CO)_5OR^- + {}^{13}CO \rightleftharpoons M(CO)_{5-n}({}^{13}CO)_nOR^- + CO \qquad (14)$$

The CO-labilizing property of alkoxide and aryloxide ligands is attributed to the ability of these ligands to π-donate electrons to the metal centre, and thus stabilize the developing intermediate resulting from CO dissociation (Fig. 8.8). Consistent with this account aryloxide derivatives with more electron-withdraw-

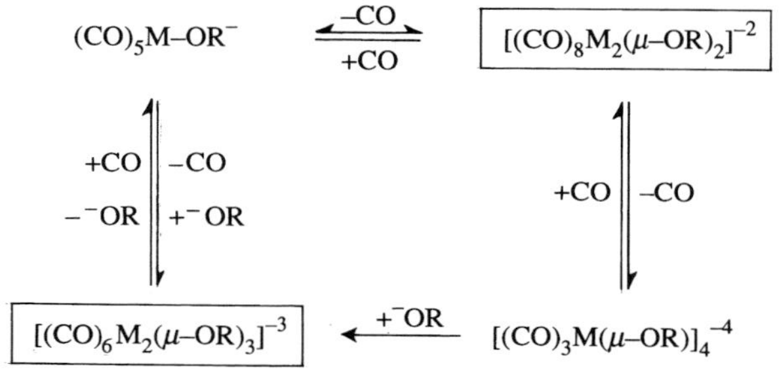

$$(CO)_5M\text{–}OR^- \underset{+CO}{\overset{-CO}{\rightleftharpoons}} \boxed{[(CO)_8M_2(\mu\text{–}OR)_2]^{-2}}$$

$(CO)_5M\text{–}OR^-$ ⇅ (+CO, $-{}^-OR$ / −CO, $+{}^-OR$) $\boxed{[(CO)_6M_2(\mu\text{–}OR)_3]^{-3}}$

$[(CO)_8M_2(\mu\text{–}OR)_2]^{-2}$ ⇅ (+CO / −CO) $[(CO)_3M(\mu\text{–}OR)]_4^{-4}$

$$\boxed{[(CO)_6M_2(\mu\text{–}OR)_3]^{-3}} \xleftarrow{+{}^-OR} [(CO)_3M(\mu\text{–}OR)]_4^{-4}$$

Scheme 8.4

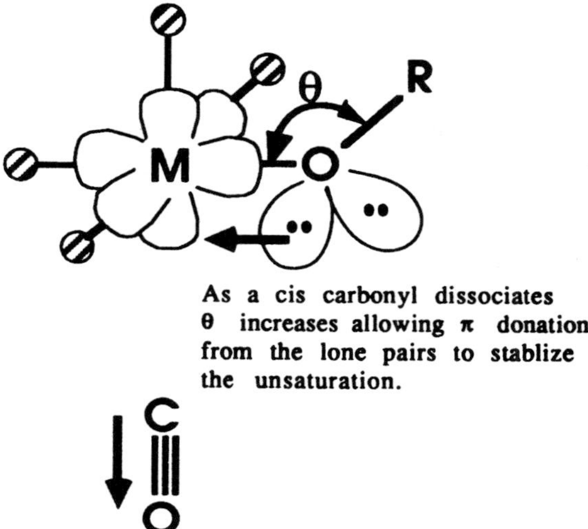

R
θ
M
O
C
O
As a cis carbonyl dissociates
θ increases allowing π donation
from the lone pairs to stablize
the unsaturation.

Fig. 8.8 Role of oxygen π-donor ligands in the labilization of CO ligands.

ing substituents are less prone to carbonyl dissociation. In addition the π-donor interaction of alkoxide ligands has been demonstrated in the ground-state structure of the formally 16-electron Ir(III) derivative, $IrH_2(OR)(PCy_3)_2$ (Lunder *et al.* 1991). In this complex the Ir–O–C angle is 138.0(11)°, with an Ir–O bond length of 2.032(10) Å Upon reacting this derivative with CO to provide the 18-electron complex, $IrH_2(OR)(CO)(PCy_3)_2$, the Ir–O distance lengthens to 2.169(7) Å and the Ir–O–C angle closes to 118.4(7)° which supports strong O → Ir π donation in the former complex.

Further indication of significant O → W π donation in $W(CO)_5OR^-$ derivatives is seen in the reactivity of these species in the presence of the corresponding alcohol, where the complexes are greatly stabilized towards CO dissociation. Alcohols have been shown both spectroscopically and structurally to hydrogen bond via the lone pairs of the –OR ligand (Kegley *et al.* 1987; Darensbourg *et al.* 1989). Hence, in the presence of alcohol less π-electron density on the –OR ligand is available for π donation to the metal centre. Figure 8.9 illustrates the strong hydrogen-bonding interaction generally observed in these derivatives by way of the structurally defined $W(CO)_4O_2C_6H_4^{-2}$ complex.

In addition to the interaction of alcohols with the lone pairs of metal-bound alkoxide or aryloxide moieties, other electrophiles such as alkali metal ions can undergo a similar interaction, **D**. Hence, it would be anticipated that the electrophilic site in carbon dioxide initially form an adduct with the M–OR functionality prior to insertion. Scheme 8.5 summarizes the insertion reaction of CO_2 with the $W(CO)_5OPh^-$ anion. As indicated in Scheme 8.5, when wet carbon dioxide is employed in the reaction sequence or when water is added to the phenylcarbonate complex, irreversible production of the stable carbonate

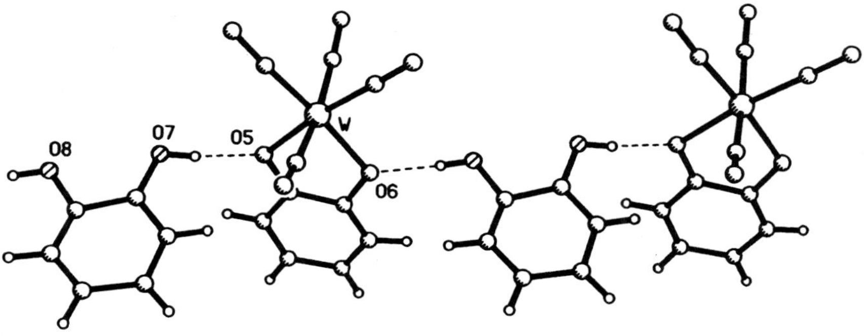

Fig. 8.9 X-ray structure of the $W(CO)_4(O_2C_6H_4)^{-2}$ anion, illustrating strong intermolecular hydrogen bonding with *m*-dihydroxybenzene.

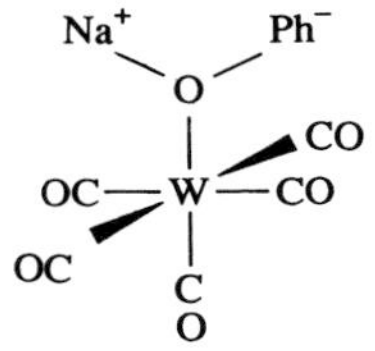

D

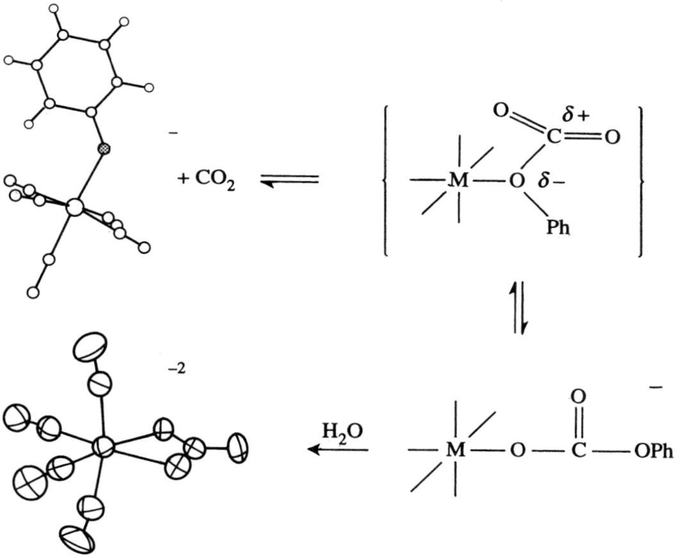

Scheme 8.5

derivatives results. Consistent with this nondissociative view of the CO_2 insertion process, the addition of carbon monoxide does not retard the rate of CO_2 insertion nor does it lead to production of $W(CO)_6$. Moreover, the insertion reaction proceeds at a reduced rate in the presence of alcohols which competively bind at the M–OR site.

The reaction sequence in Scheme 8.5 is not significantly different from the mechanism of action proposed for CO_2 hydration (eqn (15)) catalysed by the zinc metalloenzyme, carbonic anhydrases (Brown 1990). The catalytic site consists of a tetrahedral coordination environment in which Zn^{+2} is bound to three histidine imidazole groups and a water molecule, $[His_3Zn\text{-}OH_2]^{+2}$ (His = Histidine). The catalytic cycle for hydration of CO_2 is skeletally presented in Scheme 8.6.

$$CO_2 + H_2O \rightleftharpoons HCO_3^- + H_3O^+ \tag{15}$$

The question of whether the zinc bicarbonate intermediate is a four- or five-coordinate species is still under discussion; however, recent studies on model complexes, including investigations involving the isoelectronic NO_3^- anion, indicate HCO_3^- is probably bound in a monodentate manner to Zn^{+2} (Alsfasser *et al.* 1991; Han and Parkin 1991). It is also of importance to note that the active site of carbonic anhydrase is a conical, hydrophobic cavity some 15 Å deep. Hence, carbon dioxide concentration is enhanced in the cavity, i.e. CO_2 is more soluble in organic solvents than in water, and there is most likely less hydrogen bonding to the Zn–OH functionality in carbonic anhydrase versus model

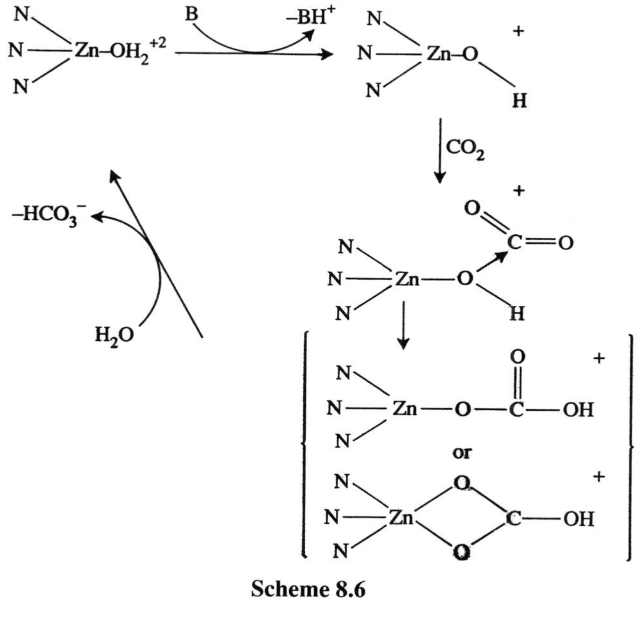

Scheme 8.6

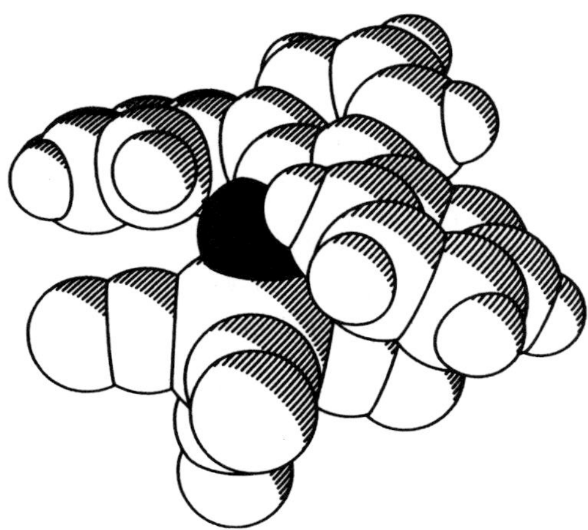

Fig. 8.10 Space-filling model of $W(CO)_5O\text{-}2,6\text{-}Ph_2C_6H_3^-$ utilizing van der Waals radii, demonstrating the crowded environment around the metal–oxygen centre.

complexes. This in turn may account in part for the much slower rates of CO_2 hydration observed in model complexes as compared with the enzyme-catalysed reactions.

Whereas CO_2 insertion into the W–O bond of $(CO)_5W\text{-}OPh^-$ occurs in a very facile manner at ambient temperature and atmospheric pressure of CO_2, the electronically quite similar $W(CO)_5O\text{-}2,\ 6\text{-}Ph_2C_6H_3^-$ anion (see Fig. 8.10 for a space filling model) is unreactive towards CO_2 even at pressures of carbon dioxide sufficient to induce its condensation at room temperature (Darensbourg *et al.* 1991*d*). This lack of CO_2 reactivity in the diphenyl derivative is presumed to be due to the steric interference of the phenyl rings around the metal–oxygen bond. On the other hand this complex reacts rapidly with 1 atm of COS to yield the tungsten–aryl thiocarbonate insertion product (eqn (16)).

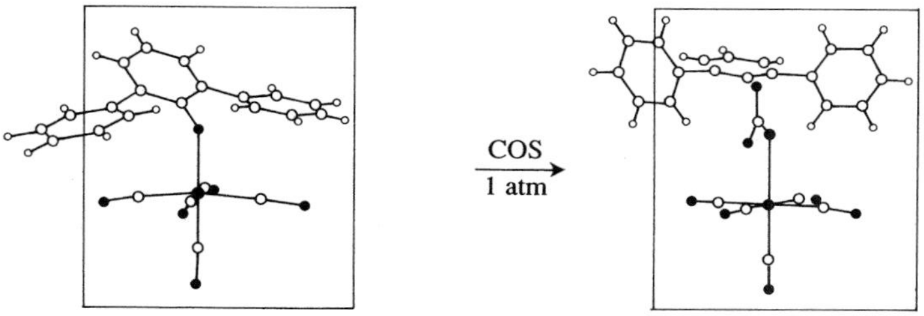

$$\xrightarrow[\text{1 atm}]{\text{COS}} \qquad (16)$$

Steric hinderance of the W-OPh reaction centre is seen also when sterically encumbering phosphine ligands are substituted *cis* to the phenoxide ligand. That is, phosphine substituted derivatives, *cis*-$W(CO)_4(L)(OPh)^-$, react more slowly

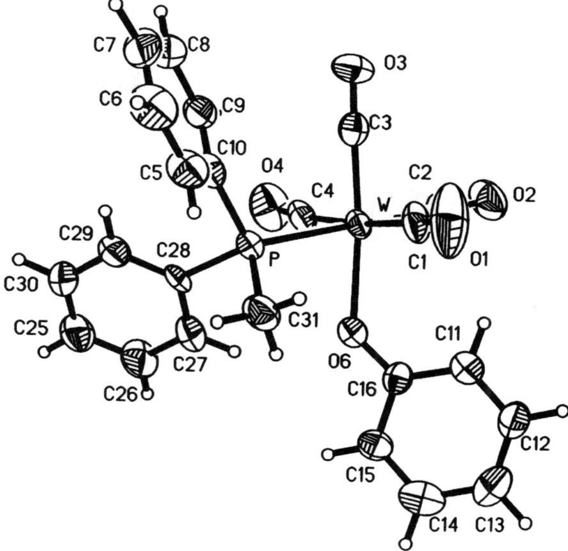

Fig. 8.11 X-ray structure of the anion, *cis*-$W(CO)_4(PPh_2Me)(OPh)^-$.

with CO_2 than the parent $W(CO)_5OPh^-$ complex. The relative rate of reaction decreases with an increase in the L ligand's cone angle, L = CO>>$P(OMe)_3$>PMe_3>PPh_3. The solid-state structure of one of the phosphine substituted complexes, namely *cis*-$W(CO)_4[PPh_2Me](OPh)^-$, was determined (Fig. 8.11). The structure clearly indicates steric interaction in that the phenoxide ligand adopts a conformation which places its lone paris in the direction of the phosphine ligand. This reduced reactivity towards CO_2 is counter to what would be anticipated based on increased electron donation from phosphines versus carbonyls. The more donating phosphines should increase the electron density on the tungsten centre thus making the phenoxide ligand more nucleophilic and reactive towards CO_2. This is what is observed for the analogous CO_2 insertion reactions involving tungsten-alkyl derivatives, where substitution of phosphines for CO significantly accelerate the rate of CO_2 insertion into the W–R bond. Thus, in the case of metal aryloxides where there is directed reactivity of the CO_2 towards the lone pairs on the oxygen, steric influences are important but in the instance where there is no directionality to the CO_2 approach for insertion, as in metal alkyls, steric hinderance is less important and electronic effects determine the energetics of the process.

The observation contained in eqn (16) is accounted for in the transition state for the reaction detailed in eqn (17). The most significant change which occurs in the transition state in going from CO_2 to COS as a substrate is the much greater metal–sulphur interaction (X = S) as compared to the corresponding metal–oxygen interaction (X = O). This stronger interaction in the case of COS overcomes the steric retardation experienced by CO_2 upon reacting with bulky aryloxide ligands. Consistent with these arguments the COS insertion process is irreversible.

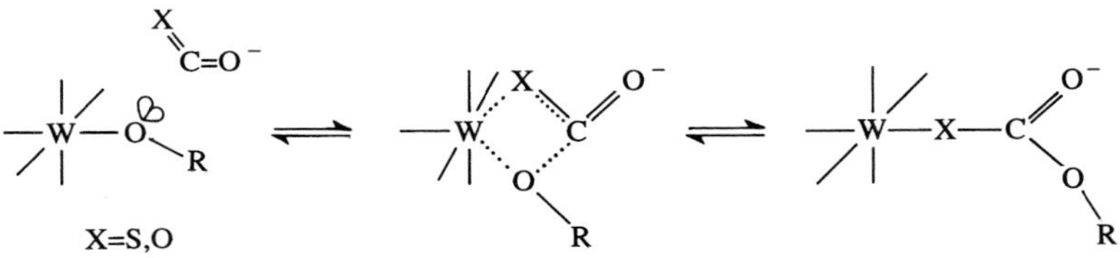

X
C=O$^-$
—W—O—R ⇌ —W(X)(O–R)C—O$^-$ ⇌ —W—X—C(O$^-$)—O—R (17)
X=S,O

Related to the subject of CO_2 insertion into M–O bonds is the copolymerization reaction of carbon dioxide with epoxides in the presence of organozinc or organoaluminium metal catalysts (eqn (18)) (Inoue 1976; Rokicki and Kuran 1981; Aida *et al.* 1986). The working hypothesis for the reaction pathway catalysed by organozinc reagents is outlined in Scheme 8.7, where CO_2 insertion into a zinc alkoxide bond is a key propagation step in the process.

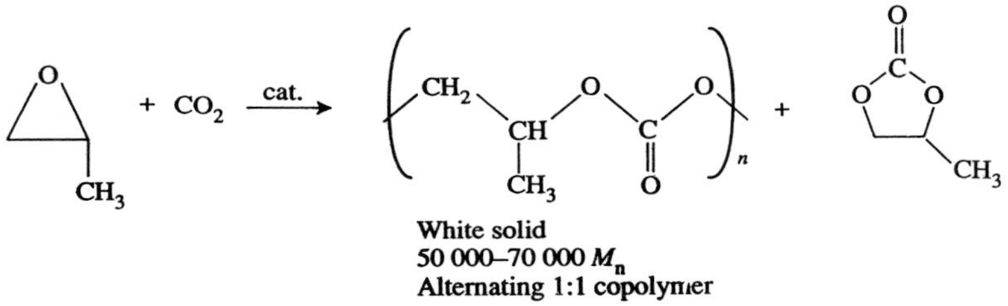

propylene oxide (CH$_3$) + CO_2 $\xrightarrow{\text{cat.}}$ (CH$_2$–CH(CH$_3$)–O–C(=O)–O)$_n$ + cyclic carbonate (CH$_3$) (18)

White solid
50 000–70 000 M_n
Alternating 1:1 copolymer

As indicated in eqn (18) an accompanying reaction which consumes a sizeable quantity of monomer is the formation of cyclic carbonate. The $W(CO)_5OCR^-$ anions discussed herein have been found to be homogeneous catalysts for this latter process, probably proceeding via a pathway similar to that of

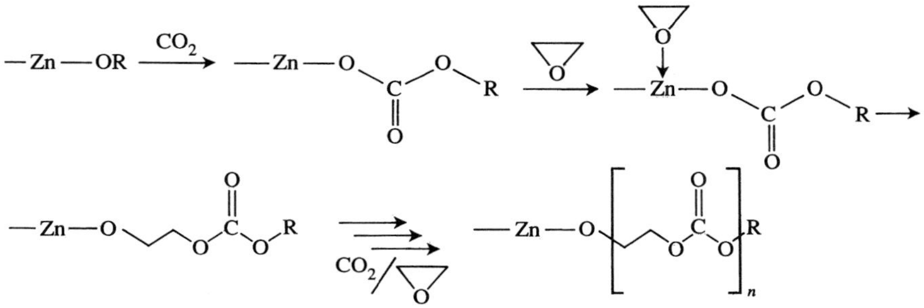

—Zn—OR $\xrightarrow{CO_2}$ —Zn—O–C(=O)–O–R $\xrightarrow{\text{epoxide}}$ (epoxide)→Zn—O–C(=O)–O–R →

—Zn—O~~O–C(=O)–O–R $\xrightarrow{CO_2/\text{epoxide}}$ —Zn—O[~~O–C(=O)–O]$_n$R

Scheme 8.7

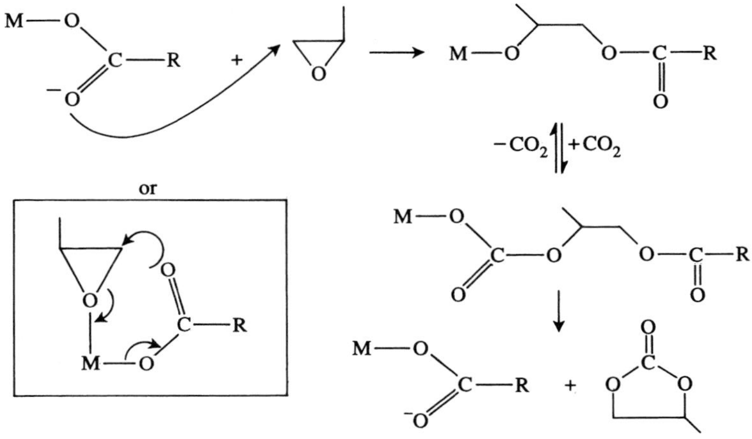

Scheme 8.8

the Zn(II) catalysts. A proposed mechanism is suggested in Scheme 8.8 which is currently being scrutinized in our laboratories. If this pathway is better understood it might be feasible to slow the progress of cyclic carbonate production relative to the desired copolymerization process. Scheme 8.8 is to be contrasted with the reactivity of epoxides with transition metal derivatives which readily undergo oxidative-addition reactions, i.e. providing oxametallocylobutane products which subsequently can react with CO_2 to afford metallocarbonates (Scheme 8.9) (Bäckvall *et al.* 1980; Klein *et al.* 1988; Aye *et al.* 1990).

The copolymerization of carbon dioxide and epoxides represents an important process for the utilization of carbon dioxide, for polycarbonates are useful

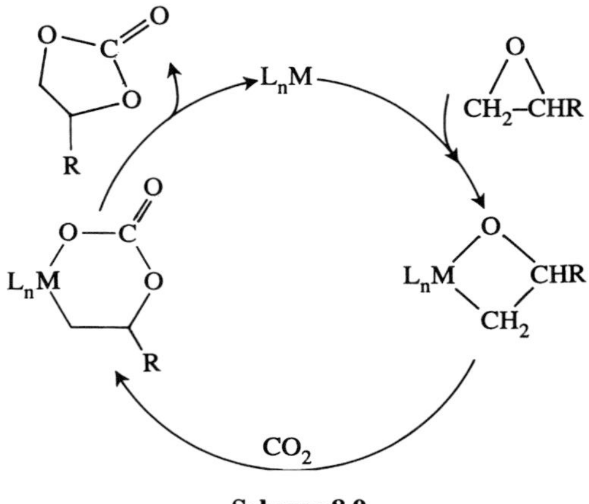

Scheme 8.9

materials with generally long lifetimes (Aresta 1990). Furthermore, an alternative method for their production, the reaction of alcohols with phosgene, has additional environmental risks.

Acknowledgements

The author is most grateful to the National Science Foundation, whose support has made possible his contributions to the research described herein. He is likewise extremely appreciative to all of his students and postdoctorals mentioned in the references, whose many original contributions have made this such an exciting area of research in which to work.

References

Abrahamson, D.E. (1989). In *The challenge of global warming* (ed. D.E. Abrahamson), pp. 3–34. Island Press, Washington, DC.

Aida, T., Ishakawa, M., and Inoue, S. (1986). *Macromolecules*, **19**, 8–13.

Alsfasser, R., Trofimenko, S., Looney, A., Parkin, G., and Vahrenkamp, H. (1991). *Inorg. Chem.*, **30**, 4098–100.

Aresta, M. (1990). In *Enzymatic and model carboxylation and reduction reactions for carbon dioxide utilization,* NATO ASI Series C314 (ed. M. Aresta and J.V. Schloss), pp. 1–18. Kluwer, Dordrecht, The Netherlands.

Aye, K.-T., Gelmini, L., Payne, N.C., Vital, J.J., and Puddephatt, R.J. (1990). *J. Am. Chem. Soc.*, **112**, 2464–5.

Bäckvall, J.-E., Karlsson, O., and Ljunggren, S.O. (1980). *Tetrahedron Lett.*, **21**, 4985–8.

Bo, C. and Dedieu, A. (1989). *Inorg. Chem.*, **28**, 304–9.

Brown, R.S. (1990). In *Enzymatic and model carboxylation and reduction reactions for carbon dioxide utilization*, NATO ASI Series C314 (ed. M. Aresta and J.V. Schloss), pp. 145–80. Kluwer,Dordrecht, The Netherlands.

Darensbourg, D.J. and Rokicki, A. (1982*a*). *Organometallics*, **1**, 1685–93.

Darensbourg, D.J. and Rokicki, A. (1982*b*). *J. Am. Chem. Soc.*, **104**, 349–50.

Darensbourg, D.J. and Kudaroski, R. (1984). *J. Am. Chem. Soc.*, **106**, 3672–3.

Darensbourg, D.J. and Grotsch, G. (1985). *J. Am. Chem. Soc.*, **107**, 7474–6.

Darensbourg, D.J. and Pala, M. (1985). *J. Am. Chem. Soc.*, **107**, 5687–93.

Darensbourg, D.J., Baldwin, B.J., and Froelich, J.A. (1980). *J. Am. Chem. Soc.*, **102**, 4688–94.

Darensbourg, D.J., Rokicki, A., and Darensbourg, M.Y. (1981). *J. Am. Chem. Soc.*, **103**, 3223–4.

Darensbourg, D.J., Kudaroski, R., Bauch, C.G., Pala, M., Simmons, D., and White, J.N. (1985). *J. Am. Chem. Soc.*, **107**, 7463–73.

Darensbourg, D.J., Bauch, C.G., and Rheingold, A.L. (1987). *Inorg. Chem.*, **26**, 977–80.

Darensbourg, D.J., Sanchez, K.M., Reibenspies, J.H., and Rheingold, A.L. (1989). *J. Am. Chem. Soc.*, **111**, 7094–103.

Darensbourg, D.J., Wiegreffe, H.P., and Wiegreffe, P.W. (1990*a*). *J. Am. Chem. Soc.*, **112**, 9252–7.

Darensbourg, D.J., Wiegreffe, P., and Riordan, C.G. (1990*b*). *J. Am. Chem. Soc.*, **112**, 5759–62.

Darensbourg, D.J., Muller, B.L., Bischoff, C.J., Johnson, C.C., Sanchez, K.M., and Reibenspies, J.H. (1990*c*). *Isr. J. Chem.*, **30**, 369–76.

Darensbourg, D.J., Joyce, J.A., Bischoff, C.J., and Reibenspies, J.H. (1991*a*). *Inorg. Chem.*, **30**, 1137–42.

Darensbourg, D.J., Longridge, E.M., Atnip, E.V., and Reibenspies, J.H. (1991*b*). *Inorg. Chem.*, **30**, 357–8.

Darensbourg, D.J., Joyce, J.A., and Rheingold, A. (1991*c*). *Organometallics*, **10**, 3407–10.

Darensbourg, D.J., Muller, B.L., Bischoff, C.J., Chojnacki, S.S., and Reibenspies, J.H. (1991*d*). *Inorg. Chem.*, **30**, 2418–24.

Deacon, G.B., Faulks, S.J., and Pain, G.N. (1986). *Adv. Organomet. Chem.*, **25**, 237–76.

Dedieu, A., Bo, C., and Ingold, F. (1992). *Metal ligand interactions: from atoms, to clusters, to surfaces* (ed. D. Salahub), NATO ASI Series. Kluwer, Dordrecht, The Netherlands.

Ford, P.C. and Rokicki, A. (1988). *Adv. Organomet. Chem.*, **28**, 139–217.

Han, R. and Parkin, G. (1991). *J. Am. Chem. Soc.*, **113**, 9707–8.

Idso, S.B. (1988). *Carbon dioxide and global change: earth in transition*, pp. 1–15. IBR Press, Tempe, Arizona.

Inoue, S. (1976). *Chemtech*, **6**, 588–94.

Kegley, S.E., Schaverein, C.J., Freudenberger, J.H., Bergman, R.J., Nolan, S.P., and Hoff, C.D. (1987). *J. Am. Chem. Soc.*, **109**, 6563–5.

Klein, D.P., Hayes, J.C., and Bergman, R.J. (1988). *J. Am. Chem. Soc.*, **110**, 3704–6.

Lunder, D.M., Lobkovsky, E.B., Streib, W.E., and Caulton, K.G. (1991). *J. Am. Chem. Soc.*, **113**, 1837–8.

McNeese, T.J., Cohen, M.B., and Foxman, B.M. (1984). *Organometallics*, **3**, 552–6.

McNeese, T.J., Mueller, T.E., Wierda, D.A., Darensbourg, D.J., and Delord, T.J. (1985). *Inorg. Chem.*, **24**, 3465–8.

Mehrota, R.C. and Bohra, R. (1983). *Metal carboxylates*. Academic Press, New York.

Rokicki, A. and Kuran, W. (1981). *J. Macromol. Sci-Rev. Macromol. Chem.*, **C21**, 135–86.

Sakaki, S. and Ohkubo, K. (1989). *Inorg. Chem.*, **28**, 2583–90.

Slater, S.G., Lusk, R., Schumann, B.F., and Darensbourg, M.Y. (1982). *Organometallics*, **1**, 1662–6.

Sullivan, B.P. and Meyer, T.J. (1984). *J. Chem. Soc., Chem. Commun.*, 1244–5.

Tooley, P.A., Ovalles, C., Kao, S.C., Darensbourg, D.J., and Darensbourg, M.Y. (1986). *J. Am. Chem. Soc.*, **108**, 5465–77.

Tsuda, T., Nakatsuka, T., Hirayama, T., and Saegusa, T. (1974). *J. Chem. Soc., Chem. Commun.*, 557–8.

Tsuda, T., Chujo, Y., and Saegusa, T. (1976*a*). *J. Chem. Soc., Chem. Commun.*, 415–16.

Tsuda, T., Sanda, S., Ueda, K., and Saegusa, T. (1976*b*). *Inorg. Chem.*, **15**, 2329–32.

Tsuda, T., Washita, H., Watanobe, K., Miwa, M., and Saegusa, T. (1978*a*). *J. Chem. Soc., Chem. Commun.*, 815–16.

Tsuda, T., Chujo, Y., and Saegusa, T. (1978*b*). *J. Am. Chem. Soc.*, **100**, 630–2.

Tsuda, T., Chujo, Y., and Saegusa, T. (1980). *J. Am. Chem. Soc.*, **102**, 431–3.

9

Catalytic additions of carbon dioxide adducts to alkynes: selective synthesis of carbamates, ureas, and carbonates

Christian Bruneau and Pierre H. Dixneuf

Introduction

CO_2 is the meeting point of several idealistic criteria for the functionalization of organic substrates: abundance, low cost, nontoxicity, and reusable carboxylation carbon unit. Its potential wide-range utilization is actually limited by its high stability, but motivates the search for general activation processes. One of the best uses of CO_2 is the carboxylation of organic halides via Grignard reagents. However, the overall reaction cost implies loss of the halide and oxidation of magnesium (eqn (1)). In contrast, new generation carboxylation processes should involve the direct coupling with hydrocarbons and the use of a more easily reducible metal catalyst.

$$\text{R–H} \longrightarrow \text{R–X} \xrightarrow{\text{Mg(0)}} \text{R–MgX} \xrightarrow{CO_2} \text{R–}CO_2\text{ MgX} \longrightarrow \text{R}CO_2\text{H} + \text{Mg(II)}X_2 \quad \text{HX} \qquad (1)$$

CO_2 has also the natural vocation to become the key reagent for access to functional carbamates and carbonates, molecules of interest for the preparation of agricultural chemicals or polymers and as intermediates for organic synthesis. CO_2 easily gives stable nucleophilic adducts on reactions with amines and alcoholate derivatives. This property is commonly used for the temporary protection of nucleophilic centres such as hydroxy or amino groups (Katritzky *et al.* 1987). The most convenient way to produce carbamates and carbonates is based on the use of phosgene derivatives, isocyanates, chloroformates or carbamoyl halides, despite their toxicity. In addition, on reaction with classical nucleophiles they release an acid halide that has to be neutralized, but the use of basic amines is no longer suitable for pharmaceutical chemical syntheses (eqn (2)). As far as the carbon source is concerned, the overall transformation involves that of CO into CO_2.

The environmental and safety requirements for modern chemistry suggest looking for new processes resulting from the use of nontoxic reagents without the release of by-products. One can expect that carboxylated derivatives should be produced directly from CO_2 and on decomposition would release reusable CO_2. In this direction our objective is to find new routes for the access to functional carbamates, ureas, and carbonates directly from CO_2 as a substitute of phosgene derivatives.

$CO + Cl_2$
ROH
$COCl_2$
RO Cl
HCl
O
$R'NH_2$
$R'NH_2$
2 HCl
HCl
O=C=NR′ ROH RO NHR′
H_2O
O
CO_2
+
ROH, $R'NH_2$

(2)

Electrophilic activation approach

The catalytic incorporation of CO_2 into hydrocarbons has been achieved at electron-rich metal centres:

1. The oxidative addition of hydrocarbons at a palladium(0) or nickel(0) centre, followed by coupling with CO_2 and reductive elimination has allowed the formation of lactones from butadiene (Inoue *et al.* 1978; Musco *et al.* 1978; Braunstein *et al.* 1988) or methylidenecyclopropane derivatives (Inoue *et al.* 1979; Binger and Weintz 1984).
2. The oxidative coupling of alkynes or allenes, directly with CO_2 followed by reductive elimination has given rise to pyrones and lactones (Walther *et al.* 1987; Inoue *et al.* 1980; Aresta *et al.* 1985; Sasaki 1989; Tsuda *et al.* 1990). The generation of nickel(0) moities can be achieved electrochemically and releases acrylic acids in the presence of Mg(II) salts (Duñach *et al.* 1989; Dérien *et al.* 1991).

Another concept is based on the nucleophilicity of adducts of CO_2 with amines, alcohols, hydroxide or even hydride to give carbamates, carbonates, monocarbonate, and formate, respectively. The formation of these intermediates is essential in enzymatic processes involving ribulose-bisphosphate carboxylase/oxygenase in the photosynthetic fixation of CO_2 (Schneider 1992), biotin in CO_2 fixation (Kluger 1992), or methanofuran, for the fixation of CO_2 in methanogenesis (Wolfe 1991). Nucleophilic adducts of CO_2 have also been used in metal-assisted reactions via electrophilic activation of oxiranes to produce 2-hydroxycarbamates (Kajima *et al.* 1986; Yoshida *et al.* 1988) or to afford urethanes in the presence of copper (Tsuda *et al.* 1978).

In order to produce unsaturated and functional carbamates and carbonates we have used a strategy based on the concept of electrophilic activation of hydrocarbons. Our interest in the catalytic activation of alkynes by ruthenium(II) complexes has led us to study the coupling of alkynes with nucleophilic adducts of CO_2. This study is connected with the catalytic addition of carboxylic acids to alkynes in the presence of $Ru_3(CO)_{12}$ (Rotem and Shvo 1983), $Ru(C_8H_{11})_2$ (Mitsudo *et al.* 1987*a*), and (arene)$RuCl_2(PR_3)$ (Bruneau *et al.* 1991) as catalyst precursors.

Synthesis of vinyl carbamates

Acetylene is an industrially available feedstock and the functional vinyl compounds resulting from addition reactions are important industrial products, especially for access to a wide variety of polymers. Vinylcarbamates are used for the manufacture of transparent polymers and varnishes; they are commonly produced from amines and vinylchloroformate generated from phosgene.

Vinylcarbamates can be considered as resulting from the formal addition of carbamic acids to acetylene. Based on the fact that triple bonds can be activated by ruthenium complexes (Bruneau *et al.* 1991) and that ammonium carbamates are easily obtained from amines and CO_2, we investigated the 'one pot' synthesis of vinylcarbamates from CO_2, amine, and acetylene in the presence of ruthenium derivatives.

When an amine and acetylene, dissolved in acetonitrile, were heated for 20 h at 80–100 °C under a 2 MPa CO_2 pressure, in the presence of catalytic amounts of ruthenium derivatives, vinylcarbamates **1** were formed (30–40 per cent) (eqn (3)).

$$R_2NH + CO_2 + H{-}C{\equiv}C{-}H \xrightarrow{[Ru]} R_2N{-}C({=}O){-}O{-}CH{=}CH_2 \quad \mathbf{1} \tag{3}$$

The best result was obtained with pyrrolidine in the presence of 0.4 mmol of $[(norbornadiene)RuCl_2]_n$ as catalyst precursor affording 63 per cent of vinylcarbamate, but $RuCl_3{\cdot}3H_2O$ and more elaborated complexes of the type (arene)$RuCl_2(PR_3)$ were also efficient catalysts (Sasaki and Dixneuf 1987*a*; Bruneau *et al.* 1988).

Table 9.1 Formation of vinyl carbamates from phenylacetylene and diethylamine

Catalyst	Conversion (%)	Carbamate **2** % (Z/E)	Dimer **3** % (Z/E)
(hexamethylbenzene)$RuCl_2(PMe_3)$	98	67 (83/16)	16 (44/56)
(norbornadiene)$RuCl_2(pyridine)_2$	98	64 (77/23)	12 (42/58)

Amine (20 mmol), phenylacetylene (10 mmol), CO_2 (5 MPa), MeCN (10 ml), catalyst (0.2 mmol), 125 °C, 20 h, conversion and yields based on the initial alkyne.

The extension of this reaction to monosubstituted alkynes was of interest since methods for the synthesis of unsaturated carbamates are scarce. Enol carbamates have been prepared by dehydrohalogenation of haloalkyl carbamates obtained from carbamoyl chlorides and ketones (Franko-Filipasic and Patarcity 1969) or α (or β)-haloalkyl chloroformates and amines (Olofson 1988; Shimizu *et al.* 1987). Stang and Anderson (1981) prepared vinyl carbamates by reaction of an isocyanate with an alkylidene carbene generated from silylvinyltriflate, whereas Olofson *et al.* (1978) added amines to vinyl chloroformates obtained from mercury derivatives.

Monosubstituted alkynes reacted with amines at 125 °C for 20 h under a 5 MPa CO_2 pressure, in the presence of ruthenium complexes to regioselectively afford the vinylcarbamates **2** corresponding to the addition of ammonium carbamate to the terminal carbon of the alkyne (eqn (4)) (Table 9.1) (Mahé *et al.* 1986, 1989; Sasaki and Dixneuf 1986). Similar results were obtained using (hexamethylbenzene)$RuCl_2(PMe_3)$ or (norbornadiene)$(pyridine)_2RuCl_2$, which led to a complete conversion of the alkyne (>98 per cent), good yields of vinylcarbamates **2** (64–67 per cent) with the Z carbamate as the major isomer (*Z/E* ~ 80/20) and formation of (*Z*) and (*E*)-1,4-diphenylbut-3-en-1-ynes **3** (12–16 per cent) as by-products, resulting from the dimerization of phenylacetylene.

$$\text{Ph–C}\equiv\text{C–H} + CO_2 + HNEt_2 \xrightarrow[125\,^\circ C]{[Ru]} \left[\begin{array}{c} \text{PhCH=CH–O–C(=O)–NEt}_2 \\ \mathbf{2}\ (Z,E) \\ + \\ \text{PhCH=CH–C}\equiv\text{CPh} \\ \mathbf{3}\ (Z,E) \end{array} \right] \quad (4)$$

The main characteristics of this reaction were (Mahé *et al.* 1989):

— The optimum temperature was located around 100 °C, higher temperatures leading to the preferential formation of alkyne oligomers at the expense of the carbamate formation.

— Acetonitrile and tetrahydrofuran were found to be the most appropriate solvents.
— The initial pressure of CO_2 was an important factor, too low a pressure favouring the formation of polymers of the alkyne.
— Mononuclear ruthenium(II) complexes especially (arene)$RuCl_2(PR_3)$ appeared to be the most efficient catalysts, compared to polynuclear ruthenium species such as $Ru_3(CO)_{12}$.
— This regioselective addition was specific of secondary aliphatic amines.

Mitsudo *et al.* (1987*b*) reported that Ru(0)(COD)(COT) and bis-(η^5-cyclooctadienyl)Ru(II) were also efficient catalysts precursors for this type of reaction.

(Arene)$RuCl_2(PR_3)$ complexes as catalyst precursors allowed the synthesis of various vinylcarbamates from phenylacetylene and secondary amines such as dimethylamine, morpholine, piperidine, *N*-methyl-*N*-butylamine and also from hex-1-yne and diethylamine.

However, these complexes were inactive for the addition to propyne. Only some complexes in the series $(Ph_2P(CH_2)_nPPh_2)$(2-methylallyl)$_2$Ru made possible this triple bond activation to regioselectively produce 1-[(morpholinocarbamoyl)oxy]prop-1-ene **4** with a high regio- and stereoselectivity (80 per cent) but in moderate yield (31 per cent) (eqn (5)).

$$\text{O}(\text{CH}_2\text{CH}_2)_2\text{N—H} + CO_2 + \text{HC}\equiv\text{C—CH}_3 \longrightarrow \text{O}(\text{CH}_2\text{CH}_2)_2\text{N—C(=O)—O—CH=CH—Me} \quad \mathbf{4} \qquad (5)$$

Synthesis of dienyl carbamates

(*E*)-dienyl carbamates have a potential as Diels–Alder reaction substrates or as precursors for polymerization (De Cusati and Olofson 1990). These 1-dienyl carbamates were expected to be prepared in one step via electrophilic activation of alkenyl acetylenes in the presence of a secondary amine and CO_2. Diethylamine or *N*-methyl-*N*-butylamine, CO_2 and isopropenylacetylene led to 37 per cent and 50 per cent yield of the corresponding carbamates **5** in the presence of $RuCl_3{\cdot}3H_2O$ as catalyst, in CH_3CN at 100 °C for 20 h (Dixneuf *et al.* 1990). Unfortunately, this catalytic precursor was not active with cyclic secondary amines. The use of $(Ph_2P(CH_2)_2PPh_2)$(2-methylallyl)$_2$Ru as catalyst precursor containing labile hydrocarbon ligands and a chelating phosphine ligand made possible the activation of isopropenylacetylene toward cyclic ammonium carbamates (eqn (6)) Höfer *et al.* 1992). This complex gave higher yields of 1-dienyl carbamates with similar regio- and stereoselectivities, as $RuCl_3{\cdot}3H_2O$ with noncyclic amines (Table 9.2).

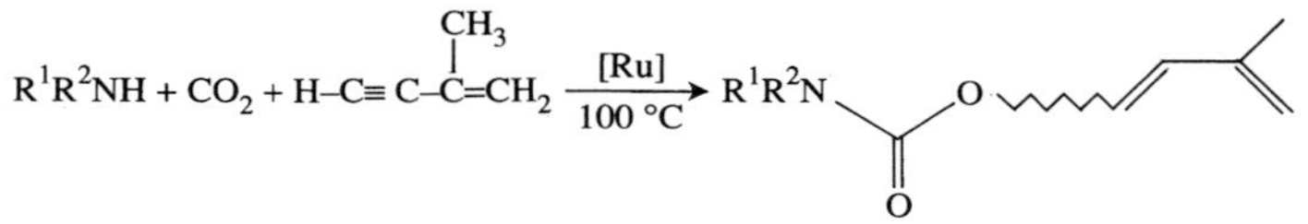

$$R^1R^2NH + CO_2 + H\text{–}C{\equiv}C\text{–}C(CH_3){=}CH_2 \xrightarrow[100\ ^\circ C]{[Ru]} R^1R^2N\text{–}C({=}O)\text{–}O\text{∿∿}CH{=}CH\text{–}C(CH_3){=}CH_2 \qquad (6)$$

5 (*Z,E*)

Table 9.2 Synthesis of 1-dienyl carbamates from isopropenylacetylene

Catalyst	Amine	Carbamate **5** yield (%)	Distribution *Z/E*
$RuCl_3.3H_2O$	Et_2NH	37	75/25
"	MeBuNH	50	70/30
"	 NH	28	
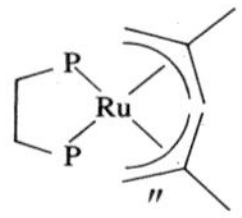 P Ru P	 NH	36	76/24
"	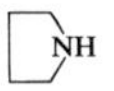 NH	50	94/6
"	 O NH	62	84/16

Amine (20 mmol), isopropenylacetylene (10 mmol), CO_2 (5 MPa), MeCN (10 ml), catalyst (0.2 mmol), 100 °C, 20 h, yields based on the initial alkyne.

Synthesis of ureas

Ureas are important compounds widely used for agriculture, medicine and industry. They are usually prepared by reaction of amines with isocyanates, chloroformates, formamides, carbamates or carbonates. Starting from CO_2 and amines, ureas can be prepared on a laboratory scale in the presence of dicyclohexylcarbodiimide (Ogura *et al.* 1978), *N*-phosphonium salt derivatives (Yamazaki *et al.* 1975) or triphenylstibine oxide (Nomura *et al.* 1987).

We have found an efficient method for the direct access to symmetrical ureas directly from CO_2 and primary amines (eqn (7)) (Fournier *et al.* 1991). This reaction required the presence of both a ruthenium catalyst precursor and a terminal alkyne.

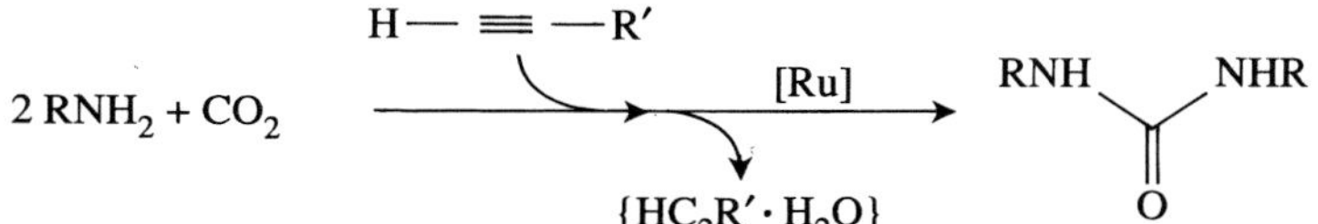

$$2\,RNH_2 + CO_2 \xrightarrow[\{HC_2R'\cdot H_2O\}]{H\text{—}{\equiv}\text{—}R'\quad [Ru]} RNH\text{–}C({=}O)\text{–}NHR \qquad (7)$$

6

When *n*-decylamine and 2-methylbut-3-yn-2-ol were reacted for 20 h at 140 °C under a 5 MPa CO_2 pressure, in the presence of $RuCl_3{\cdot}3H_2O$ and PBu_3, di-*n*-decylurea was formed in 68 per cent yield. Propargyl alcohols, especially 2-methylbut-3-yn-2-ol, used as solvent, were found to be the terminal alkynes leading to the best results, and expected to play the role of water trapping reagent. (Arene)$RuCl_2(PR_3)$ complexes and the system $RuCl_3{\cdot}3H_2O/2PBu_3$(A) exhibited similar catalytic activities, since under standard conditions (hexamethylbenzene)$RuCl_2(PMe_3)$,(*p*-cymene)$RuCl_2(PPh_3)$ and A gave 61 per cent, 56 per cent and 61 per cent yields of dicyclohexylurea, respectively.

Several symmetrical ureas were obtained from various amines (40–70 per cent) (Fournier *et al.* 1991), but aromatic amines did not react, so the reaction appears to be specific of aliphatic and araliphatic amines.

Catalytic synthesis of cyclic carbonates

Carbonates are industrially used as solvents and precursors of transparent polycarbonate glasses (Keim 1987). They are also efficient protecting groups for alcohols and diols and intermediates in organic synthesis (Aresta and Quaranta, 1991).

Cyclic carbonates have been obtained by phosgenation of α-difunctional substrates such as diols or α-hydroxy ketones, insertion of CO_2 into a C–O bond of an oxirane. The most suitable methods used for the synthesis of α-methylene carbonates start from propargyl alcohol derivatives and are based on the intramolecular cyclization of an adduct resulting from the addition of CO_2 to a propargyl alcohol derivative (Dimroth *et al.* 1963; Iritani *et al.* 1986; Sasaki 1986; Schneider 1986; Inoue *et al.* 1988, 1990).

We have shown that this reaction can be efficiently catalysed by phosphines (Fournier *et al.* 1989; Joumier *et al.* 1991) without any metal catalyst (eqn (8)). PBu_3 was the most active phosphine for the formation of cyclic carbonates **7** at low temperature (50 °C, 20 h), whereas PPh_3 or P(cyclohexyl)$_3$ required a higher temperature (140 °C).

$$\mathrm{H{-}C{\equiv}C{-}C(R^1)(R^2){-}OH} + CO_2 \xrightarrow[100\,°C]{PBu_3(cat.)} \mathbf{7} \qquad (8)$$

The formation of cyclic carbonates **7** is specific of tertiary α-propynyl alcohols and is also limited to terminal prop-2-ynols except when the substituent is an unsaturated conjugated group such as an ethynyl or vinyl group, leading to carbonates **7a** and **7b** in 54 per cent and 47 per cent yields, respectively.

The introduction of an unsaturated substituent at the C1 carbon of the propargyl alcohol makes possible the synthesis of allylic carbonates **7c**.

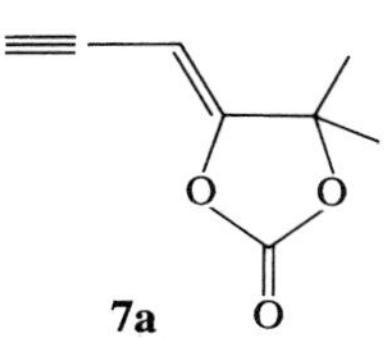

We have shown that these cyclic carbonates react very easily with alcohols in the presence of Et_3N or KCN to afford β-oxopropyl carbonates (**8**) in good yields (67–91 per cent) (eqn (9)) (Joumier *et al.* 1991).

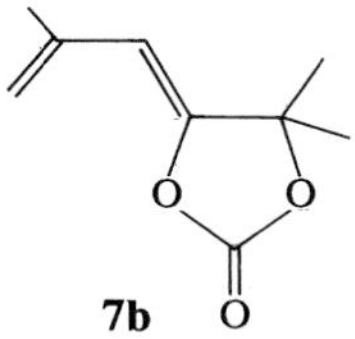

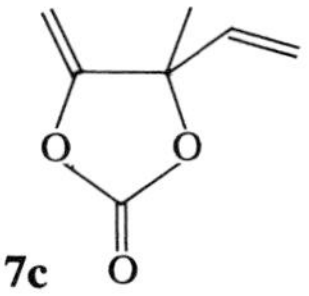

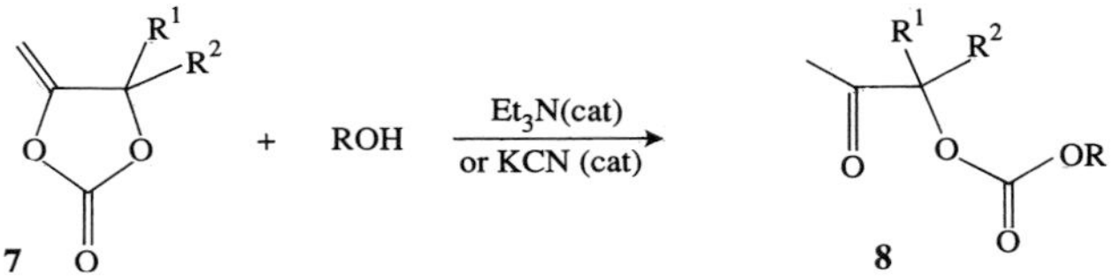

(9)

This reaction gives access to a variety of polyfunctional carbonates, useful as intermediates for organic synthesis. The presence of an exocyclic double bond in carbonates (**7**) favoured the cleavage of the C–O bond, which gave an enolate intermediate affording an acetyl group on protonation.

Synthesis of β-oxopropyl carbamates

The use of propargyl alcohol, as simple functional terminal alkyne, for the synthesis of hydroxylated vinyl carbamates, led to the unexpected formation of β-oxopropyl carbamates (**9**) (eqn (10)) (Bruneau and Dixneuf 1987; Sasaki and Dixneuf, 1987*b*).

$$R_2NH + CO_2 + H{-}C{\equiv}C{-}C(R^1)(R^2){-}OH \xrightarrow{[Ru]} R_2N{-}C(=O){-}O{-}C(R^1)(R^2){-}C(=O){-}Me \quad (10)$$

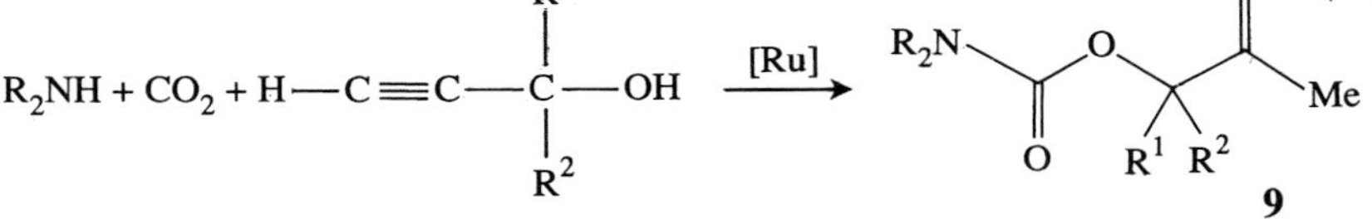

In the presence of a catalytic amount of $Ru_3(CO)_{12}$, diethylamine reacted with propargyl alcohol under a 5 MPa CO_2 pressure at 80 °C for 20 h to specifically produce β-oxopropyl-*N,N*-diethylcarbamate in 54 per cent yield. The reaction was possible with cyclic secondary amines such as morpholine, piperidine or pyrrolidine, and also with various α-alkynyl alcohols such as but-3-yn-2-ol (R^1 = H, R^2 = Me), or 2-methylbut-3-yn-2-ol (R^1 = R^2 = Me). Recently, Kim *et al.*

(1990) have shown that iron complexes are also good catalysts for this reaction.

A mechanism based on an intramolecular transcarbamatation involving the intermediates (I) and (II) can be proposed to account for the formation of derivatives (**9**).

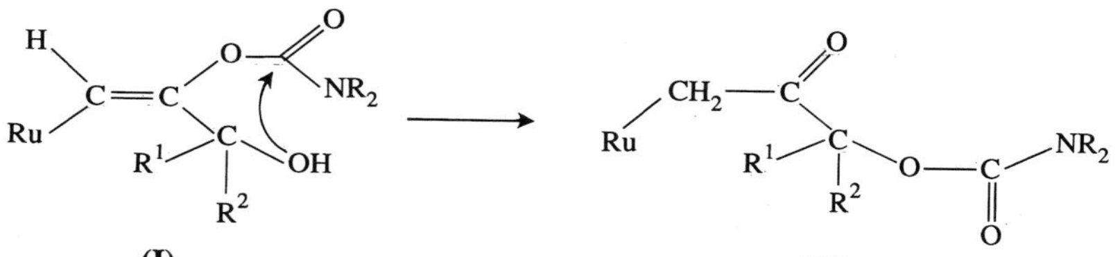

(I) (II)

Another possibility is the intermediate formation of a cyclic carbonate, which on reaction with a secondary amine would give the β-oxopropyl carbamate.

Sasaki (1986) has reported that the formation of α-methylene cyclic carbonates was catalysed by ruthenium complexes. We have shown that secondary amines readily reacted at room temperature with α-methylene cyclic carbonates to afford β-oxopropyl carbamates in good yields (65–85 per cent (eqn (11)).

This second mechanism is supported by the experimental results obtained with tertiary propargyl alcohols, but the fact that no cyclic carbonate can be obtained from propargyl alcohol itself makes the mechanism unprobable in the case of secondary and primary alcohols.

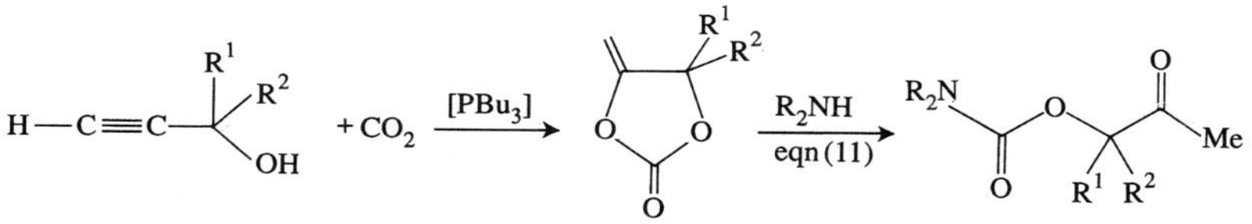

As a consequence of the above results, β-oxopropyl carbamates can be obtained from secondary amines either from propargyl alcohols via ruthenium catalysed reactions in one step (route 1 – eqn (10)), or from tertiary propargyl alcohols via cyclic carbonates in two steps (route 2 – eqns (8), (11)).

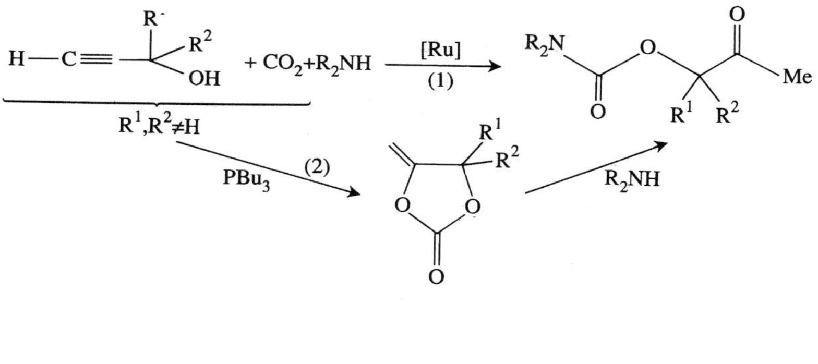

Synthesis of oxazolidinones

4-methylene oxazolidin-2-ones

Oxazolidinones are an important class of heterocyclic organic compounds used in organic synthesis. They are obtained from α, β-difunctional substrates such as amino alcohols or oxiranes or aziridines in the presence of phosgene, carbonate, isocyanate, and also CO_2 (Saito *et al.* 1986).

5-Methylene oxazolidinones have been prepared in one step from CO_2 and secondary propargyl amines in the presence of ruthenium complexes (Mitsudo *et al.* 1987*b*) or copper derivatives (Dimroth and Pasedach 1964) as catalysts. 4-methylene oxazolidinones have been prepared by Shachat and Bagnell (1963) by intramolecular cyclization of a propynyl carbamate, and by Dimroth *et al.* (1964) directly from CO_2, propargyl alcohols, and primary amines in the presence of copper salts.

The authors have reported that the direct synthesis of 4-methylene oxazolidin-2-ones (**10**) from propargyl alcohols, CO_2, and primary amines can also be catalysed by phosphines (eqn (12)) (Fournier *et al.* 1990).

$$H{-}C{\equiv}C{-}C(R^1)(R^2)OH + CO_2 + RNH_2 \xrightarrow[\text{cat.}]{PR_3} \mathbf{10} \qquad (12)$$

As was the case for the synthesis of cyclic carbonates, the reaction was specific to tertiary α-propynyl alcohols but required more severe conditions (110–140 °C). The reaction of aromatic amines to yield aromatic oxazolidinones was possible in good yields (aniline 63 per cent, p-toluidine 72 per cent), which was an advantage as compared to the low yields obtained by the copper catalysed process (Dimroth *et al.* 1964).

4-hydroxy oxazolidinones

Bulky primary amines such a *t*-butylamine or cyclohexylamine reacted at low temperature (0–20 °C) with α-methylene cyclic carbonates to afford the corresponding linear β-oxopropyl carbamates, as was the case with secondary amines (eqn (11)). But less hindered primary amines readily afforded 4-hydroxy oxazolidines **11** in one step and very good yields (Table 9.3).

These 4-hydroxy oxazolidinones (**11**) were probably intermediates in the formation of 4-methylene oxazolidin-2-ones (**10**) since heating of **11** at 110 °C in the presence of a catalytic amount of PBu_3 gave **10** in quantitative yield.

Table 9.3 Formation of 4-hydroxy-4-methyl-5,5-dimethyl oxazolidin-2-ones **11** from 4,4-dimethyl-5-methylene-1,3-dioxolan-2-one

Oxazolidinone **11**		Isolated yield (%)
	R = Ph–CH(CH_3)–	88
	Ph–CH_2–	89
	Ph–$(CH_2)_2$–	92
	CH_2 = CH-CH_2–	76
	$(CH_3)_2$CH-	88

Carbonate **7** (10 mmol), amine (10 mmol), toluene (5 ml), 20 °C, 15 h.

Both formations of oxazolidinones **10** and **11** are specific of tertiary α-propynyl alcohols and can be summarized as indicated in the following scheme.

H—C≡C—CR_2OH + CO_2 $\xrightarrow[110\ °C]{PBu_3\ (cat.)}$

R′NH_2/PBu_3(cat.) 110 °C

R′NH_2

20 °C

110 °C PBu_3(cat.)

10 **11**

Conclusion

The variety of syntheses of organic compounds that we have presented above shows that CO_2 can be used directly for access to new derivatives or to functional products which were previously made via multi-step syntheses from phosgene. It can be expected that the most successful of them will be used in the future for reasons of low cost, nontoxicity, and performance easiness.

References

Aresta, M. and Quaranta, E. (1991). *Tetrahedron*, **47**, 9489–502.

Aresta, M., Quaranta, E., and Ciccarese, A. (1985). *C_1 Mol. Chem.*, **1**, 283–95.

Binger, P. and Weintz, J.H. (1984). *Chem. Ber.*, **117**, 654–65.

Braunstein, P., Matt, D., and Nobel, D. (1988). *J. Am. Chem. Soc.*, **110**, 3207–12.

Bruneau, C. and Dixneuf, P.H. (1987). *Tetrahedron Lett.*, **28**, 2005–8.

Bruneau, C., Dixneuf, P.H., Lécolier, S. (1988). *J. Mol. Catal.*, **44**, 175–8.

Bruneau, C., Neveux, M., Kabouche, Z., Ruppin, C., and Dixneuf, P.H. (1991). *Synlett.*, **11**, 755–63.

De Cusati, P.F. and Olofson, R.A. (1990). *Tetrahedron Lett.*, **31**, 1405–12.

Dérien, S., Duñach, E., and Périchon, J. (1991). *J. Am. Chem. Soc.*, **113**, 8447–54.

Dimroth, P. and Pasedach, H. (1964). Ger. Pat. 1 164 411.

Dimroth, P., Schefczik, E., and Pasedach, H. (1963). Ger. Pat. DPB 1 145 632.

Dimroth, P., Pasedach, H., and Schefczik, E. (1964). Ger. Pat. 1 151 507.

Dixneuf, P.H., Bruneau, C., and Fournier, J. (1990). Catalytic incorporation of CO_2 for the synthesis of organic compounds. In *Enzymatic and model carboxylation and reduction reactions for carbon dioxide utilization*, NATO ASI Series, Vol. 314, (ed. M. Aresta and J.V. Schloss), pp.65–77. Kluwer, Dordrecht.

Duñach, E., Dérien, S., and Périchon, J. (1989). *J. Organomet. Chem.*, **364**, C33–6.

Fournier, J., Bruneau, C., and Dixneuf, P.H. (1989). *Tetrahedron Lett.*, **30**, 3981–2.

Fournier, J., Bruneau, C., and Dixneuf, P.H. (1990). *Tetrahedron Lett.*, **31**, 1721–2.

Fournier, J., Bruneau, C., Dixneuf, P.H., and Lécolier, S. (1991). *J. Org. Chem.*, **56**, 4456–8.

Franko-Filipasic, B.R. and Patarcity, R. (1969). *Chem. Ind.*, **8**, 166–7.

Höfer, J., Doucet, H., Bruneau, C., and Dixneuf, P.H. (1991). *Tetrahedron Lett.* **32**, 7409–10.

Inoue, Y., Sasaki, Y., and Hashimoto, H. (1978). *Bull. Chem. Soc. Jpn*, **51**, 2375–8.

Inoue, Y., Hibi, T., Satake, M., and Hashimoto, H. (1979). *J. Chem. Soc., Chem. Commun.*, 982.

Inoue, Y., Itoh, Y., Hazama, H., and Hashimoto, H. (1980). *Bull. Chem. Soc. Jpn*, **53**, 3329–33.

Inoue, Y., Ohuchi, K., and Imaizumi, S. (1988). *Tetrahedron Lett.*, **29**, 5941–2.

Inoue, Y., Itoh, Y., Yen, I.F., and Imaizumi, S. (1990). *J. Mol. Catal.*, **60**, L1–3.

Iritani, K., Yanagihara, N., and Utimoto, K. (1986). *J. Org. Chem.*, **51**, 5501–3.

Joumier, J.M., Fournier, J., Bruneau, C., and Dixneuf, P.H. (1991). *J. Chem. Soc. Perkin Trans.1*, 3271–4.

Kajima, F., Aida, T., and Inoue, S. (1986). *J. Am. Chem. Soc.*, **108**, 391–402.

Katritzky, A.R., Fan, W.-Q., Koziol, A.E., and Palenik, G.J. (1987). *Tetrahedron*, **43**, 2343–8.

Keim, W. (1987). Industrial uses of carbon dioxide. In *Carbon dioxide as a source of carbon*, NATO ASI Series, Vol. 206 (ed. M. Aresta and G. Forti), pp. 23–31. D. Reidel, Dordrecht.

Kim, T.J., Kwon, K.H., Kwon, S.C., Baeg, J. O., Shim, S.C., and Lee, D.H. (1990). *J. Organomet. Chem.*, **389**, 205–17.

Kluger, R. (1992). In *Carbon dioxide fixation and reduction in biological and model systems* (ed. C.-I. Bränden and G. Schneider). Oxford University Press, Oxford.

Mahé, R., Dixneuf, P.H., and Lécolier, S. (1986). *Tetrahedron Lett.*, **27**, 6333–6.

Mahé, R., Sasaki, Y., Bruneau, C., and Dixneuf, P.H. (1989). *J. Org. Chem.*, **54**, 1518–23.

Mitsudo, T., Hori, Y., Yamakawa, Y., and Watanabe, Y. (1987*a*). *J. Org. Chem.*, **52**, 2230–9.

Mitsudo, T., Hori, Y., Yamakawa, Y., and Watanabe, Y. (1987*b*). *Tetrahedron Lett.*, **28**, 4417–8.

Musco, A., Perego, C., and Tartiari, V. (1978). *Inorg. Chim. Acta*, **28**, L147–8.

Nomura, R., Yamamoto, M., and Matsuda, H. (1987). *Ind. Eng. Chem. Res.*, **26**, 1056–9.

Ogura, H., Takeda, K., Tokue, R., and Kobayashi, T. (1978). *Synthesis*, 394–6.

Olofson, R.A. (1988). *Pure & Appl. Chem.*, **60**, 1715–24.

Olofson, R.A., Bauman, B.A., and Wancowicz, D.J. (1978). *J. Org. Chem.*, **43**, 752–4.

Rotem, M. and Shvo, Y. (1983). *Organometallics*, **2**, 1689–91.

Saito, N., Hatakeda, K., Ito, S., Asano, T., and Toda, T. (1986). *Bull. Chem. Soc. Jpn*, **59**, 1629–31.

Sasaki, Y. (1986). *Tetrahedron Lett.*, **27**, 1573–4.

Sasaki, Y. (1989). *J. Mol. Catal.*, **54**, L9–12.

Sasaki, Y. and Dixneuf, P.H. (1986). *J. Chem. Soc., Chem. Commun.*, 790–1.

Sasaki, Y. and Dixneuf, P.H. (1987*a*). *J. Org. Chem.*, **52**, 314–5.

Sasaki, Y. and Dixneuf, P.H. (1987*b*). *J. Org. Chem.*, **52**, 4389–91.

Schneider, G. (1992). In *Carbon dioxide fixation and reduction in biological and model systems* (ed. C.-I. Bränden and G. Schneider). Oxford University Press, Oxford.

Schneider, K. (1986). Ger. Pat. DE 3 433 403 A1.

Shachat, N. and Bagnell, J.J. (1963). *J. Org. Chem.*, **28**, 991–5.

Shimizu, M., Tanaka, E., and Yoshioka, H. (1987). *J. Chem. Soc., Chem. Commun.*, 136–7.

Stang, P.J. and Anderson, G.H. (1981). *J. Org. Chem.*, **46**, 4585–6.

Tsuda, T., Washita, H., Watanabe, K., Miwa, M., and Saegusa, T. (1978). *J. Chem. Soc., Chem. Commun.*, 815–6.

Tsuda, T., Morikawa, S., Hasegawa, N., and Saegusa, T. (1990). *J. Org. Chem.*, **55**, 2978–81.

Walther, D., Schönberg, H., Dinjus, E., and Sieler, J. (1987). *J. Organomet. Chem.*, **334**, 377–88.

Wolfe, R.S. (1991). *Ann. Rev. Microbiol.*, **45**, 1–35.

Yamazaki, N., Iguchi, T., and Higashi, F. (1975). *Tetrahedron*, **31**, 3031–4.

Yoshida, Y., Ishii, S., Kawato, A., Yamashita, T., Yano, M., and Inoue, S. (1988). *Bull. Chem. Soc. Jpn*, **61**, 2913–16.

10

Activation and reduction of carbon dioxide to formylmethanofuran – a unique method of carbon dioxide fixation in the archaebacteria (archaea)

Alain Wasserfallen and Ralph S. Wolfe

Methanogens constitute a taxonomic group of strict anaerobes characterized by their ability to generate methane from the oxidation of molecular hydrogen and the reduction of carbon dioxide or the metabolism of other simple compounds such as formate, acetate, methanol, methylamines (Jones *et al.* 1987). Methanogenic organisms are currently grouped into three orders (Methanobacteriales, Methanococcales, Methanomicrobiales) in what is now referred to as the domain of archaea (Woese *et al.* 1990). Methanogens grow in environments where the redox potential is lower than –330 mV (Smith and Hungate 1958) and have been found in nearly every anaerobic habitat. Various methanogens have been found to use the same pathway of carbon dioxide fixation (Jones *et al.* 1985). As shown in Fig. 10.1 the oxidation of hydrogen and reduction of carbon dioxide to methane is a thermodynamically favourable reaction. Methanogenesis from this reaction readily takes place in nature (i.e. in sediments or the rumen) where the partial pressure of hydrogen may be quite low (10^{-4} atm). The C_1 intermediates at the formyl, formaldehyde, and methyl levels of oxidation are coenzyme bound, and are not detectable as free compounds.

The first evidence that a new coenzyme was involved in carbon dioxide reduction in these organisms was obtained by following methane formation by treated cell extracts under a hydrogen and carbon dioxide atmosphere. As seen

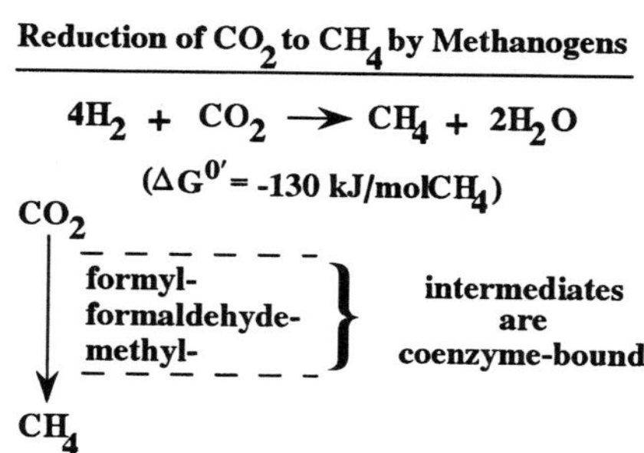

Fig. 10.1 Reduction of CO_2 to CH_4 by methanogens.

Table 10.1 Discovery of methanofuran in cell extracts of *Methanobacterium*

Reaction vial	Additions	CO_2 reduced to CH_4 (nmol)
1	Cell extract proteins passed through a Sephadex G-25 column (G-25 protein)	85
2	G-25 protein + small compounds eluted from G-25 column	1600
3	G-25 protein + boiled cell extract	3750

Assays were performed as described by Romesser and Wolfe (1982*a*).

in Table 10.1, the protein pass-through fraction from an anaerobic Sephadex-G-25 column possessed little ability to reduce carbon dioxide. However, when the eluate from the column was added to the protein fraction, the amount of carbon dioxide reduced to methane increased 18-fold, indicating a factor of low molecular weight was involved. When boiled cell extract was added to the protein fraction, the amount of carbon dioxide reduced to methane increased 44-fold. This factor was named the carbon dioxide reduction factor (CDR factor) (Romesser and Wolfe 1982*a*). Purification of the factor proved to be very difficult, since it had no absorption that was useful spectrophotometrically. By use of the laborious methanogenic assay to test each fraction during purification the compound was purified to homogeneity, and its structure was determined (Leigh *et al.* 1984). The formula is shown in Fig. 10.2. The compound was named methanofuran due to the furan moiety which also possessed a primary amine. The linear nature of the molecule is intriguing, as is the nature of the tetracarboxylic acid structure which forms the highly polar end of the molecule. The function of the molecule was not readily apparent, and it was not until methanogenesis from $^{14}CO_2$ was inhibited that evidence for radio label was

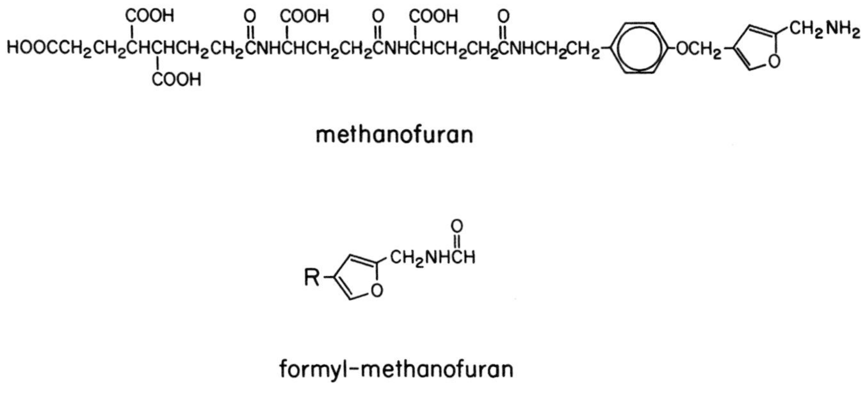

HOOCCH$_2$CH$_2$CH(COOH)CH(COOH)CH$_2$CH$_2$C(=O)NHCH(COOH)CH$_2$CH$_2$C(=O)NHCH(COOH)CH$_2$CH$_2$C(=O)NHCH$_2$CH$_2$–C$_6$H$_4$–OCH$_2$–(furan)–CH$_2$NH$_2$

methanofuran

R–(furan)–CH$_2$NHC(=O)H

formyl-methanofuran

Fig. 10.2 Structural formulas for methanofuran and formyl-methanofuran.

found to accumulate on methanofuran. Analysis of the labelled compound revealed that the label was carried on the primary amine (Fig. 10.2), formyl-methanofuran being the product (Leigh *et al.* 1985). This new method of carbon dioxide fixation was found in all methanogens which grew by the reaction shown in Fig. 10.1. In addition, when [^{14}C]-formyl-methanofuran was added to extracts under a nitrogen atmosphere (in the absence of hydrogen), $^{14}CO_2$ was evolved, indicating that a formyl-methanofuran dehydrogenase existed in methanogens (Leigh *et al.* 1985).

During the fractionation of methanofuran from *Methanosarcina* the compound exhibited unusual chromatographic behaviour. Analysis of the structure (Fig. 10.3) revealed that the polar end of the molecule was modified so that two glutamyl moieties replaced the tetracarboxylic structure (Bobik *et al.* 1987). This modification of the molecule was named methanofuran b. Another modification of the polar end (Fig. 10.3) from *Methanobrevibacter smithii* is named methanofuran c (White 1988). Methanofuran is found in certain non-methanogenic members of the archaea. Although other modifications in the polar end of the molecule are indicated (White 1988), no modifications at the furan end of the molecule have been found.

Formyl-methanofuran is the first stable product of carbon dioxide fixation to be detected in the pathway of methanogenesis. It is assumed, in analogy to the spontaneous reaction of furfurylamine with carbon dioxide, that the initial reaction involves the formation of a carbamate followed by enzymically catalysed

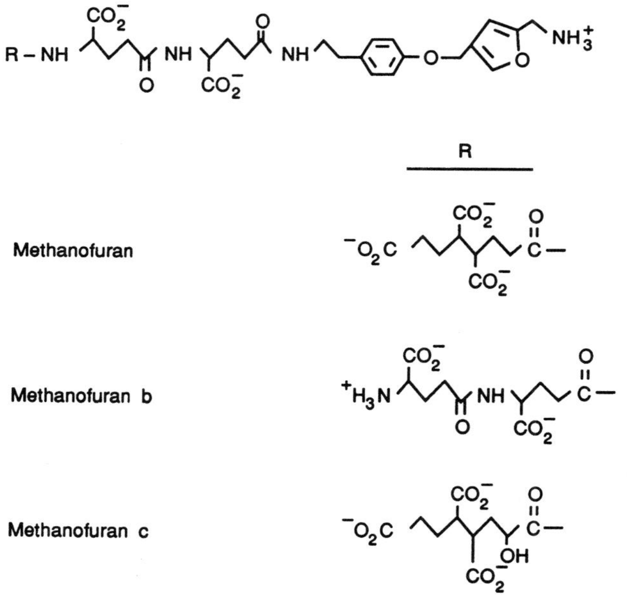

Methanofuran

Methanofuran b

Methanofuran c

Fig. 10.3 Variations in the structure of methanofuran. (By permission of *Annual Review of Biochemistry.*)

reduction to the formyl level. As shown in Fig. 10.4, the reduction of carbon dioxide to methane requires eight electrons and six unusual coenzymes (DiMarco *et al.* 1990). Methanogens employ three of these coenzymes as C1 carriers, methanofuran, tetrahydromethanopterin, and coenzyme M. Methanofuran carries a C1 group at the formyl level (reaction 1). Tetrahydromethanopterin accepts the formyl group (reaction 2) which is sequen-

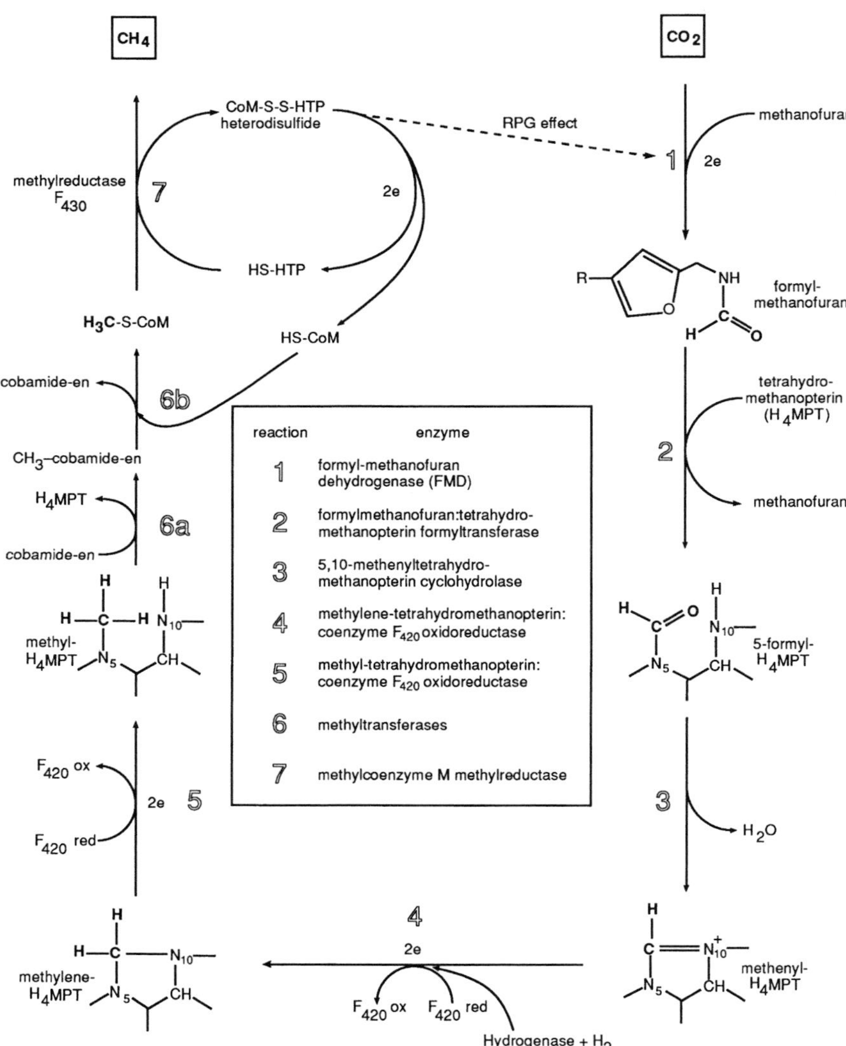

Fig. 10.4 Coenzymes and enzymes which carry out the reduction of CO_2 to CH_4. Only that portion of methanofuran and tetrahydromethanopterin (H_4MPT) which is involved in carrying a C1 group is shown.

tially modified from the formyl to the methenyl (reaction 3), to the methylene (reaction 4), to the methyl (reaction 5). The methyl-transferases (reactions 6a, 6b) transfer the methyl group to HS-coenzyme M, 2-mercaptoethanesulphonic acid, to form CH_3–S–CoM, 2-[methylthio]ethanesulphonic acid. CH_3–S–CoM is the substrate for the terminal methylreductase (reaction 7) which evolves methane as a product. This reaction involves two other unusual coenzymes, the electron donor 7-mercaptoheptanoylthreonine phosphate (HS-HTP) and the nickel tetrapyrrole, F430, a tetrahydrocorphin (DiMarco *et al.* 1990). Another unique electron carrier, the deazaflavin F420, is involved in reactions 4 and 5 (DiMarco *et al.* 1990). The precise electron donor for the formation of formyl-methanofuran, reaction 1, is uncertain.

In certain methanogens there is a dramatic coupling of the methylreductase, reaction 7, with the activation and reduction of carbon dioxide. This phenomenon was discovered by R.P. Gunsalus and was named the RPG effect (Gunsalus and Wolfe 1977). When cell extract from the thermophile, *Methanobacterium thermoautotrophicum*, was incubated under a hydrogen and carbon dioxide atmosphere in the presence of CH_3–S–CoM the rate of methane formation increased 30-fold and the amount of methane formed exceeded by 11fold that formed CH_3–S–CoM under a hydrogen atmosphere. Only a negligible amount of methane was formed under hydrogen and carbon dioxide in the absence of added CH_3–S–CoM. This coupling had no definitive stoichiometry and was believed to involve an unstable product of the methylreductase reaction, since the reaction could be repeatedly initiated in the same reaction mixture by sequential addition of CH_3–S–CoM. The magnitude of the RPG effect in various methanogens varied greatly as shown in Table 10.2 (Romesser and Wolfe 1982*b*), and formaldehyde, serine, pyruvate or ethyl-coenzyme M were initiators of the RPG effect; so it appeared that any compound which could donate a C1 group or participate in the methylreductase reaction could initiate the effect. However, malate and fumarate (Table 10.3) also initiated the RPG effect, and we made the erroneous assumption that these compounds in some manner gave rise to a C1 group which entered the methanogenic pathway (Romesser,

Table 10.2 Presence of the RPG effect in various methanogens

	Additional CO_2 reduced by RPG effect (nmol)
Methanobacterium bryantii	2000
Methanospirillum hungatei	708
Methanobacterium formicicum	94
Methanobrevibacter ruminantium	24
Methanosarcina barkeri	0

Assay conditions were as described by Romesser and Wolfe (1982*b*).

Table 10.3 Induction of the RPG effect in dialysed cell extracts by intermediates of the citric acid cycle

	Methane formed (nmol)		
Substrate	From substrate	From CO_2	CO_2 reduced per mol of substrate
L-malate	14	724	52
fumarate	14	1405	100

Assay conditions were as described by Romesser (1978).

unpublished thesis, 1978). In addition thiols were found to inhibit the RPG effect (Romesser and Wolfe 1982*b*). At that time these observations made little sense.

With the discovery of methanofuran and the characterization of formyl-methanofuran (Leigh *et al.* 1985) it became possible to study the RPG effect in a simplified system where the formation of formyl-methanofuran from carbon dioxide could be measured rather than the complete reduction of carbon dioxide to methane. However, it was not until the methylreductase reaction was defined that further progress could be made. With the identification of the electron donor as 7-mercaptoheptanylthreonine phosphate, HS-HTP (Noll *et al.* 1986), the stoichiometry of the terminal reaction was defined (Bobik *et al.* 1987; Ellermann *et al.* 1988), the heterodisulphide of the two coenzymes being a product:

$$CH_3\text{–S–CoM} + \text{HS–HTP} \rightarrow \text{CoM–S–S–HTP} + CH_4$$

It now became possible to explore the effect of the heterodisulphide on the activation and reduction of carbon dioxide to methane. As shown in Fig. 10.5 the heterodisulphide substituted for CH_3–S–CoM in stimulating the activation and reduction of carbon dioxide to methane (Bobik and Wolfe 1988). Similar results were obtained, when only the synthesis of formyl-methanofuran was followed. Thus, CH_3–S–CoM was an indirect stimulator of the RPG effect, the heterodisulphide being the initiator. In addition inhibitors of the methylreductase were found to inhibit the RPG effect from CH_3–S–CoM but not from the heterodisulphide. This was the first demonstration that the RPG effect could be uncoupled from the methylreductase reaction.

Some of the curious observations about the RPG effect such as the transient nature of the intermediate and the inhibition of the RPG effect by thiols, could now be explained. The former was explained by the finding of a specific heterodisulphide reductase in cell extracts (Bobik and Wolfe 1988; Hedderich and Thauer 1988), the latter by the finding that the heterodisulphide could be reduced by dithiothreitol (Fig. 10.6) (Bobik and Wolfe 1988). In addition, the curious initiation of the RPG effect by fumarate was finally explained when an unusual fumarate reductase was discovered in extracts of *Methanobacterium*

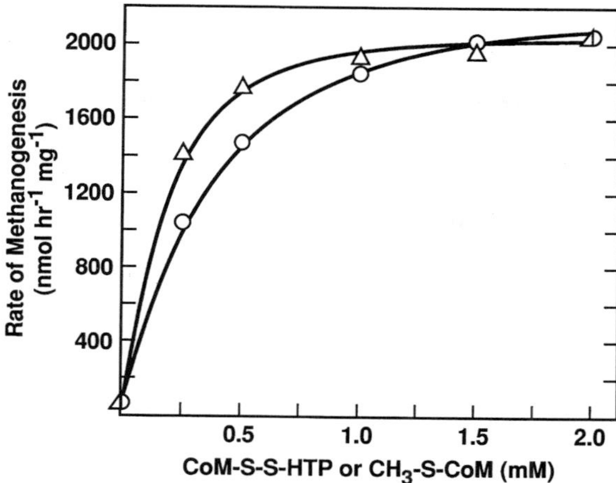

Fig. 10.5 Comparison of CoM–S–S–HTP and CH_3–S–CoM dependent stimulation of methanogenesis from CO_2 by cell extracts. Rate of methanogenesis: Δ CH_3–S–CoM; ○ CoM–S–S–HTP. Conditions were as described by Bobik and Wolfe (1988). (With permission of the National Academy of Sciences USA.)

(Fig. 10.7) (Bobik and Wolfe 1989*a*). The enzyme is highly specific for HS–CoM and HS–HTP as coelectron donors! The products of the reduction are succinate and the heterodisulphide CoM–S–S–HTP. This alternative mechanism

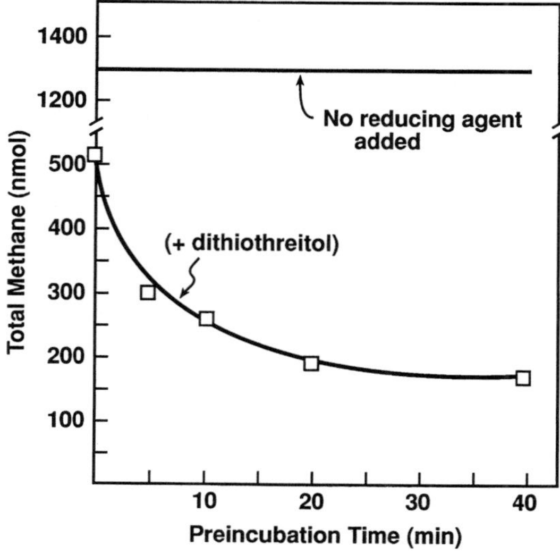

Fig. 10.6 Effect of dithiothreitol on the stimulation of CO_2 reduction to methane by CoM–S–S–HTP. Conditions were as described by Bobik and Wolfe (1988). (With permission from the National Academy of Sciences USA.)

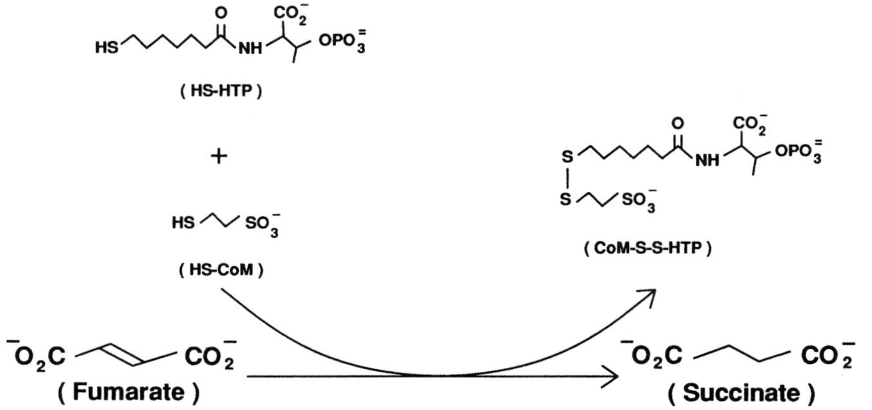

(HS-HTP)

+

(HS-CoM)

(CoM-S-S-HTP)

(Fumarate)

(Succinate)

Fig. 10.7 The unusual fumarate reductase from *Methanobacterium thermoautotrophicum* which uses HS–CoM and HS–HTP as coelectron donors with the formation of succinate and CoM–S–S–HTP as products. (From Bobik and Wolfe 1989*a*.) (Reproduced with permission from the *Journal of Biological Chemistry*.)

for producing the heterodisulphide came as a complete surprise. Thus, there is a common thread that links the curious observations of the RPG effect, that thread being CoM–S–S–HTP. The stimulatory effect of formaldehyde was shown to arise from the spontaneous reaction of formaldehyde with tetrahydro–methanopterin to form methylene-H_4MPT, and intermediate in the pathway of carbon dioxide reduction to methane (Escalante-Semerena and Wolfe 1984).

The synthesis of formylmethanofuran in the presence of the heterodisulphide is only observed under an atmosphere of hydrogen and carbon dioxide (nitrogen cannot substitute). Since the reduction of CO_2 to the formyl level requires electrons, it is likely that electrons are generated by a hydrogenase and, in the presence of the heterodisulphide, relayed in some way to formylmethanofuran dehydrogenase. Evidence for the presence of an intermediate in that electron transport chain was obtained when using metronidazole (Bobik and Wolfe 1989*b*). Reduction of metronidazole was observed in cell extracts under a hydrogen atmosphere to be dependent upon addition of CoM–S–S–HTP, Fig. 10.8. Metronidazole is commonly used in enzymic assays for ferredoxins and flavodoxins where it is rapidly reduced. In contrast it is only slowly reduced by hydrogenases. This new enzyme activity was termed HAMRA, for heterodisulphide-activating, metronidazole-reducing activity. A ferredoxin would be a logical candidate for the electron carrier in HAMRA, perhaps the polyferredoxin (Reeve *et al.* 1989) could be involved.

Recently, it was shown that HAMRA could be resolved from formylmethanofuran dehydrogenase by a simple hydrophobic interaction chromatographic step, and that both enzyme fractions were required to reconstitute *in vitro* formyl-methanofuran synthesis from CO_2 and the heterodisulphide under

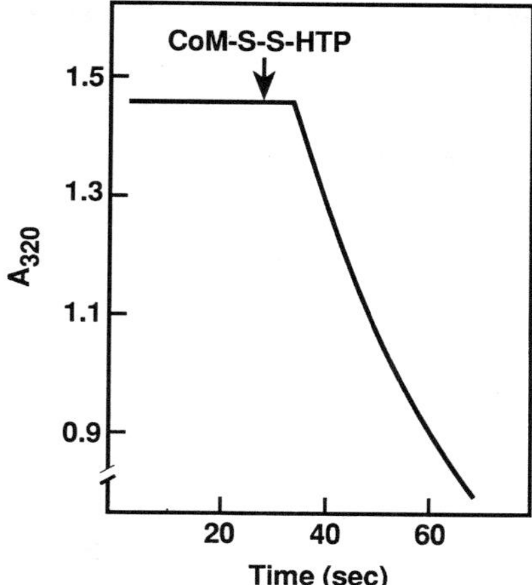

Fig. 10.8 Activation of metronidazole reduction by CoM–S–S–HTP. Assay conditions were as described by Bobik and Wolfe (1989*b*). (Reproduced with permission from the *Journal of Bacteriology.*)

hydrogen (Bobik, unpublished thesis, 1990). The available evidence made it possible to propose that the minimal structural requirements to observe HAMRA define a protein complex between hydrogenase (since molecular hydrogen is required in the assay) and a postulated intermediate electron carrier (IEC, which passes electrons on to metronidazole), as indicated in Fig. 10.9. In this model, (redrawn after Bobik *et al.* 1990), the heterodisulphide is assumed to promote in some way the activation of a hydrogenase, thus mediating the RPG effect, and the active hydrogenase subsequently passes electrons on to formyl-methanofuran dehydrogenase via an intermediate electron carrier, or to metronidazole.

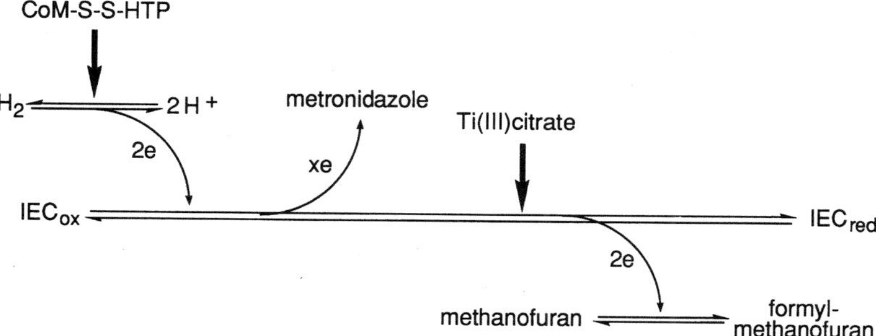

Fig. 10.9 Model for the activation of formyl-methanofuran synthesis by CoM–S–S–HTP. Titanium(III) citrate bypasses hydrogenase and donates electrons directly to IEC, the proposed intermediate electron carrier which can reduce metronidazole or CO_2 to formylmethanofuran.

HAMRA is a very labile activity, and its purification could not be attempted until the detergent CHAPS was found to stabilize the enzyme complex. So far, HAMRA has not been resolved into distinct components (Wasserfallen and Wolfe, unpublished).

An interesting finding by Bobik indicated that analysis of the RPG effect could be further simplified by use of titanium(III) citrate under an atmosphere of nitrogen and carbon dioxide. Low-potential electrons from titanium(III) citrate functionally replaced the heterodisulphide and hydrogen in the synthesis of formyl-methanofuran (Bobik and Wolfe 1989*b*). This discovery was an important simplification as the need for hydrogenase was bypassed. Titanium(III) citrate may not interact directly with FMD, but with its physiological electron donor, which is functionally replaced by metronidazole in the assay for HAMRA (Fig. 10.9). Therefore, both the HAMRA assay and titanium(III) citrate provide a handle to screen cell extracts for potential electron donors for FMD (Wasserfallen and Wolfe, unpublished data).

The heterodisulphide might conceivably act (i) as a regulator, or (ii) by reductively activating one protein component. In support of the regulatory role is the fact that the rate of methanogenesis from CO_2 was stimulated 42-fold by the heterodisulphide in a typical assay for the RPG effect. Reductive activation is more difficult to rationalize, as the heterodisulphide has been shown to undergo reduction to HS–HTP and HS–CoM by a soluble enzyme (Hedderich *et al.* 1990). The heterodisulphide reductase is able to use reduced benzyl viologen (E_0' = –360 mV) as the electron donor. Thus, it is likely that the heterodisulphide is reducible by electrons from molecular hydrogen. In enzymic terms, this reaction has been proposed to involve hydrogen (the electron source), a hydrogenase, an intermediate electron carrier, and the heterodisulphide reductase. In contrast, only low-potential electron donors such as $Cr(II)Cl_2$ and titanium(III) citrate (E_0' = –480 mV) were thus far found to substitute for the heterodisulphide in the activation of formylmethanofuran synthesis. Formyl-methanofuran dehydrogenase has been calculated to have a midpoint potential of –497 mV (Keltjens and van der Drift 1986), and in cell extracts the equilibrium favours the oxidation of formyl-methanofuran to CO_2 (Leigh *et al.* 1985).

Although considerable progress has been made in locating the components involved in the RPG effect and the activation and reduction of carbon dioxide to formyl-methanofuran, the biochemical details of this process are elusive at present.

Acknowledgements

This work was supported by grants from National institute of Allergies and infectious Diseases, The National Science Foundation, and the Department of Energy. A. Wasserfallen was supported by a postdoctoral fellowship from the Swiss National Science Foundation.

References

Bobik, T.A. (1990). Ph.D. Thesis. University of Illinois at Urbana-Champaign.

Bobik, T.A. and Wolfe, R.S. (1988). *Proc. Nat. Acad. Sci. USA*, **85**, 60–3.

Bobik, T.A. and Wolfe, R.S. (1989*a*). *J. Biol. Chem.*, **264**, 18 714–18.

Bobik, T.A. and Wolfe, R.S. (1989*b*). *J. Bacteriol.*, **171**, 1423–7.

Bobik, T.A., Donnelly, M.I., Rinhehart, K.L., Jr, and Wolfe, R.S. (1987). *Arch. Biochem. Biophys.*, **254**, 430–6.

Bobik, T.A., DiMarco, A.A., and Wolfe, R.S. (1990). *FEMS Microbiol. Rev.*, **87**, 323–6.

DiMarco, A.A., Bobik, T.A., and Wolfe, R.S. (1990). *Ann. Rev. Biochem.*, **59**, 355–94.

Ellermann, J., Hedderich, R., Böcher, R., and Thauer, R.K. (1988). *Eur. J. Biochem.*, **172**, 669–77.

Escalante-Semerena, J.C., and Wolfe, R.S. (1984). *J. Bacteriol.*, **158**, 721–6.

Gunsalus, R.P. and Wolfe, R.S. (1977). *Biochem. Biophys. Res. Commun.*, **76**, 790–5.

Hedderich, R. and Thauer, R.K. (1988). *FEBS Lett.*, **234**, 223–7.

Hedderich, R., Berkessel, A., and Thauer, R.K. (1990). *Eur. J. Biochem.*, **193**, 255–61.

Jones, W.J., Donnelly, M.I., and Wolfe, R.S. (1985). *J. Bacteriol.*, **163**, 126–31.

Jones, W.J., Nagle, D.P., J, and Whitman, W.B. (1987). *Microbiol. Rev.*, **51**, 135–77.

Keltjens, J.T. and van der Drift, C. (1986). *FEMS Microbiol. Rev.*, **39**, 259–303.

Leigh, J.A., Rinehart, K.L., Jr, and Wolfe, R.S. (1984). *J. Am. Chem. Soc.*, **106**, 3636–40.

Leigh, J.A., Rinehart, K.L., Jr, and Wolfe, R.S. (1985). *Biochemistry*, **24**, 995–9.

Noll, K.M., Rinehart, K.L., Jr, Tanner, R.S., and Wolfe, R.S. (1986). *Proc. Nat. Acad. Sci. USA*, **83**, 4238–42.

Reeve, J.N., Beckler, G.S., Cram, D.S., Hamilton, P.T., Brown, J.W., Krzycki, J.A. (1989). *Proc. Nat. Acad. Sci. USA*, **86**, 3031–5.

Romesser, J.A. (1978). Ph.D. Thesis. University of Illinois at Urbana-Champaign.

Romesser, J.A. and Wolfe, R.S. (1982*a*). *Zbl. Bakt. Hyg., I Abt. Orig.*, **C3**, 271–6.

Romesser, J.A. and Wolfe, R.S. (1982*b*). *J. Bacteriol.*, **152**, 840–7.

Smith, P.H., and Hungate, R.E. (1958). *J. Bacteriol.*, **75**, 713–18.

White, R.H. (1988). *J. Bacteriol.*, **170**, 4594–7.

Woese, C.R., Kandler, O., and Wheelis, M.L. (1990). *Proc. Nat. Acad. Sci. USA*, **87**, 4576–9.

11

Hydrogenation of carbon dioxide by catalysts

Yoshie Souma and Masahiro Fujiwara

Introduction

Huge amounts of CO_2 have been emitted by the consumption of fossil fuel, accompanied by the development of industry. The total amounts of emitted CO_2 from fossil fuels are assumed to be 19×10^9 ton year^{-1}, of which 60 per cent are emitted from thermal power stations or factories. Although CO_2 should be recovered and fixed in order to prevent global warming, at the same time, CO_2 is an important carbon source to be reused as a chemical or fuel. Various methods are known for the fixation of CO_2. However, a rapid conversion rate of CO_2, which is comparable to the combustion rate of fossil fuel, is indispensable. From this viewpoint the hydrogenation of CO_2 is expected to be the best method to fix huge amounts of CO_2. Carbon dioxide is emitted in comparatively high concentrations from power stations, factories, and ironworks and is therefore suitable for recovery and recycling use. The concept of recycling technology for CO_2 by hydrogenation is shown in Fig. 11.1. Although many reports of CO hydrogenation have been published for the synthesis of an alternative fuel after the oil crisis of the last decade (Fujimoto *et al.* 1988; Inui *et al.* 1985; Liu *et al.* 1984), the study of CO_2 hydrogenation is by no means sufficient. The research on CO_2 hydrogenation is expected to be developed rapidly by referring to the results of CO hydrogenation. Reviews (Denise and Sneeden 1982), books (Komiyama 1990) and investigation reports (NEDO 1991) for the hydrogenation of CO_2

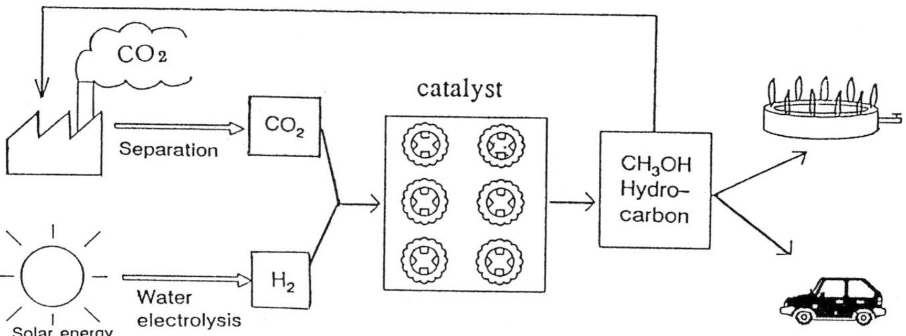

Fig. 11.1 Outline of the research on the recycling of CO_2.

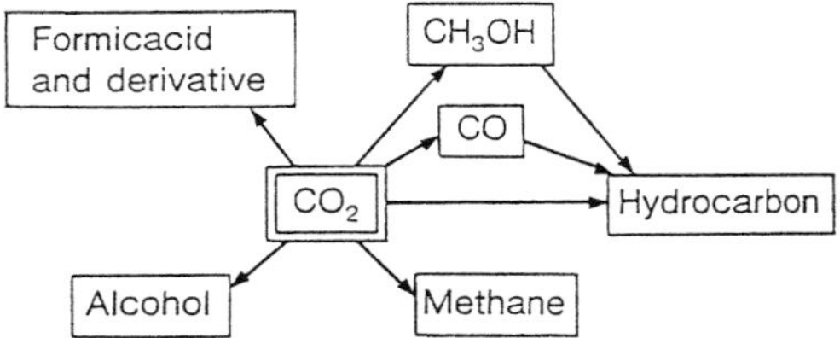

Fig. 11.2 Chemicals obtained from the hydrogenation of CO_2.

have been published. The main products obtained from CO_2 hydrogenation are shown in Fig. 11.2. Carbon monoxide, methanol, methane, hydrocarbons (C_2–C_6), higher alcohols (ethanol, propanol etc.), and formic acid and its derivatives were obtained by using various catalysts. There are many reports on methanation and methanol synthesis among them. In this chapter, recent studies on CO_2 hydrogenation will be summarized.

Methanol synthesis

Methanol is an important chemical and a transportable clean fuel. Presently methanol is commercially prepared from CO and H_2. Synthesis of methanol by the hydrogenation of CO_2 is shown in eqn (1). The exothermal heat of the reaction is almost half of that from syngas (eqn (2)).

$$CO_2 + 3H_2 \rightleftharpoons CH_3OH + H_2O \qquad \Delta H = -11.70 \text{ kcal mol}^{-1} \qquad (1)$$
$$CO + 2H_2 \rightleftharpoons CH_3OH \qquad \Delta H = -21.54 \text{ kcal mol}^{-1} \qquad (2)$$

In the hydrogenation of CO_2, reverse shift reaction (eqn (3)), and methanation (eqn (4)) take place at the same time. Therefore, an increase in the selectivity of methanol is an important subject.

$$CO_2 + H_2 \rightleftharpoons CO + H_2O \qquad \Delta H = 9.84 \text{ kcal mol}^{-1} \qquad (3)$$
$$CO_2 + 4H_2 \rightleftharpoons CH_4 + 2H_2O \qquad \Delta H = -39.42 \text{ kcal mol}^{-1} \qquad (4)$$

The relation between the equilibrium conversion of CO_2 in these four reactions and temperature is shown in Fig. 11.3. The reaction of methanol synthesis is advantageous at lower temperature and higher pressure from the viewpoint of equilibrium. Therefore, the development of an active catalyst at low temperature is desired. Although methanol synthesis from CO_2 seems to be disadvantageous relative to the synthesis from CO in equilibrium, Bardet *et al.* (1981) reported that much more methanol was obtained from CO_2–$4H_2$ in comparison to CO–$4H_2$ around 200 °C at atmospheric pressure in a low conversion region using a Cu–Zn–Al catalyst. A similar tendency was reported by Arakawa *et al.* (1992) using a Cu–ZnO/SiO_2 catalyst. A typical result is shown Fig. 11.4.

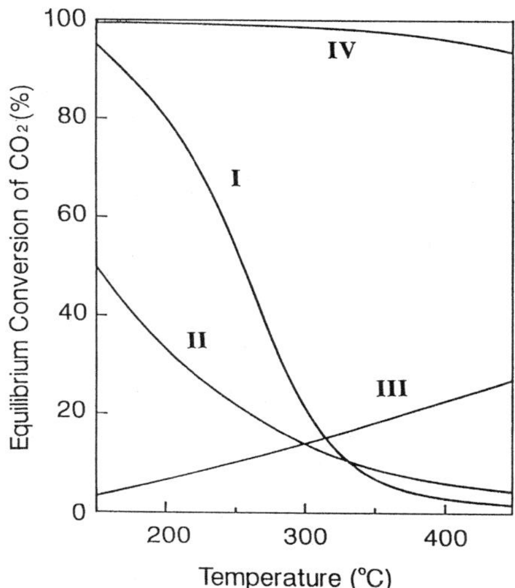

Fig. 11.3 Relation between equilibrium conversion and temperature in CO_2 hydrogenation (50 kg cm^{-2}).

I	$CO + 2H_2$	CH_3OH
II	$CO_2 + 3H_2$	$CH_3OH + H_2O$
III	$CO_2 + H_2$	$CO + H_2O$
IV	$CO_2 + 4H_2$	$CH_4 + 2H_2O$

The research on methanol synthesis from CO_2 has been mainly developed using a Cu–ZnO catalyst which is used in methanol synthesis from syngas. In research during an early period, Ipatieff and Monroe (1945) reported that methanol was obtained in high yield using a Cu/Al_2O_3 catalyst at 282 °C under 409 atm.

Detailed research has been developed in the 1980s. (Kieffer *et al.* 1981; Amenomiya 1987). Ramaroson *et al.* (1982) found that 16.4 per cent of CO_2 was converted to methanol at 300 °C, and 110 bar; the order of the activity of supports was $ThO_2>La_2O_3>Nd_2O_3>In_2O_3$, $Y_2O_3>MgO$, $Al_2O_3>SiO_2$. On the other hand, precious metal catalysts such as Pd (Ramaroson *et al.* 1982), Pt (Inoue and Iizuka 1986), Re (Iizuka *et al.* 1983), and Rh were used on various supports. Inoue and Iizuka (1986) carried out the hydrogenation of CO_2 using Pt catalysts at 240 °C, under 10 atm in rather mild conditions, and revealed that the activity of supports for hydrogenation was in the following order; ZrO_2, $Nb_2O_5>MgO>TiO_2>SiO_2$, and the order of selectivity for methanol was $MgO>TiO_2>ZrO_2>Nb_2O_5>SiO_2$.

Inui and Takeguchi (1991) summarized the activity of various catalysts for methanol synthesis as shown in Table 11.1. Inui *et al.* (1985) developed the active methanol synthesis catalyst Cu–Zn–Cr–Al by the uniform gelation method to improve the uniformity of multicomponent catalyst. The activity of

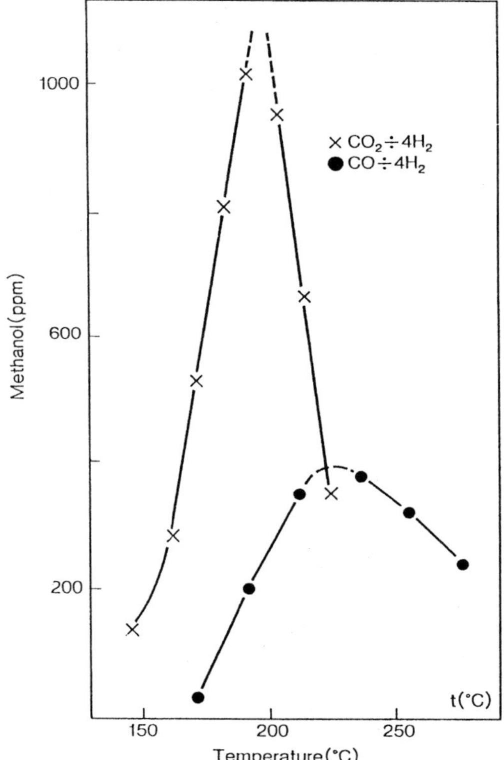

Fig. 11.4 Methanol synthesis from CO_2/H_2 and CO/H_2.

the catalyst was 1.5 times higher compared to the catalyst obtained by coprecipitation. Moreover, precious metals were added to the Cu–Zn–Cr–Al catalyst expecting the effect of hydrogen spill over, and it was found that the addition of Pd and Ag was markedly effective in methanol synthesis from CO_2, and the STY of methanol was as high as 356 g lh^{-1}, which was the highest value among those reported previously.

Concerning the mechanism of methanol synthesis from CO_2, there are two opinions. One is the direct hydrogenation of CO_2, and the other is the hydrogenation via CO. Denise *et al.* (1982) proposed a mechanism which has both reaction pathways.

Arakawa *et al.* (1991) observed IR absorption of reaction intermediates on a Cu, Zn/SiO_2 catalyst under various temperatures and pressures at rather low yield conditions. They proposed direct hydrogenation pathway of CO_2 as follows:

Table 11.1 Catalysts for methanol synthesis

Catalyst	Conditions	Conversion to MeOH (%)	MeOH STY (g/lh)	Reference
CuO–ZnO/La_2O_3	225 °C 40 atm	2.7		Beguin *et al.* 1980
Cu/ZrO_2	260 °C	4.1	36.3	Gasser and Baiber 1989
Pd/La_2O_3	350 °C 120 atm	6.1	87.1	Ramaroson *et al.* 1982
Pt/La_2O_3	240 °C 10 atm	0.3>		Inoue and Iizuka 1986
Cu–Zn–Cr–Al	250 °C 50 atm	10.5	176	Inui and Takeguchi 1991
Cu–Zn–Cr–Al–Pd	250 °C 50 atm	21.2	356	Inui and Takeguchi (1991)

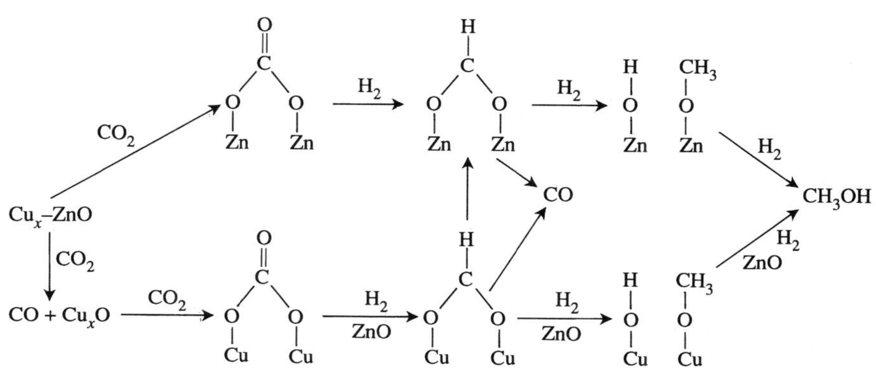

The surface conditions seem to be different among various catalysts, and further detailed research is expected to clarify these differences.

Methanation

Methanation is the easiest reaction for the hydrogenation of CO_2. The representative catalysts for hydrocarbon synthesis including methanation are shown in Table 11.2.

Methanation was studied using Group VIII metal catalysts. Weatherbee and Bartholomew (1984) reported that methane was obtained on Fe, Co, Ru, Ni catalyst in 9–11 per cent yield at 205–291 °C under 1–11 atm. The order of the activity

Table 11.2 Catalysts for fuel synthesis from CO_2

Catalyst	Conditions	CO_2 conversion(%)	Selectivity of products (%)	Reference
Co, Fe, Ru/SiO_2	205–290 °C, 1 atm	10–11	CH_4 (89–99)	Weatherbee and Bartholomew 1984
Ni–Ru–La_2O_3	227 °C		CH_4	Inui *et al.* 1980
Fe_2O_3–CuO–TiO_2-Al_2O_3–KCl	350 °C, 15 atm	10	CH_4(12), C_2(7), C_3(7), CO(47), C_4^+(27)	Barrault *et al.* 1981
Cu/SiO_2,Rh/Nb_2O_5	340 °C, 1 atm	20	C_2- C_3 (40)	Nozaki *et al.* 1987
Cu–Cr–Zn–Al H-Fe-silicate	250 °C, 50 atm 300 °C, 1 atm	25	CH_3OH (56) C_1-C_7 (100)	Inui and Takeguchi 1991
Zn-Cr + DAY	320 °C, 21 atm	20.4	C_1–C_6 (55)	Fujimoto and Shikada 1987

of supports was Co>Ru>Ni>Fe. In addition, Iizuka *et al.* (1982) studied Rh catalyst on various supports and found that hydrogenation of CO_2 occurred at a lower temperature in comparison with CO and that the order of activity of supports was ZrO_2>Al_2O_3>SiO_2>MgO.

Inui *et al.* (1980) developed an active composite catalyst Ni–La_2O_3–Ru, dispersed on the support having a meso-macro bimodal pore structure, which enabled rapid transformation of CO_2 to methane. When the three components,

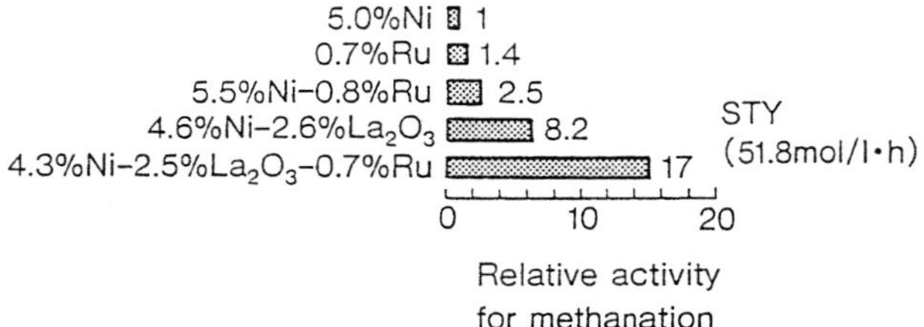

Fig. 11.5 Composition of catalyst and activity for methanation. CO_2 6 per cent, H_2 18 per cent, N_2 76 per cent, 227 °C, SV = 1000 h^{-1}.

Ni, La_2O_3, and Ru, were combined in proper ratios, a prominent synergistic effect, which was far from the sum of the activities of single components, was obtained as shown in Fig. 11.5. The La_2O_3 part would contribute to the increase of CO_2 adsorption capacity by its basicity, and the Ru part would play the port-hole for hydrogen spillover. As for the support structure, a porous material having meso pores (diameter 6 nm) and macro pores (diameter 760 nm) was suitable. Catalytic hydrogenation took place at the meso pores and reactant and products are rapidly transported through the macro pores.

Hydrocarbon synthesis

Higher hydrocarbons were prepared from the hydrogenation of CO_2 directly by using Fisher–Tropsch type Fe catalysts (Barrault *et al.* 1981; Kiennemann *et al.* 1981) or via CO (Nozaki *et al.* 1987) or methanol (Inui *et al.* 1986; Fujimoto and Shikada 1987) by using hybrid catalysts.

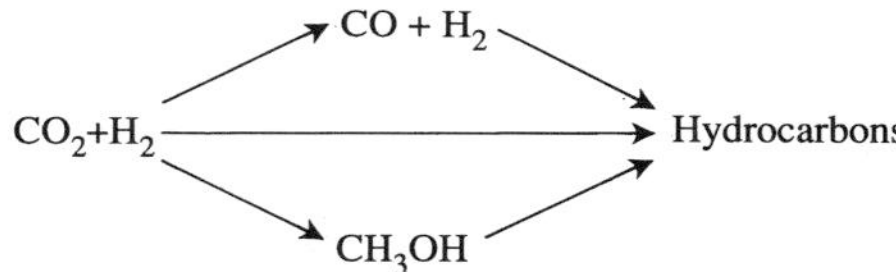

When a Fe_2O_3–CuO–TiO_2–Al_2O_3–KCl catalyst was used at 350 °C under 15 atm, the conversion of CO_2 into hydrocarbons was about 10 per cent, and the selectivity of C_2–C_7 hydrocarbons was 41 per cent. A hybrid catalyst, composed of Cu/SiO_2 and Rh/Nb_2O_5 which is useful for the conversion of CO_2 into CO and CO into hydrocarbons, was effective for CO_2 hydrogenation. The conversion of CO_2 was about 20 per cent, and the selectivity for C_2 and C_3 hydrocarbons was 40 per cent (Nozaki *et al.* 1987) at 340 °C under atmospheric pressure.

Inui and Takeguchi (1991) succeeded in obtaining gasoline from CO_2 via methanol by placing a H–Fe–silicate catalyst in the second reactor just after the first methanol synthesis reactor. The conversion of CO_2 to hydrocarbons was 24.9 per cent, the selectivity of gasoline was 45 per cent, and most of the C_2–C_4 hydrocarbons were intermediate olefins. When propylene was added in order to accelerate the conversion of olefins to gasoline, the selectivity of gasoline increased to 60 per cent (Fig. 11.6). Fujimoto and Shikada (1987) studied the hydrogenation of CO_2 by hybrid catalysts composed of Zn–Cr and DAY (de-aluminated Y type zeolite) and reported that the conversion of CO_2 and the yield of hydrocarbons were 20 per cent and 11.2 per cent, respectively, at 320 °C under 21 atm.

The formation of hydrocarbons with a high yield which exceeds the thermo-dynamic limit of the methanol formation is expected by the use of hybrid catalysts composed of methanol synthesis catalysts and zeolite. The partial

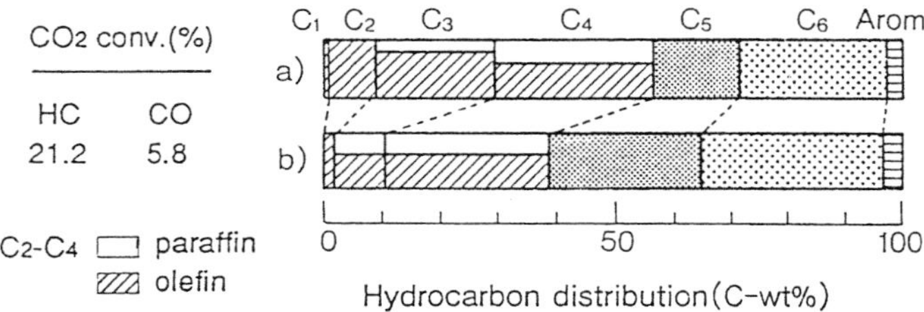

Fig. 11.6 Gasoline synthesis from CO_2 using a series reactor. First reactor: Cu–Zn–Cr–Al, 250 °C, 50 atm, SV = 4700 h^{-1}. Second reactor: H–Fe–silicate, 300 °C, 1 atm, CO_2 converted to HC 24.9 per cent. (a) Without propylene. (b) Propylene added.

pressure of methanol will be lowered by the conversion of intermediate methanol into hydrocarbons by zeolite catalyst. This will make possible further conversion of CO_2 to methanol by exceeding the thermodynamic conversion in a flow type reactor. The remarkable synergistic effect of hydrocarbon formation from CO_2 by hybrid catalysts was reported by Souma (1991*a*), and Fujiwara and Souma (1992*a*). Hydrogenation of CO_2 was carried out using hybrid catalysts composed of Cu–Zn–Cr and zeolite HY at 400 °C under 50 atm. The conversion of CO_2 was 37.1 per cent, and the yield of hydrocarbons was 8.4 per cent. Because the yield of methanol was 1.8 per cent at 400 °C using a Cu–Zn–Cr catalyst, 5 times the synergistic effect was observed by combining zeolite with Cu-Zn-Cr catalyst as shown in Fig. 11.7. Furthermore, the productivity of

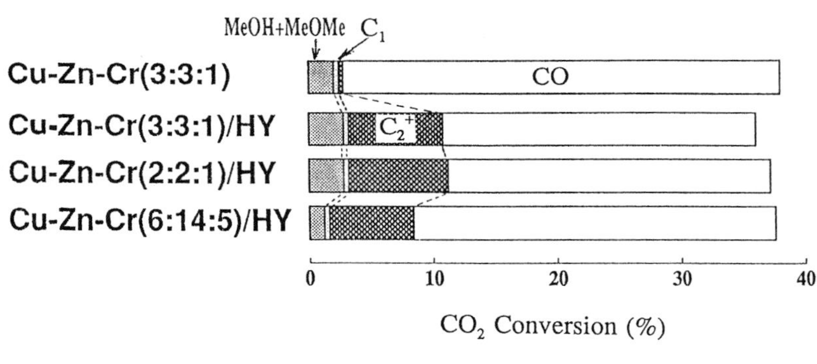

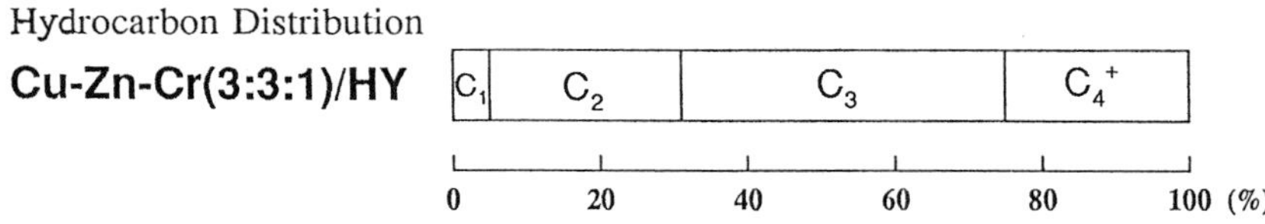

Fig. 11.7 Hydrocarbon synthesis using hybrid catalysts. Conditions: 400 °C, 50 kg cm^{-2}, SV = 3 l gh^{-1}, H_2/CO_2 = 3, HY(Si/Al = 4.8).

hydrocarbons exceeded the equilibrium conversion of CO_2 into methanol (6.5 per cent) under the same conditions. In this catalyst, the use of CrO_3 as a chromium source was important.

As the mechanism of hydrocarbon formation from methanol, dimethyl ether was formed by the dehydration of two molecules of methanol, and successive addition of methyl cation gave ethylene and longer hydrocarbons.

Other compounds

Ethanol and propanol were obtained by the hydrogenation of CO_2 using KCl–Mo–SiO_2 catalyst (Tatsumi *et al.* 1985). The yield of these alcohols was low, and further research is expected to develop this field.

Formic acid and its derivatives were obtained using complex catalysts such as $Pd(dpm)_2$, W, Cr, and Ru in organic solvent (Denise and Sneeden 1981; Darensbourg and Ovalles 1984). Potassium formate was obtained in KOH aqueous solution using Pd/C catalysts in 98 per cent yield at 20 °C under 20 atm of mixed gases of CO_2 and H_2 (Fujiwara and Souma 1992*b*).

$$CO_2 + H_2 \xrightarrow[\text{KOH}]{\text{Pd/C}} HCOOK$$

Supply of hydrogen

A source of cheap hydrogen is essential to realize the hydrogenation of CO_2. For the time being, by-product hydrogen, obtained from coke formation for the steel industry or from the steam reforming of natural gas, is available.

$$CH_4 + H_2O \longrightarrow H_2 + CO$$

However, in the twenty-first century, it is desirable that hydrogen is obtained from water electrolysis by solar cells or atomic energy, not from fossil sources. The generation of electric power by silicon solar cells has been studied energetically. The study of water electrolysis by solid polymer electrolyte has been developed, and 2 Nm^3 of hydrogen could be produced by the test plant of the Osaka National Research Institute, AIST (Takenaka 1991).

NEDO estimated the cost of methanol production using recovered CO_2 from a fixed emission source and hydrogen obtained by water electrolysis using a solid polymer electrolyte. As shown in Fig. 11.8, the cost of methanol production is proportional to the price of electricity, and the cost of hydrogen production amounts to 80 per cent of the total cost. Thus, the development of cheap hydrogen production is indispensable to the realization of the transformation of CO_2 to fuels.

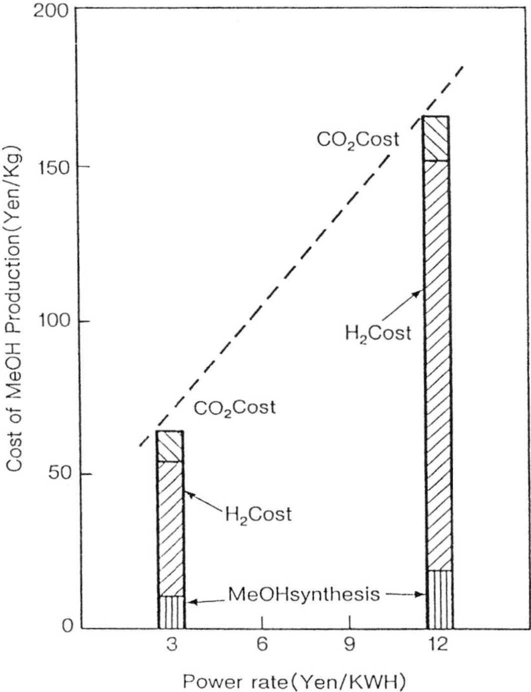

Fig. 11.8 Relation between the cost of methanol synthesis and power rate.

Conclusion

The study of CO_2 hydrogenation, which has received much attention recently, includes many interesting subjects such as the development of novel active catalysts and reaction mechanisms. Research on the production of cheap hydrogen is also very important if the hydrogenation of CO_2 is to be realized. The success of the hydrogenation of CO_2 will solve not only the global environmental problem but also the shortage of fossil fuels in the twenty-first century.

References

Amenomiya, Y. (1987). *Appl. Catalyst,* **30,** 57–68.

Arakawa, H., Sayama, K., Okabe, K., Shimomura, K., and Hagiwara, H. (1991). *Catalyst,* **33,** 103–6.

Bardet, R., Thivolle-Cazat, J., and Trambouze, Y. (1981). *J. Chim. Phys.,* **78,** 135–8.

Barrault, J., Forguy, C., Menezo, J.C., and Maurel, R. (1981). *Reac. Kin. Catal. Lett.,* **17,** 373–8.

Beguini, B., Denise, B., and Sneeden, R.P.A. (1980). *Reac. Kin. Catal. Lett.,* **14,** 9–14.

Darensbourg, D.J. and Ovalles, C. (1984). *J. Am. Chem. Soc.,* **106,** 3750–4.

Denise, B. and Sneeden, R.P.A. (1981). *J. Organomet. Chem.*, **221,** 111–16.

Denise, B. and Sneeden, R.P.A. (1982). *Chemtech,* 108–12.

Fujimoto, K. and Shikada, T. (1987). *Appl. Catal.*, **31,** 13–23.

Fujimoto, K., Saima, H., and Tominaga, H. (1988). *Ind. Eng. Chem. Res.*, **27,** 920–6.

Fujiwara, M. and Souma, Y. (1991*a*). *Proceeding of the international symposium on chemical fixation of carbon dioxide* (ed. K. Ito), pp. 373–8. Nagoya, Japan.

Fujiwara, M. and Souma, Y. (1992*a*). *J. Chem. Soc., Chem. Commun.*, 767–8.

Fujiwara, M. and Souma, Y. (1992). *Chemistry Express,* **2**, 353–6.

Gassa, D. and Baiber, A. (1989). *Appl. Catal.*, **48,** 279–94.

Iizuka, T., Tanaka, Y., and Tanabe, K. (1982). *J. Catal.*, **76,** 1–8.

Iizuka, T., Kojima, M., and Tanaka, K. (1983). *J. Chem. Soc., Chem. Commun.*, 638–9.

Inoue, T. and Iizuka, T. (1986). *J. Chem. Soc., Faraday Trans.*, **82,** 1681–6.

Inui, T. and Takeguchi, T. (1991). *Catalysis Today,* **10,** 95–106.

Inui, T., Funabiki, M., and Takegami, Y. (1980). *Ind. Eng. Chem. Res., Prod. Res. Div.*, **19,** 385–8.

Inui, T., Hagiwara, T., Yamase, O., and Kitagawa, K. (1985). *J. Jap. Petrol. Inst.*, **28,** 225–33.

Inui, T., Kitagawa, K., and Hagiwara, T. (1986). *Chemistry Exp.*, **1**, 107–10.

Ipatieff, V.N. and Monroe, G.S. (1945). *J. Am. Chem. Soc.*, **67,** 2168–71.

Kieffer, R., Ramaroson, E., Deluzarche, A., and Trambouze, Y. (1981). *Reac. Kin. Catal. Lett.*, 16, 207–12.

Kiennemann, A., Kieffer, R., and Chornet, E. (1981). *Reac. Kin. Catal. Lett.*, **16,** 371–6.

Komiyama, H. (1990). *Handbook for the protection of global warming,* IPC.

Liu, G., Willcox, D., Garland, M., and Kung, H. H. (1984). *J. Catal.*, **90,** 136–46.

NEDO (New Energy and Industrial Technology, Development Organization, Japan) (1990). *Survey report on new energy technology for global environment.* NEDO-p-8914, pp. 162–9.

NEDO (1991). *Survey report on global warming problem.* NEDO-ITE-9002-4.

Nozaki, F., Sodesawa, T., Satoh, S., and Kimura, K. (1987). *J. Catal.*, **104,** 339–46.

Ramaroson, E., Kieffer, R., and Kiennemann, A. (1982). *Appl. Catal.*, **4,** 281–6.

Souma, Y. (1991). *IEA international conference on technology responses to global environmental challenges.* No. 1 (IEA Kyoto Conference '91), pp. 475–8. Inter Group Corporation, Japan.

Takenaka, H. (1991). 1990 Annual summary of hydrogen energy R & D (Japan's Sunshine Project), pp. 1–10. AIST, MITI, Japan.

Tatsumi, T., Muramatsu, A., and Tominaga, H. (1985). *Chem. Lett.*, 593–4.

Weatherbee, G.D. and Bartholomew, C.H. (1984). *J. Catal.*, **87,** 352–62.

12

Carbon dioxide as an organic building block. Mechanisms of its activation by electron transfer and transition-metal complexes

Christian Amatore, Anny Jutand, and Merete F. Nielsen

Introduction

A large amount of work has been devoted to the reductive activation of carbon dioxide (for earlier reviews see Darensbourg and Kudarovski 1983; Silvestri 1987; Behr 1988; Braustein *et al.* 1988; Silvestri *et al.* 1990), since this molecule represents an abundant and low-cost potential source of carbon for the production of organic chemicals. Under its natural form, *viz.* CO_2, this is a rather stable molecule, inert to most organic reagents except strong nucleophiles. Conversely, its reductive activation, either by direct electron transfer (*outer-sphere activation*) or by ligation to an electron-rich metal centre (*inner-sphere activation*) considerably increases its reactivity, especially in view of allowing a facile creation of carbon–carbon bonds. Since several other chapters of this book (see for example Chapters 7–9, 13) are already devoted to the discussion of several of these synthetic perspectives, here the focus is on some mechanistic aspects relevant to the reductive activation of carbon dioxide, in view of its use as a C1 building block in organic chemistry. The first section of this chapter is devoted to the mechanism of electron-transfer activation of CO_2, as it occurs upon its electrochemical reduction at inert electrodes. In the second section we will discuss some of the mechanistic subtleties introduced by the combined use of electrochemical and transition metal activation of carbon dioxide, with respect to the formation of carbon–carbon bonds with organic substrates.

Mechanism of electrochemical reduction of carbon dioxide at inert electrodes in media of low proton availability

The direct electrochemical reduction of carbon dioxide occurs at rather negative potentials (more negative than –2 V in most aprotic solvents). Depending on the exact experimental conditions, CO_2 reduction at inert electrodes affords oxalate, formate, and equimolar amounts of carbonate and carbon monoxide (Gressin *et al.* 1979)

$$2CO_2 + 2e \rightarrow C_2O_4^{2-} \quad (1)$$
$$2CO_2 + 2e \rightarrow CO + CO_3^{2-} \quad (2)$$
$$CO_2 + H_2O + 2e \rightarrow OH^- + HCO_2^- \quad (3)$$

This product range may be widened (for example, so as to yield methanol and higher condensed products than oxalate) and the chemical selectivity possibly oriented by the use of specific electrode materials (see for example Silvestri 1987; Silvestri *et al.* 1990; and Chapter 13 for a precise account and discussion of such results). Yet, to the best of our knowledge, the mechanisms of such electrode-catalysed processes are not properly understood, except for the fact that they certainly involve strong interactions between the intermediate reduction products and each particular electrode material. Therefore, in the following, we will restrict our presentation to the mechanism of CO_2 reduction at inert electrodes, such as mercury. At such electrodes carbon dioxide reduction proceeds via the intermediate formation of its reactive anion radical, $CO_2^{\bullet -}$

$$CO_2 + e \Leftrightarrow CO_2^{\bullet -} \quad (4)$$

at a standard potential $E^0 = -2.21$ V versus SCE (saturated calomel electrode) in dimethylformamide (DMF) (Lamy *et al.* 1977). The strong reorganization of the molecule upon this electron transfer is reflected by the rather low magnitude of its rate constant of electron transfer ($k_0 = 6 \times 10^{-3}$ cm s^{-1}) (Lamy *et al.* 1977); this may be rationalized by considering that CO_2 is a linear molecule while its anion radical is bent, with localization of the anionic charge on one oxygen and of the unpaired electron at the sp^2 carbon. From such a structure one may expect $CO_2^{\bullet -}$ to undergo several competitive routes which afford eventually the products in eqns (1)–(3). Thus its protonation by residual or purposively added water leads to the oxidized form of formate which is reduced under these conditions to yield a formate ion (Roberts and Sawyer 1965)

$$CO_2^{\bullet -} + H_2O \rightarrow OH^- + HCO_2^{\bullet} \qquad (k_{H_2O}) \quad (5)$$
$$HCO_2^{\bullet} + e \rightarrow HCO_2^- \quad (6)$$

Another nucleophilic reaction may involve an attack of the negatively charged oxygen centre on an unreduced carbon dioxide molecule, leading to the formation of a carbon–oxygen bond; reduction of the transient undelocalized anion radical thus formed (see Scheme 12.1), and fragmentation of the corresponding dianion, yields a carbon monoxide molecule together with a carbonate ion

$$CO_2^{\bullet -} + CO_2 \rightarrow {}^{\bullet}C(=O)\text{–}O\text{–}C(=O)O^- \qquad (k_{CO_2}) \quad (7)$$
$${}^{\bullet}C(=O)\text{–}O\text{–}C(=O)O^- + e \rightarrow CO + CO_3^{2-} \quad (8)$$

Finally, owing to the radical character of the carbon centre of $CO_2^{\bullet -}$, a carbon–carbon bond may also be created by condensation of two anion radicals, thus yielding an oxalate ion

$$2CO_2^{\bullet -} \rightarrow C_2O_4^{2-} \qquad (k_{dim}) \quad (9)$$

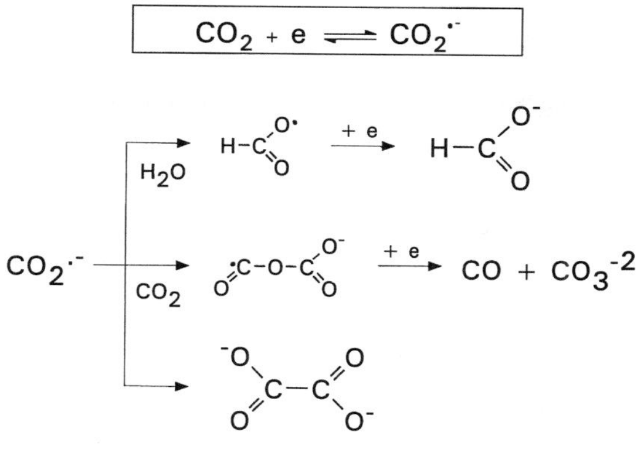

$$CO_2 + e \rightleftharpoons CO_2^{\bullet-}$$

$$CO_2^{\bullet-} \xrightarrow{H_2O} H\text{-}C(O^{\bullet})\text{=}O \xrightarrow{+e} H\text{-}C(O^-)\text{=}O$$

$$CO_2^{\bullet-} \xrightarrow{CO_2} O\text{=}\dot{C}\text{-}O\text{-}C(O^-)\text{=}O \xrightarrow{+e} CO + CO_3^{-2}$$

$$CO_2^{\bullet-} \longrightarrow {}^-O(O\text{=})C\text{-}C(\text{=}O)O^-$$

Scheme 12.1

The distribution of the products in eqns (1)–(3) depends strongly (Gressin *et al.* 1979) upon operational factors such as the current density, the initial carbon dioxide or water concentrations or the hydrodynamic regime of electrolysis, which influence the competition between the three steps in eqns (5) (formate), (7) (carbon monoxide), and (9) (oxalate).

The variation of the yield of formate versus that of carbon monoxide upon changing the respective concentrations of water and carbon dioxide is easily rationalized owing to the individual rate laws pertaining to eqns (5) and (7); indeed, their instant relative chemical yield is given at any time of electrolysis by (Amatore and Savéant 1981)

$$(d[HCO_2^-]/d[CO])_{chem} = (k_{H_2O}/k_{CO_2}) \times ([H_2O]/[CO_2]) \qquad (10)$$

The experimental validity of the expression in eqn (10) is confirmed by the plot in Fig. 12.1(a); furthermore, from the slope of the correlation line, the ratio of the rate constants (k_{H_2O}/k_{CO_2}) is found to be equal to 0.25. The respective yields of carbon monoxide and oxalate are more difficult to predict owing to the fact that both reactions (7) and (9) involve a second-order dependence on carbon dioxide concentration. However, Table 12.1 shows that their relative yields are extremely dependent on the electrolysis current density; when the latter is only a fraction of the limiting current density, i.e. when $(i/i_{lim}) \ll 1$, a significant amount of carbon monoxide is produced; conversely when the current density is closer to the limiting current density, the yield is oxalate is nearly quantitative. Such a result can be easily rationalized by considering the concentration profiles of CO_2 and $CO_2^{\bullet-}$ in the diffusion layer, *viz.* at a distance of a few micrometres from the electrode surface. Such concentration profiles are in Fig. 12.2 for two different electrolysis regimes: $(i/i_{lim}) = 0.5$, in (a), and $(i/i_{lim}) \approx 1$, in (b). The large current density in (b) imposes a low concentration of unreduced carbon dioxide in the region near the electrode surface where the $CO_2^{\bullet-}$ concentration is not null; conversely, in (a), at the electrode surface, $[CO_2]$ is approximately only

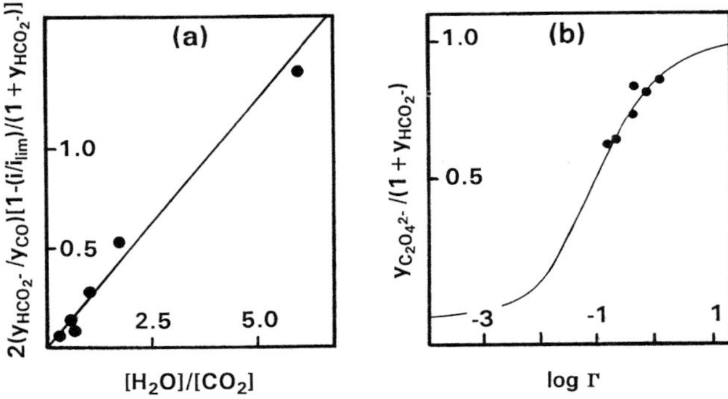

Fig. 12.1 Correlation between the yields of formate, carbon monoxide, and oxalate during the electrolysis of carbon dioxide at a mercury electrode in DMF. The points correspond to the experimental data (Gressin *et al.* 1979) and the solid lines to the theoretical predictions for the mechanism in Scheme 12.1 (Amatore and Savéant 1981) with: $(k_{H_2O}/k_{CO_2}) = 0.25$, in (a), and $(k_{dim}/k_{CO_2}{}^{3/2}) = 55\ M^{1/2}s^{1/2}$, in (b). (a) Correlation between the yields of formate and carbon monoxide. (b) Correlation between the yields of oxalate and carbon monoxide (note that a correction for the simultaneous production of formate is included); $\Gamma = (D/2[CO_2])^{1/2}\delta^{-1}(i/i_{lim})\ (1 + 2y_{HCO_2^-}/y_{CO})\ [1-(i/i_{lim})(1 + y_{HCO_2^-})]^{-3/2}$.

Table 12.1 Direct electrochemical reduction of CO_2 in DMF at a mercury electrode. Variations of the yields in formate, carbon monoxide, and oxalate as a function of electrolysis conditions. (Data from Gressin *et al.* (1979))

i/i_{lim}	δ (μm)*	$[CO_2]$ (M)	$[H_2O]$ (M)	Formate (%)†	Carbon monoxide (%)†	Oxalate (%)†
0.27	15	0.166	0.045	0.5	13	86
0.28	15	0.130	0.233	4	11	85
0.32	30	0.039	0.236	5	6	88
0.12	15	0.124	0.080	2	24	74
0.12	30	0.130	0.156	5	30	65
0.06	15	0.124	0.090	1	35	64

* Diffusion layer thickness.
† Chemical yields in per cent.

half of its value in the bulk of the solution, i.e. greatly exceeds the local concentration of $CO_2^{\bullet-}$. Since the instant relative chemical yield of oxalate (eqn (9)) versus that of carbon monoxide (eqn (7)) is given by (Amatore and Savéant 1981)

$$(d[C_2O_4^{2-}]/d[CO])_{chem} = (k_{dim}/k_{CO_2}) \times ([CO_2^{\bullet-}]/[CO_2]) \quad (11)$$

one understands why increasing the current density results in increasing the

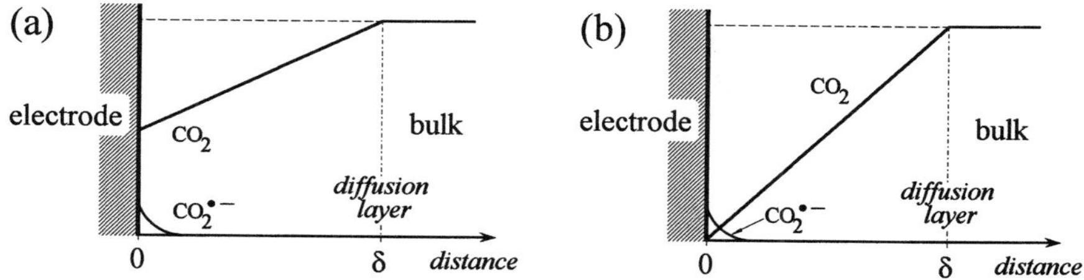

Fig. 12.2 Schematized concentration profiles of CO_2 and $CO_2^{\bullet-}$ as a function of the electrode distance during electrolysis of carbon dioxide, for two different current densities: $i = i_{lim}/2$; in (a), and $i = i_{lim}$ in (b).

yield of oxalate, because the available amount of carbon dioxide near the electrode surface decreases. This qualitative rationalization is further confirmed by the plot in Fig. 12.1(b) (Amatore and Savéant 1981), which compares the experimental yields given in Table 12.1 to those predicted on the basis of eqn (11) (solid line). Furthermore this plots allows the determination of the rate parameter ($k_{dim}/k_{CO_2}^{3/2}$), which is found to be equal to 55 $M^{1/2}s^{1/2}$. The good agreement between the experimental data and the theory allows also the ruling out of another possible mechanism for the formation of oxalate (Aylmer-Kelly *et al.* 1973), *viz.*

$$CO_2^{\bullet-} + CO_2 \rightarrow {}^{\bullet}O(O=)C{-}C(=O)O^- \quad (k'_{CO_2}) \qquad (12)$$

$$^{\bullet}O(O=)C{-}C(=O)O^- + e \rightarrow C_2O_4^{2-} \qquad (13)$$

Indeed, in eqn (7) we have considered that reaction of $CO_2^{\bullet-}$ with CO_2 necessarily involved a nucleophilic attack of the anionic oxygen of $CO_2^{\bullet-}$ on the positive carbon of CO_2. However, an alternative reaction between the same reactants may involve a radical attack of the carbon-centred radical of $CO_2^{\bullet-}$ on one π bond of CO_2, with formation of a carbon–carbon bond, as in eqn (12), rather than a carbon–oxygen bond as in eqn (7). Such a possibility would amount to replacing eqn (11) by eqn (14)

$$(d[C_2O_4^{2-}]/d[CO])_{chem} = (k_{dim}/k_{CO_2}) \times ([CO_2^{\bullet-}]/[CO_2]) + (k'_{CO_2}/k_{CO_2}) \qquad (14)$$

which is found to give a poor fit with the experimental data shown in Table 12.1 (Amatore and Savéant 1981), thus showing that the sequence in eqns (12) and (13) cannot be responsible for a large yield of oxalate, if any.

Since ($k_{dim}/k_{CO_2}^{3/2}$) = 55 $M^{1/2}s^{1/2}$, and (k_{H_2O}/k_{CO_2}) = 0.25 are known, knowledge of one of these rate constants allows the determination of all three. An estimate of $k_{dim} \approx 10^7$ $M^{-1}s^{-1}$, can be obtained from the limit of chemical reversibility of the cyclic voltammogram reported previously (Lamy *et al.* 1977) for the reduction of carbon dioxide in DMF; indeed, based on the above values of ($k_{dim}/k_{CO_2}^{3/2}$) and (k_{H_2O}/k_{CO_2}) it can be shown that under the conditions where these cyclic voltammetric experiments have been performed, the main route for $CO_2^{\bullet-}$

consisted of its dimerization in eqn (9). This allows us to evaluate (Amatore and Savéant 1981) the rate constants of the main reaction steps controlling the reactivity of $CO_2^{\bullet-}$, when electrogenerated by reduction of CO_2 at an inert electrode (Scheme 12.1): carbon–carbon self-coupling (eqn (9)) to afford oxalate, $k_{dim} = 10^7 M^{-1}s^{-1}$; carbon–oxygen coupling with CO_2 (eqn (7)) to afford equimolar amounts of carbon monoxide and carbonate ion, $k_{CO_2} = 3 \times 10^3\ M^{-1}s^{-1}$, protonation by water (eqn (5)) to afford formate, $k_{H_2O} = 8 \times 10^2\ M^{-1}s^{-1}$.

The mechanism presented in Scheme 12.1 involves only homogeneous chemical steps, except for the electron transfers which occur at the electrode surface (note that the intermediate radicals formed in eqns (5) and (7) can be reduced in solution by $CO_2^{\bullet-}$ rather than by the electrode). As such, its domain of validity is restricted to those electrodes at which adsorption of any intermediate in Scheme 12.1 is unimportant. This has been elegantly confirmed to be the case for mercury, in a recent work (Isse *et al.*, unpublished, 1991). Indeed, the data in the first row of Table 12.2, present the result of a direct electrolysis of carbon dioxide in DMF at a mercury electrode; it is seen that these data compare satisfactorily with those in the second row, in which the electrolysis was performed indirectly with benzonitrile (BN) as an electron-transfer mediator (*i.e.* without any possible interference with the electrode surface), under otherwise identical conditions

$$\mathrm{BN} + \mathrm{e} \Leftrightarrow \mathrm{BN}^{\bullet-} \quad \text{(electrode)} \quad (15)$$

$$\mathrm{BN}^{\bullet-} + \mathrm{CO_2} \rightarrow \mathrm{BN} + \mathrm{CO_2}^{\bullet-} \quad \text{(solution)} \quad (16)$$

$$\mathrm{CO_2}^{\bullet-} \rightarrow \text{etc.} \quad (17)$$

In agreement with the formulation in eqns (15)–(17), similar results, shown in the third row, were obtained using a platinum electrode, the electrolysis still being performed indirectly using the same mediator. Contrastingly, the data in the fourth row show that a direct electrolysis at the platinum electrode results in an almost complete inversion of the yields of oxalate and carbon monoxide, with respect to the average product distribution in rows 1–3. This effect emphasizes that some electrodes, such as platinum, may actively participate in the reduction

Table 12.2 Comparison between direct and indirect* electrolysis of CO_2 at mercury or platinum electrodes. (Data from Isse *et al.* (1991))

Electrode material	$[CO_2]$ (M)	[Mediator] (mM)*	Formate (%)†	Carbon monoxide (%)†	Oxalate (%)†
Mercury	0.101	none	1	10	76
Mercury	0.101	1.8	1.5	1	80
Platinum	0.101	2.0	1.5	10	60
Platinum	0.101	none	6	91	3.5

* Mediator: benzonitrile.

† Faradic yields in per cent.

mechanism of carbon dioxide and may then be used to orientate the selectivity of the reaction toward one target product (compare for example Chapters 11 and 13).

Combined use of transition metals and electron transfer in the activation of carbon dioxide. Application to the electrocarboxylation of organic halides

In the organic or organometallic fields the importance of carbon dioxide coordination by metal complexes has been well recognized (see for example Chapters 7–9 as well as Darensbourg and Kudarovski 1983; Behr 1988; Braustein *et al.* 1988), with a particular emphasis on the creation of carbon–carbon bonds. For example, CO_2 insertion in a carbon–metal bond (eqn (18)) results in the formation of carboxylic acids or their derivatives (Kolomnikov *et al.* 1974; Darensbourg and Rokicki 1982; Behr *et al.* 1983, 1984; Darensbourg and Kudarovski 1984; Darensbourg and Grötsch 1985; Darensbourg *et al.* 1985*a*,*b*, 1987) in a way reminiscent of the reaction of a Grignard reagent with carbon dioxide

$$\mathcal{M}\text{–R} + CO_2 \rightarrow \mathcal{M}\text{–}O_2C\text{–R} \rightarrow \text{R–}CO_2H \tag{18}$$

In these reactions, the reducing agent may consist of the metal centre, $\mathcal{M}$, itself, the anionic organic ligand or possibly be an external reagent, for example such as hydrogen in eqn (19) (Alvarez *et al.* 1985)

$$\{\eta_2 CH_2 = CHR, \mathcal{M}\} + CO_2 + H_2 \rightarrow CH_2RCH_2\text{–}CO_2H \tag{19}$$

In an attempt to activate simpler organic substrates, in the presence of only catalytic amounts of readily available transition-metal complexes, several authors have investigated the transposition of such reactions to electrochemical conditions, where a cathode may play the rôle of the reducing reagent (for reviews see Silvestri 1987; Silvestri *et al.* 1990; and Chapter 13)

$$\text{R–X} + CO_2 + 2e\text{–}\{\epsilon \text{ per cent } MX'_2L_n\} \rightarrow \text{R–}CO_2^- + X^- \tag{20}$$

The synthetic utility of this reaction is best illustrated by the electrosynthesis of anti-inflammatory drugs by electrocarboxylation of benzylic halides catalysed by $NiCl_2$(dppe), dppe: 1,2-(diphenylphosphino)ethane (Fauvarque *et al.* 1986) or by the electrocarboxylation of aromatic halides (Fauvarque *et al.* 1984), the latter reaction being also efficiently catalysed by palladium complexes, for example $PdCl_2(PPh_3)_2$ (Torii *et al.* 1986).

Mechanism of $NiCl_2$(dppe)-catalysed electrocarboxylation of aromatic halides

In the absence of carbon dioxide and transition-metal catalyst, reduction of aromatic halides, ArX, results in the consumption of two electrons per mole and the quantitative production of the corresponding hydrocarbon, ArH (see for example Weinberg 1975). Conversely, the same electrolysis, performed in the presence of carbon dioxide and catalytic amounts (*ca* 10 per cent) of a nickel catalyst such as $NiCl_2(dppe)$, occurs at a potential considerably less negative than those required for electrolysis of ArX or CO_2 alone, and yields quantitatively the corresponding carboxylate, $ArCO_2^-$, according to the stoichiometry in eqn (20) (Fauvarque *et al.* 1984). Of interest for our mechanistic purpose here, is the fact that under identical conditions but in the absence of carbon dioxide, electrolysis proceeds at the same potential to afford stoichiometrically the corresponding biaryl (Amatore and Jutand 1988), with a net consumption of one electron *per* mole of ArX

$$\mathrm{ArX} + \mathrm{e}\text{–}\{10 \text{ per cent } \mathrm{NiCl_2(dppe)}\} \rightarrow {}^1/_2\, \mathrm{Ar\text{–}Ar} + \mathrm{X}^- \qquad (21)$$

The process in eqn (21) has been shown (Amatore and Jutand 1988) to proceed smoothly via an efficient chain mechanism involving a rapid succession of one-electron transfer steps and chemical reactions involving diamagnetic and paramagnetic nickel-centred complexes, as outlined in Scheme 12.2. In the presence of carbon dioxide, the efficient catalytic cycle in Scheme 12.2 is totally deactivated, less than a few tenths of 1 per cent of biaryl being produced under these conditions. This indicates that in the presence of carbon dioxide, one of the transient nickel species involved in the catalysis is trapped efficiently by carbon dioxide, with the net result of interrupting the propagation of the biaryl catalytic cycle and initiating a concurrent cycle leading to the quantitative formation of arylcarboxylate. In the following we want to show how the qualitative use of cyclic voltammetry and steady-state voltammetry may allow us to identify which of the intermediates in Scheme 12.2 is intercepted by CO_2.

The fact that carbon dioxide is inert, within our experimental conditions and our time-scale, *vis-à-vis* $Ni^{II}Cl_2(dppe)$, $Ni^{I}Cl(dppe)$ or $Ni^{0}(dppe)$ is established by the invariance of the cyclic voltammograms in Fig. 12.3 obtained either in the presence or in the absence of saturated carbon dioxide. Moreover, it is noted

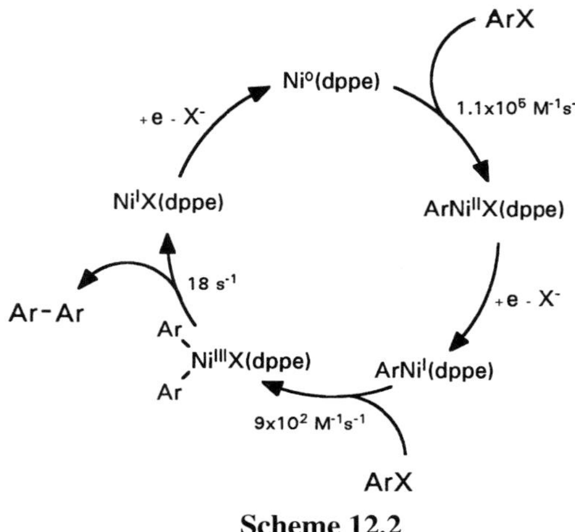

Scheme 12.2

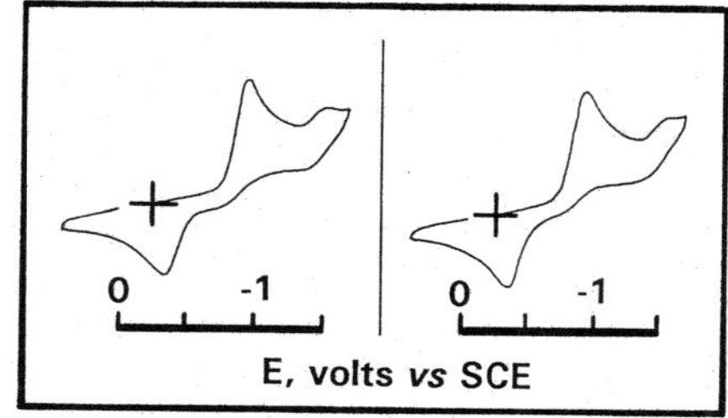

Fig. 12.3 Cyclic voltammetry of $Ni^{II}Cl_2$(dppe), 2mM in THF/HMPA (2:1 v/v) + 0.1 M NBu_4BF_4 at 25 °C in the absence (*left*) or in the presence (*right*) of saturated CO_2 (Amatore and Jutand 1991*a*).

in Fig. 12.4(a) that the reduction wave of Ar–Ni^{II}X(dppe) is still observed in the presence of CO_2, and furthermore that its limiting current increases upon increasing the carbon dioxide concentration. This, and the fact that the electrolysis potential must be set on this wave in order for the electrocarboxylation to proceed, shows that the formation of Ar–Ni^{I}(dppe) by one-electron transfer reduction of Ar–Ni^{II}X(dppe) (Amatore and Jutand 1988) is required for allowing the formation of arylcarboxylate. At this point, one is then led to the conclusion that the deactivation of the biaryl catalytic cycle occurs at the level of either Ar–Ni^{I}(dppe) or $(Ar)_2Ni^{III}$X(dppe). Let us then investigate each of these possibilities.

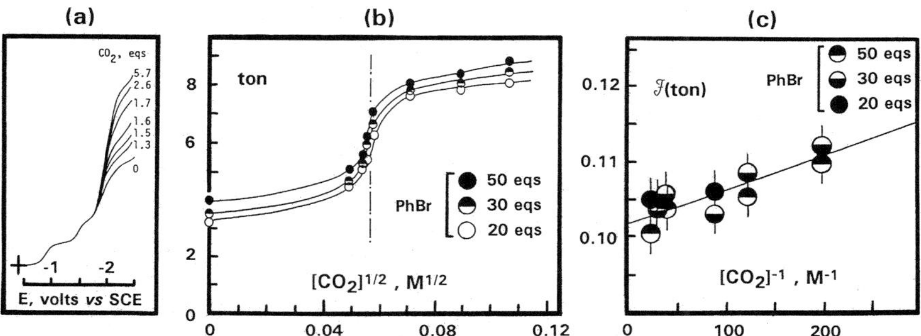

Fig. 12.4 Steady state voltammetry of $Ni^{II}Cl_2$(dppe), 2 mM in THF/HMPA (2:1 v/v) + 0.1 M NBu_4BF_4 at 25 °C in the presence of various excesses of bromobenzene and carbon dioxide (as indicated in equivalents on each figure). (a) Steady-state voltammograms in the presence of 10 equivalents of PhBr. (b) Variations of the turnover number (ton) as a function of $[CO_2]$ and [PhBr]. (c) Variations of $\mathcal{J}(\text{ton}) = (2\delta/D^{1/2})/(1 + \text{ton})-(k_0\,[\text{ArX}])^{-1/2}$, as a function of $[CO_2]^{-1}$ (see text and eqns. (27) and (28). (Data from Amatore and Jutand 1991*a*)

A reaction at the level of $(Ar)_2Ni^{III}X(dppe)$, necessarily implies that the propagation rate of the carboxylation chain cannot exceed the rate of oxidative addition of ArX to $Ar–Ni^{I}(dppe)$, *viz.* that of the step affording $(Ar)_2Ni^{III}X(dppe)$. In other words this implies that the catalytic current observed in Fig. 12.4(a), at the level of the $Ar–Ni^{II}X(dppe)$ reduction wave, should be independent of the concentration of carbon dioxide. Indeed, we have established previously (Amatore and Jutand 1988) that under the conditions of Fig. 12.4(a), the catalytic current observed at this wave is almost entirely controlled by the rate of oxidative addition of ArX to $Ar–Ni^{I}(dppe)$. Therefore, the net increase of this current in Fig. 12.4(a), in the presence of CO_2, establishes that the biaryl chain is deactivated before $(Ar)_2Ni^{III}X(dppe)$ is produced, which in turns demonstrates that the chain is intercepted at the level of $Ar–Ni^{I}(dppe)$.

This may occur by a direct insertion of carbon dioxide into the aryl-nickel (compare Darensbourg *et al.* (1987) for a discussion in the context of Rh(I)) bond to afford an arylcarboxylate-nickel(I) intermediate, as in eqn (22)); one-electron reduction of the latter (eqn (23)) should then close the catalytic cycle by regeneration of $Ni^{0}(dppe)$ and expulsion of the arylcarboxylate, as is observed (Amatore and Jutand 1991*a*) during the two-step reduction mechanism of an authentic sample of $(Ar–CO_2)_2Ni^{II}(dppe)$

$$Ar–Ni^{I}(dppe) + CO_2 \rightarrow Ar–CO_2–Ni^{I}(dppe) \qquad (22)$$
$$Ar–CO_2–Ni^{I}(dppe) + e \rightarrow Ar–CO_2^{-} + Ni^{0}(dppe) \qquad (23)$$
$$Ni^{0}(dppe) + ArX \rightarrow Ar–Ni^{II}X(dppe), \text{ etc.} \qquad (k_0) \qquad (24)$$

However, such a mechanism would imply that the propagation rate of the carboxylation chain is either limited by that of CO_2 insertion (eqn (22)) or by that of the oxidative addition of ArX to $Ni^{0}(dppe)$ (eqn (24)). In other words the catalytic current observed for the reduction of $Ar–Ni^{II}X(dppe)$ in Fig. 12.4(a) should always increase upon increasing either $[CO_2]$ (rate determining step: eqn (22)) or [ArX] (rate-determining step: eqn (24)), but neither saturate upon an increase of both concentrations. The fact that this conclusion disagrees with the experimental observations is illustrated by the plots in Fig. 12.4(b), where a saturation of the chain turnover, *viz.* of the catalytic current in Fig. 12.4(a), is observed with both parameters. In order to reconcile these experimental data with our previous conclusion that the carboxylation chain is initiated by trapping $Ar–Ni^{I}(dppe)$ with CO_2, one must consider that the formation of the arylcarboxylate–nickel(I) species occurs in a stepwise fashion, rather than in a concerted one as suggested by the formulation in eqn (22). A possible scenario (compare Chapter 7 for related cases) involves first a coordination of carbon dioxide by $Ar–Ni^{I}(dppe)$ (eqn (25)), in a similar fashion as it occurs with several electron-rich zerovalent nickel complexes (Aresta and Nobile 1977; Döhring *et al.* 1985; Collin *et al.* 1988), followed by an intramolecular formation of the carbon–carbon bond (eqn (26))

$$Ar–Ni^{I}(dppe) + CO_2 \rightarrow Ar–Ni^{III}(CO_2)(dppe) \qquad (k_1) \qquad (25)$$
$$Ar–Ni^{III}(CO_2)(dppe) \rightarrow Ar–CO_2–Ni^{I}(dppe) \qquad (k_2) \qquad (26)$$

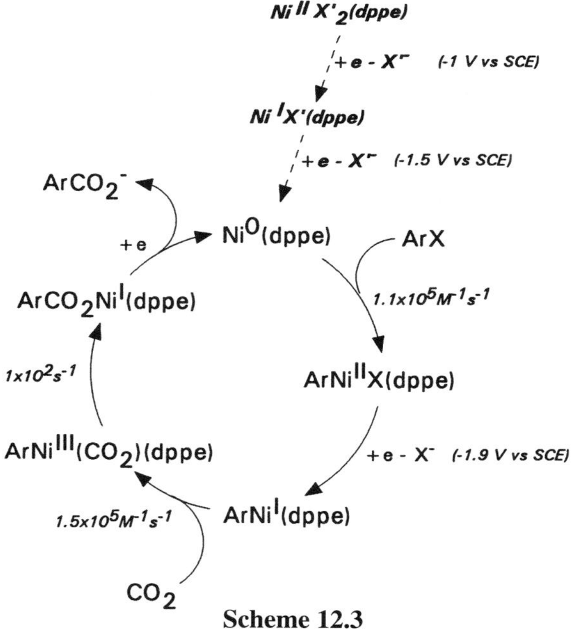

Scheme 12.3

Kinetic control of the chain propagation rate by the latter reaction is expected to occur upon increasing both concentrations of carbon dioxide and organic halide, i.e. upon making the rates of the reactions in eqns (24) and (25) increasingly large with respect to that in eqn (26). Since the rate of eqn (26) is independent of $[CO_2]$ or [ArX], it is deduced that under such conditions the propagation rate of the carboxylation chain should become independent of both of these parameters, in agreement with the saturation effects observed in Fig. 12.4(b). A quantitative kinetic treatment (Amatore and Jutand 1991*a*,*b*) of the propagation of the whole carboxylation chain outlined in Scheme 12.3 shows that the chain turnover (ton), i.e. the apparent number of electrons corresponding to the catalytic current observed for the reduction of Ar–Ni^{II}X(dppe) (compare for example Fig. 12.4(a)) is

$$\mathcal{A}(\text{ton}) = k_2^{-1/2} + (k_2^{1/2}/k_1) \times [CO_2]^{-1} \quad (27)$$

where $\mathcal{A}$(ton) is defined as

$$\mathcal{A}(\text{ton}) = (2\delta/D^{1/2})/(1 + \text{ton}) - (k_0 \times [\text{ArX}])^{-1/2} \quad (28)$$

and is determined directly from the experimental values of ton, since $(2\delta/D^{1/2}) = 1.09\ s^{1/2}$ and $k_0 = 1.1 \times 10^5\ M^{-1}s^{-1}$ are known independently (Amatore and Jutand 1988) (note that the parameter $(2\delta/D^{1/2})$, which relates the diffusion layer thickness, δ, to the average diffusion coefficient, D, of nickel containing species, reflects the role of the hydrodynamic conditions prevailing in the electrolysis cell near the electrode). The plot of $\mathcal{A}$(ton) versus $[CO_2]^{-1}$ represented in Fig. 12.4(c) shows that the experimental data in Fig. 12.4(a),(b) are consistent with the kinetic equations (27), (28) derived for the chain mechanism in Scheme 12.3. Moreover, from the intercept and slope of the correlation line in Fig. 12.4(c), k_1 (eqn (25)) and k_2 (eqn (26)) are found to be equal to $(1.5 \pm 0.5) \times 10^5 M^{-1}s^{-1}$ and $10^2 s^{-1}$ respectively.

Mechanism of $PdCl_2(PPh_3)_2$-catalysed electrocarboxylation of aromatic halides

On a strict operational basis, the $PdCl_2(PPh_3)_2$-catalysed electrocarboxylation of aromatic halides proceeds along the same lines as those described above for the nickel-catalysed reaction. Thus the divalent palladium complex is reduced in a zerovalent low-ligated palladium centre, which undergoes oxidative addition by the aryl halide, ArX, to afford a σ-arylpalladium(II) species, $Ar–Pd^{II}X(PPh_3)_2$, akin to that, $Ar–Ni^{II}X(dppe)$, involved in Schemes 12.2 and 12.3. In agreement with this mechanism, the success of the electrosynthetic carboxylation requires the cathode potential to be set on the reduction wave of $Ar–Pd^{II}X(PPh_3)_2$ (Torii *et al.* 1986). Indeed, in the presence of carbon dioxide, an overall two-electron reduction of the latter species affords the arylcarboxylate and regenerates the zerovalent palladium centre which enters a new catalytic cycle. At this stage, the only noticeable difference between the two mechanisms is related to the mechanism of electron-transfer activation of the divalent precursors of the catalysts. Indeed, whereas $Ni^{II}Cl_2(dppe)$ is reduced in a succession of two one–electron steps (Amatore and Jutand 1988)

$$Ni^{II}Cl_2(dppe) + e \rightarrow Ni^{I}Cl(dppe) + Cl^- \quad (E^{1/2} = -1.0 \text{ V versus SCE}) \quad (29)$$

$$Ni^{I}Cl(dppe) + e \rightarrow Ni^{0}(dppe) + Cl^- \quad (E^{1/2} = -1.5 \text{ V versus SCE}) \quad (30)$$

$PdCl_2(PPh_3)_2$ is reduced in a single two-electron wave (Amatore *et al.* 1990)

$$Pd^{II}Cl_2(PPh_3)_2 + 2e \rightarrow Pd^{0}(PPh_3)_2 + 2Cl^- \quad (E^{1/2} = -1.04 \text{ V versus SCE}) \quad (31)$$

This may appear *a priori* as a minor variant, since both processes (*viz.* eqns (29) and (30) on the one hand, or eqn (31) on the other) yield a zerovalent species at the potentials where electrocarboxylations are performed. However, the same difference exits also for the reductions of the σ-arylmetal complexes, $Ar–Ni^{II}X(dppe)$ and $Ar–Pd^{II}X(PPh_3)_2$. Indeed, as explained above, $Ar–Ni^{II}X(dppe)$ reduction involves a one-electron transfer (Amatore and Jutand 1988) and affords a paramagnetic intermediate, $Ar–Ni^{I}(dppe)$ whose reaction with carbon dioxide (eqn (25)) was central to the carboxylation process. Oppositely, either in the presence or in the absence of CO_2, reduction of $Ar–Pd^{II}X(PPh_3)_2$ occurs via a single two-electron overall process (Amatore *et al.* 1992).

The investigation of a series of σ-arylpalladium(II) complexes, $Ar–Pd^{II}X(PPh_3)_2$, allowed to establish that their reduction proceeds along an overall two-electron mechanism to afford an anionic σ-arylpalladium(0) intermediate, $Ar–Pd^{0}(PPh_3)_2^-$, whose reversible endergonic dissociation yields a free σ-aryl anion, Ar^-, and $Pd^{0}(PPh_3)_2$

$$Ar{-}Pd^{II}X(PPh_3)_2 + 2e \rightarrow Ar{-}Pd^{0}(PPh_3)_2^- + X^- \tag{32}$$
$$Ar{-}Pd^{0}(PPh_3)_2^- \Leftrightarrow Ar^- + Pd^{0}(PPh_3)_2 \tag{33}$$

In the absence of carbon dioxide Ar^- is protonated by the medium to give the corresponding hydrocarbon. In the presence of carbon dioxide, it undergoes a nucleophilic addition on the latter, to afford the sought arylcarboxylate. On the other hand, and in the presence of excess aryl halide, the zerovalent low-ligated palladium centre formed in eqn (33), goes through an oxidative addition by ArX (Amatore *et al.* 1991) to regenerate $Ar{-}Pd^{II}X(PPh_3)_2$ and closes the catalytic cycle represented in Scheme 12.4.

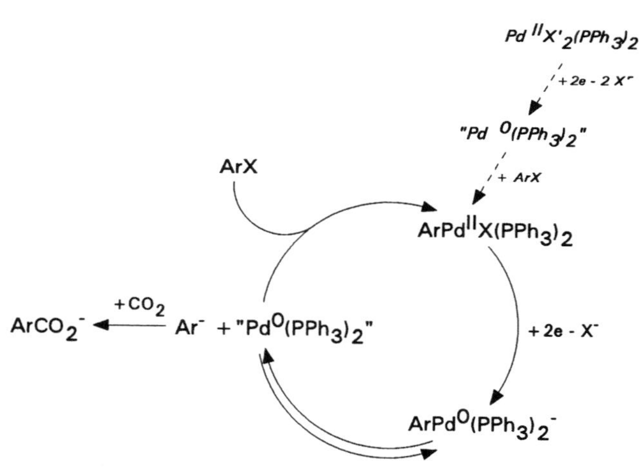

Scheme 12.4

This establishes that the palladium-catalysed carboxylation does not involve any paramagnetic intermediates, as was the case for nickel (compare Scheme 12.3) but only diamagnetic ones; thus, despite their apparent identity, the two catalytic mechanisms differ drastically. In the case of nickel the transition metal plays a double role, by allowing an activation of the aryl halide at a potential considerably less than that of its direct reduction, but also by participating in the creation of the carbon–carbon bond which occurs within its coordination sphere (eqns (25) and (26)). The palladium catalyst also allows a facile activation of the aryl halide; yet in this case its role is limited to affording an aryl anion at a potential much less than that required for the reduction of ArX, since the carbon–carbon bond creation occurs outside the coordination shell of the palladium centre, between a free aryl anion and a free carbon dioxide molecule.

Acknowledgements

This work has been supported in parts by the Centre National de la Recherche Scientifique (CNRS) and by Ecole Normale Supérieure. M.F.N. acknowledges also the award of a postdoctoral grant from the University of Copenhagen during her stay in Paris. Professor E. Vianello and his research group are cordially thanked for communicating to us the results reported in Table 12.2 before their publication.

References

Alvarez, R., Carmona, E., Cole-Hamilton, D.J., Galindo, A., Gutierez-Puebla, E., Monge A. *et al.* (1985). *J. Am. Chem. Soc.*, **107**, 5529–31.

Amatore, C. and Jutand, A. (1988). *Organometallics*, **7**, 2203–14.

Amatore, C. and Jutand, A. (1991*a*). *J. Am. Chem. Soc.*, **113**, 2819–25.

Amatore, C. and Jutand, A. (1991*b*). *J. Electroanal. Chem.*, **306**, 141–56.

Amatore, C. and Savéant, J.-M. (1981) *J. Am. Chem. Soc.*, **103**, 5021–3.

Amatore, C. Azzabi, M., Calas, P., Jutand, A., Lefrou, C., and Rollin, Y. (1990). *J. Electroanal. Chem.*, **288**, 45–63.

Amatore, C., Jutand, A., Khalil, F. and Nielsen, M.F. (1992). *J. Am. Chem. Soc.*, *114*, 7076–85.

Amatore, C., Azzabi, M., and Jutand, A. (1991). *J. Am. Chem. Soc.*, **113**, 8375–84.

Aresta, M. and Nobile, C.F. (1977). *J. Chem. Soc., Dalton Trans.*, 708–11.

Aylmer-Kelly, A.W.B., Bewick, A., Cantrill, P.R., and Tuxford, A.M. (1973). *Discuss. Faraday Soc.*, **56**, 96–107.

Behr, A. (1988). *Angew. Chem., Int. Ed. Engl.*, **27**, 661–78.

Behr, A., Keim, W., and Thelen, G. (1983). *J. Organomet. Chem.*, **249**, C38–C40.

Behr, A., Kanne, U., and Thelen, G. (1984). *J. Organomet. Chem.*, **269**, C1–C3.

Braustein, P., Matt, D., and Nobel, D. (1988). *Chem. Rev.*, **88**, 747–64.

Collin, J.P., Jouaiti, A., and Sauvage, J.P. (1988). *Inorg. Chem.*, **27**, 1986–90.

Darensbourg, D.J. and Grötsch, G. (1985). *J. Am. Chem. Soc.*, **107**, 7473–6.

Darensbourg, D.J. and Kudarovski, R.A. (1983). *Adv. Organomet. Chem.*, **22**, 129–68.

Darensbourg, D.J. and Kudarovski, R.A. (1984). *J. Am. Chem. Soc.*, **106**, 3672–3.

Darensbourg, D.J. and Rokicki, A. (1982). *J. Am. Chem. Soc.*, **104**, 349–50.

Darensbourg, D.J. Hanckel, R.K., Bauch, C.G., Pala, M., Simmons, D., and White, J.N. (1985*a*). *J. Am. Chem. Soc.*, **107**, 7463–76.

Darensbourg, D.J., Kudarovski, R.A., and Delord, T. (1985*b*). *Organometallics*, **4**, 1094–7.

Darensbourg, D.J., Grötsch, G., Wiegreffe, P., and Rheingold, A.L. (1987). *Inorg. Chem.*, **26**, 3827–30.

Döhring, A., Jolly, P.W., Krüger, C. and Romào, M.J. (1985). *Z. Naturforsch. B*, **40**, 484–8.

Fauvarque, J.F., Chevrot, C., Jutand, A., François, M., and Perichon, J. (1984). *J. Organomet. Chem.*, **264**, 273–81.

Fauvarque, J.F., Jutand, A., and François, M. (1986). *New J. Chem.*, **10**, 119–22.

Gressin, J.C., Michelet, D., Nadjo, L., and Savéant, J.-M. (1979). *Nouv. J. Chim.*, **3**, 545–54.

Isse, A.A., Gennaro, A., Severin, M.G., and Vianello, E. (1991). Unpublished work cited here with the authors' permission.

Kolomnikov, I.S., Gusev, A.O., Belopotapova, T.S., Grigoryan, M.K., Struchkov, Y.T., and Volpin, M.E. (1974). *J. Organomet. Chem.*, **69**, C10–C12.

Lamy, E., Nadjo, L., and Savéant, J.-M. (1977). *J. Electroanal. Chem.*, **78**, 403–7.

Roberts, J.L. and Sawyer, D.T. (1965). *J. Electroanal. Chem.*, **9**, 1–7.

Silvestri, G. (1987). In *Carbon dioxide as a source of carbon*, NATO ASI Series (Ser. C) (ed. M. Aresta and G. Forti), **206**, 339–69. Kluwer, Dordrecht, The Netherlands.

Silvestri, G., Gambino, S., and Filardo, G. (1990). Electrochemical syntheses involving carbon dioxide. In *Enzymatic and model carboxylation and reduction reactions for CO_2 utilization*, NATO ASI Series, **314**, 101–27. Kluwer, Dordrecht, The Netherlands.

Torii, S., Tanaka, H., Hamatani, T., Morisaki, K., Jutand, A., Pflüger, F. *et al.* (1986). *Chem. Lett.*, 169–72.

Weinberg, N.L. (1975). *Techniques of electroorganic synthesis.* Part II, p. 189. Wiley Interscience, New York.

13

Electrochemical syntheses involving carbon dioxide

Giuseppe Silvestri, Salvatore Gambino, and Giuseppe Filardo

Introduction – historical background

The early speculations on the possibility of carbon dioxide being the starting material for chemical synthesis of organic compounds are due to one of the founders of modern organic chemistry, Justus Liebig. In 1847 he suggested that in the future a laboratory mimesis of the natural pathway to complex molecules such as carboxylic acids, alcohols or sugars would be possible. In 1861 two other illustrious German chemists, Kolbe and Schmitt, described the occurrence of formates by addition of carbon dioxide and water vapour to metallic sodium or potassium, thus confirming the hypothesis Liebig had proposed 'mit prophetischem Geiste'. Following an analogous procedure, but in anhydrous conditions, Drechsel (1868) reduced carbon dioxide to oxalates. The electrochemical approach was an obvious extension of those findings: soon afterwards Beketov (1869) and Royer (1870) reported the formation of formic acid by cathodic reduction of aqueous alkaline solutions of carbon dioxide. Further investigations involved both the chemical reduction with metals of the I and II groups, free or amalgamated (Lieben 1895) and electrochemical reduction in aqueous media (Cohen and Jahn 1904; Ehrenfeld 1905; Fischer and Prziza 1914).

The questions whether the reduction involves molecular CO_2 or the anion HCO_3^-, and whether the reductive process is due to the direct activation of one of those species or to 'nascent' hydrogen (Rabinowitsch and Maschowetz 1930; Van Rysselberghe and Alkyre 1944; Van Rysselberghe 1946; Van Rysselberghe *et al.* 1946; von Stackelberg and Stracke 1949; Vlcek 1953), received a convincing answer only in 1954 by Teeter and Van Rysselberghe who demonstrated the direct involvement of molecular carbon dioxide in the electrodic process. In 1960 Jordan and Smith on the basis of polarographic data postulated that the anion radical $CO_2^{\bullet-}$ was an intermediate in the reduction of aqueous solutions of carbon dioxide. Paik *et al.* (1969) and Ryu *et al.* (1972) developed a detailed kinetic analysis of the polarographic curves of the CO_2/water system.

The chemical reduction of CO_2 by metals of the first or second group, or by their amalgams, which can be considered as an indirect two-stage electrochemical process, received attention until the beginning of the sixties. According to this procedure sodium formate in water (Chechel and Antropov 1958) and calcium oxalate in anhydrous conditions (Hohn *et al.* 1953) were obtained in high yields from the corresponding amalgams. The most interesting features of this indirect reduction are summarized in a paper by Miller *et al.* (1962): together with oxalate, and depending on the reaction conditions, carbon monoxide and carbonate, formate, acetate, glycollate, malonate, and tartronate anions were obtained, thus anticipating much of the reactivity later observed by electrochemical reductions in aprotic or partially protic systems.

The first studies on electrochemistry of carbon dioxide in aprotic media were published in the same period. From polarographic determinations in dimethysulphoxide (Dehn *et al.* 1962; Shöber 1964) a half wave potential of –2, 24 V (versus Hg pool) was measured, and the formation of oxalate anion as the result of the reduction was hypothesized. A remarkable difference of the voltammetric and chronopotentiometric behaviour of carbon dioxide between amalgamated platinum and gold cathodes was observed (Roberts and Sawyer 1965) and later confirmed in preparative electrolyses (Haynes and Sawyer 1967): on gold electrodes the reduction leads essentially to carbon monoxide and carbonate, whereas formate (probably due to adventitious water) was observed on mercury as the main product. The synthesis of oxalic acid by direct reduction of carbon dioxide in *N,N*-dimethylformamide was first published in 1972 (Tyssee *et al*).

A large expansion of the investigation into the electrochemistry of carbon dioxide took place in the last two decades involving: (a) better definition of synthetic possibilities offered by the reduction of carbon dioxide alone in various media; (b) the synthesis of complex carboxylic acids starting from CO_2 and suitable costarting organic compounds; (c) adsorption and electrode poisoning, related to fuel cells development. In this review the main developments of researches into points (a) and (b) will be presented. Other areas of interest to electrochemistry of carbon dioxide, which cannot be included in this review, involve researches on photoelectrochemistry (see Chapter 14), on electrochemical gas sensors, on electrochemical gas separation membranes, and on the use of supercritical mixtures containing carbon dioxide as media for electroanalytical investigations or for electrochemical syntheses. Some reviews on electrochemical transformations of carbon dioxide have already been published (Taniguchi 1989; Silvestri *et al.* 1990). Other reviews are focused on specific aspects, such as catalytic systems based on transition metal complexes (O'Connell *et al.* 1987; Sullivan *et al.* 1988; Collin and Sauvage 1989) or on electrocarboxylations (Silvestri 1987; Chaussard *et al.* 1990; Silvestri *et al.* 1991) and the reader is addressed to them for a deeper insight into the various aspects of these processes.

The reduction of carbon dioxide alone

Interest in the different aspects of electrochemistry and photoelectrochemistry of carbon dioxide has been growing in the last decades, parallel to the concern of the general public on the problems raised by the increase of concentration of carbon dioxide in the atmosphere, and by the prospect of a shortage of crude oil. The most recent investigations into the 'one pot' electrochemical reduction of carbon dioxide have added new derivatives, such as hydrocarbons (both saturated and unsaturated), alcohols (methanol, ethanol, and higher omologs), aldehydes, fatty acids, and various hydroxycarboxylic acids, to the list of products obtained by this procedure. Most attention has been focused on compounds of high energy content, such as saturated hydrocarbons, considered for energy transportation and storage, and on compounds which could be the starting materials for a future 'petrochemistry without crude', such as carbon monoxide, olefins, and alcohols. In this context oxalic acid could be considered as a possible building block for complex organic molecules, and formic acid as a possible hydrogen carrier, due to its easy decomposition into hydrogen and carbon dioxide.

A series of realistic considerations has to be taken into account, concerning the requisites those processes must fulfil if they are to compete with other chemical or physical approaches to using carbon dioxide for commodities, basic chemicals, and energy storage. Considering large scale processes, energy aspects have paramount importance. Rather favourable thermodynamics are observed if the whole reduction processes, leading to the various derivatives, are taken under consideration (Scheme 13.1), but uncatalysed systems still present strong kinetic hindrances resulting in very high electrode overpotentials. Optimization of the catalytic systems should therefore combine enhanced selectivities towards the desired products to high turnover numbers, and operative cathodic potentials as close to as possible to the corresponding thermodynamic values.

Electrochemical reactors always present space–time yields almost one order of magnitude lower than a common catalytic gas-phase reactor. Therefore, electrochemical reactions should proceed through fast kinetics, in order to minimize this drawback and be competitive with chemical technologies. Concerning the stability of the electrochemical system, aqueous systems with inorganic supporting electrolytes are by far the most reliable. For low costs of separation the best conditions are those in which the products are gaseous and the catalytic system is heterogeneous, or those for which there is a large difference of volatility

			$E°$ (V)
$CO_2\,(g) + 2e^- + 2H^+$	$\rightarrow$	$(COOH)_2\,(aq)$	−0.904
$CO_2\,(g) + 2e^- + 2H^+$	$\rightarrow$	$HCOOH\,(aq)$	−0.614
$CO_2\,(g) + 2e^- + 2H^+$	$\rightarrow$	$CO(g) + H_2O(l)$	−0.518
$CO_2\,(g) + 6e^- + 6H^+$	$\rightarrow$	$CH_3OH(l) + H_2O(l)$	−0.384
$CO_2\,(g) + 8e^- + 8H^+$	$\rightarrow$	$CH_4(g) + 2H_2O(l)$	−0.245

Scheme 13.1 Reduction of carbon dioxide to various derivatives. Equilibrium potentials referred to NHE at pH 7.

between solvent and products, or when the product of the synthesis can be used as solvent of the electrochemical system (for example the production of methanol in methanol).

Investigations on the electrochemical reduction of carbon dioxide alone are at present focused on the following topics:

- the use as cathodes of metals showing particular catalytic properties, which offer long range reliability and acceptable mechanical, thermal, and chemical stability. These cathodes generally show high overpotentials in operative conditions;
- the use of metal complexes as homogeneous or heterogenous catalysts, which are generally more selective and offer lower overpotentials than metals although they are less stable.

Very recent investigations into the use of electrochemically triggered enzymatic systems for the fixation of carbon dioxide into organic molecules will also briefly be mentioned.

The reduction of carbon dioxide on metals

Aqueous systems Scheme 13.2 summarizes the fundamental reaction pathways taking place at different electrode materials. Pathway 1a, the reduction to formate, has been thoroughly investigated in its mechanistic and synthetic aspects, especially the behaviour of cathodes from moderate to high hydrogen overvoltage, such as Sn, In, Cd, Pb, Hg, and amalgamated metals (Russel *et al.* 1977; Kitani *et al.* 1978; Ito *et al.* 1980; Vassiliev *et al.* 1985; Ikeda *et al.* 1987). Besides formic acid, various C2–C5 carboxylic acids were found (Kitani *et al.* 1978; Ito *et al.* 1980). A diaphragm electrochemical cell with a rotating amalgamated copper cathode was described by Udupa *et al.* (1971) by which formic acid is obtained with Faradic yields of up to 80 per cent in prolonged electrolyses. Gas diffusion electrodes bearing Pb particles embedded in the hydrophobic PTFE matrix gave current yields up to 100 per cent for the reduction to formate, at current densities of 115 mA cm^{-2}, already suitable for industrial applications (Mahmood *et al.* 1987*a*).

Related to the reduction of carbon dioxide on metals are the investigations on electrochemical systems for the disposal of carbon dioxide from closed atmospheres such as in submarines or space systems (Leitz 1967; Meller 1968; Grishaenkov *et al.* 1986; Butenko *et al.* 1987; Hooker *et al.* 1990; Sawada *et al.* 1990). The formation of carbonaceous layers on the cathode (pathway 1b) was reported for platinum at very high current densities (Srivastava and Shukla 1970), for silver (McQuillan *et al.* 1975; Mahoney *et al.* 1980) and recently for copper (Cook *et al.* 1988*a*; DeWulf and Bard 1989). In the case of copper the formation of a similar carbonaceous layer by electroreduction of aqueous solutions of formic acid (Cook *et al.* 1989) confirms the hypothesis that this acid is an intermediate of the reduction.

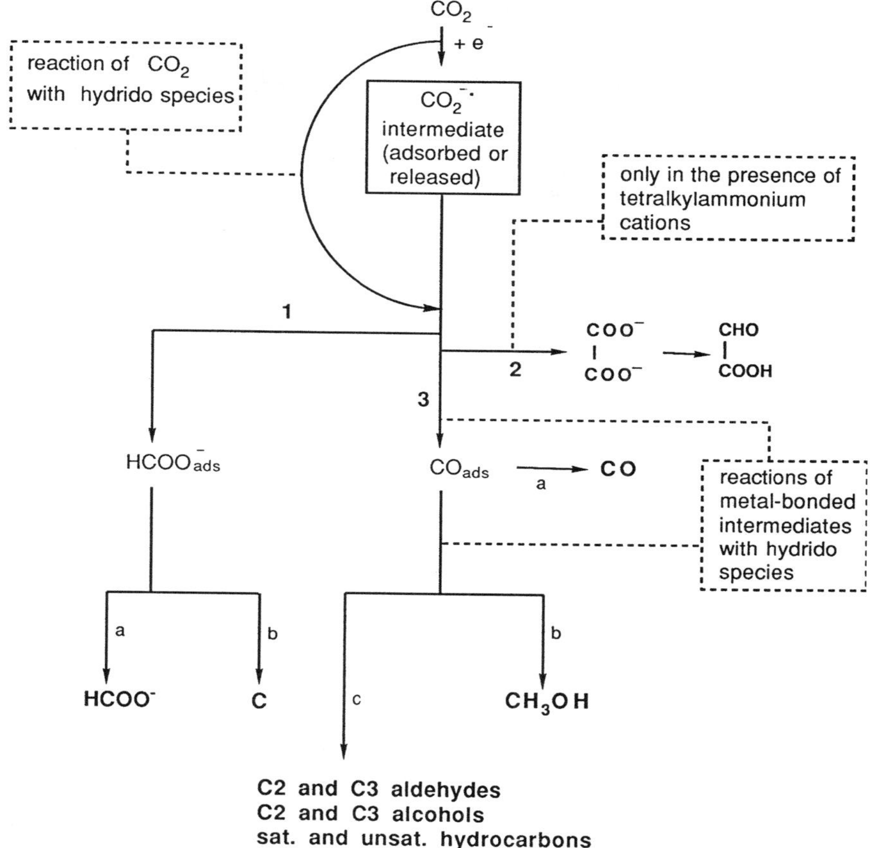

Scheme 13.2 Reaction pathways at different electrode materials.

The selectivity of the cathodic reaction is strongly influenced not only by the nature and structure of the electrode surface but also by the composition of the double layer: in the presence of quaternary ammonium salts the formation of a relatively anhydrous layer on the electrode induces the formation of different products such as oxalate or C3 and C4 carboxylic acids (Ito *et al.* 1981) or glycollate and malate on lead or mercury (Bewick and Greener 1969, 1970; Wolf and Rollin 1977). More recently Eggins *et al.* (1988) reported that in similar conditions on graphite cathodes oxalate or, at more negative potential, glyoxylate are obtained in substantial yields depending on the cathodic potential (pathway 2).

Most of the reactions schematized in pathway 3 were only recently disclosed: through them it is possible to obtain, just by a proper selection of the cathode material, carbon monoxide (pathway 3a), or more reduced products, such as methanol (pathway 3b), or (pathway 3c) aldehydes, higher alcohols, and hydrocarbons.

Adsorbed carbon monoxide is recognized as the key stable intermediate before either the release of molecular CO, or the reaction with hydrides leading to more reduced species. Evidence for its formation on copper and on nickel have been published (Islam and Kunimatsu 1989; Hori and Murata 1990*b*; Hori *et al.* 1991). The particular selectivity of gold towards formation and release of CO was observed in nonaqueous media (Haynes and Sawyer 1967) and more recently confirmed also in aqueous systems (Hori *et al.* 1985, 1987; Maeda *et al.* 1987; Noda *et al.* 1990). The same behaviour towards the release of CO is also shown by silver. The electroanalytical behaviour of gold electrodes in aqueous acidic media is indicative of the existence of interactions between carbon dioxide molecules and H adatoms (Alonso *et al.* 1988). Low yields of hydrocarbons were reported for the reduction of silver at low temperature (Azuma *et al.* 1990*a*), and for Au-solid polymer electrolyte systems (Cook *et al.* 1990*b*).

Methanol, with Faradic yields of up to 85 per cent, was obtained on molybdenum (Summers *et al.* 1986) and, together with methane and carbon monoxide, on ruthenium cathodes (Frese and Leach 1985). Amazingly low overpotentials were reported for both reactions, although checked only for very small amounts of charge ($Q < 30$ C) and at low current densities (0.3–0.6 mA cm^{-2} for Ru and 50–100 μA cm^{-2} for Mo). Disappointingly, the only check available on the behaviour of Mo and Ru cathodes at current densities of 24.6 mA cm^{-2} and 10.2 mA cm^{-2} respectively has shown that the reduction of water to hydrogen in these conditions is the main, if not the only, process taking place (Noda *et al.* 1990).

The importance of the hydride coverage of the cathode surface in the mechanism of formation of reduced products was in evidence both in the case of Mo (Summers *et al.* 1986; Koike *et al.* 1989) and of Ru (Frese and Summers 1988). A Ru surface deactivation due to the formation of a carbonaceous layer was observed at $t > 85°$.

Since the first communications (Hori *et al.* 1985, 1986) it appeared clear that systems based on copper cathodes are among the most promising for the electrochemical reduction of carbon dioxide to hydrocarbons in aqueous media, although presenting rather unfavourable energetics. Methane and ethylene are the main hydrocarbon derivatives but selectivity and current yields strongly depend on various parameters, such as the surface state of copper (Koga *et al.* 1989), the temperature (Azuma *et al.* 1989, 1990*a*), the cathode potential (Hori *et al.* 1989*a*; Noda *et al.* 1989), the pH and the supporting electrolyte (Hori *et al.* 1989a), and the history of the metal surface in the course of the electrolysis (DeWulf *et al.* 1989). Thus ethylene, ethanol, and propanol are obtained in KCl, K_2SO_4, $KClO_4$, and diluted HCO_3^- solutions, whereas methane is the main derivative in concentrated HCO_3^- or phosphate solutions (Hori *et al.* 1989*a*). Furthermore, by a proper choice of the cathodic potential, ethanal and propanal can be obtained together with alcohols (Noda *et al.* 1989). It is interesting from the applicative point of view that methane has been obtained with acceptable Faradic efficiency on commercial copper (Cook *et al.* 1987*a*) and with very good current yields on electrodeposited copper (Cook *et al.* 1987*b*, 1988*a*; Ikeda *et al.* 1988).

Long range electrolyses on copper cathodes showed progressive decrease of the selectivity towards the formation of hydrocarbons, parallel to the appearance of a carbonaceous layer on the exposed cathode surface (DeWulf *et al.* 1989). The authors propose, on the basis of blank experiments, two parallel reaction pathways, one of which goes through CO and leads to hydrocarbons and the other which involves the formation of formate, and, as already said, is preferentially reduced to carbon. Copper-based gas diffusion electrodes gave very good Faradic yields of ethylene, up to 69 per cent, with minor yields of methane, at cathodic current densities in the range of 400–600 mA cm^{-2}, at cathodic potentials ranging from –3 to –4.5 V versus Ag/AgCl (Cook *et al.* 1990*a*).

Positive, although very preliminary results have been reported on the use of systems in which the reduction is performed at the interface of the copper/solid polymer electrolyte (Cook *et al.* 1988*b*; DeWulf and Bard 1988). Of course much effort, both speculative and experimental, has been devoted to the formulation of a reasonable explanation for the exceptional properties of copper cathodes, and on the possible reaction mechanisms through which the reaction evolves to the formation of hydrocarbons. Almost all the authors agree on the hypothesis that hydrido intermediates should take part into the various steps of the process: in this context it was observed (Kim *et al.* 1988) that Cu, in comparison to other metals such as Ru, can sustain larger overpotentials without evolution of hydrogen therefore allowing the reduction of carbon dioxide. Other authors (Cook *et al.* 1988*a*) attribute the main reason for the selectivity in the hydrogenation of adsorbed CO_2 to the slow recombination rate to hydrogen of hydrido species present on copper. Analogies to the mechanism of the Fisher–Tropsch reaction have been hypothesized, with the involvement of surface carbon as precursor of methane (Kim *et al.* 1988), or through the intermediation of $Cu = CH_2$ species, precursors both of methane and ethylene (Cook *et al.* 1988*a*; Hori *et al.* 1989*a*; DeWulf *et al.* 1989). Interesting results have also been obtained in the attempt of induce some modifications of the copper activity by deposition of different metals. Thus various Cu alloys have been tested (Watanabe *et al.* 1991), although for a very small amount of charge passed and at very low current density, a very different reactivity than pure copper is observed; as an example, on Cu/Ni 6/4 alloy no hydrocarbons nor higher alcohols were formed, and only methanol and formic acid, with an overall yield of 20 per cent were obtained. The same effect has been observed in the case of Cu/Cd electrodes (Hori *et al.* 1990*a*) with a strong inhibition of the formation of hydrocarbons and a pronounced increase of the selectivity towards CO and formate. Low surface concentrations of Ni or Fe on Cu result in the increase of the selectivity of the electrodes towards methane, reasonably connected to a higher concentration of hydrido species on the electrode due to the presence of the other metals (Hori *et al.* 1989*b*).

The behaviour of nickel, palladium, and platinum has been the object of some attention, due to the well-known surface interactions of these metals with hydrogen and carbonaceous species: Pd and Pt yield mainly hydrogen, but also formic acid, carbon monoxide and hydrocarbons in low yields have been detected (Azuma *et al.* 1990*b*; Nakagawa *et al.* 1991). Formate and formaldehyde were obtained (Osetrova *et al.* 1984) in diaphragmless cells with Pt electrodes, by electrolysing a saturated solution of carbonate and carbon dioxide; it is suggested that anodically formed percarbonate dianion $C_2O_6^{2-}$, could be responsible for this unusual reactivity. Hydrogen is invariably the major derivative, also with nickel, but hydrocarbons are formed at increasing rates with increasing temperature; the intermediation of CO in these reduction processes has recently been demonstrated (Hori and Murata 1990*b*). A total Faradic yield of 3.9 per cent for hydrocarbon formation on Ni has been observed at a carbon dioxide pressure of 60 atm (Nakagawa *et al.* 1991). From a careful scanning of the properties of a large number of other metals it has been possible to detect from traces to minor amounts of hydrocarbons, methanol, or carbon monoxide from any metal tested (Azuma *et al.* 1989, 1990*a*,*b*; Cook *et al.* 1990b; Nakagawa *et al.* 1991). No valid alternatives to copper emerged from these investigations, nevertheless important indications can be deduced for the design of new catalysts based on polymetal surfaces. It is interesting to quote in this context the enhanced CO formation at Ni/Cd electrodes (Murata and Hori 1991).

Surface interactions of hydrido species with carbon dioxide and its reduction derivatives seem to play a decisive role in addressing the reaction towards the various products. The different pathways the reaction can take are related to the different strengths of binding to the cathode surface of the various intermediates, and among them carbon monoxide has a pre-eminent position as well as the hydrido availability of the surface itself. An apparent 'on–off' selectivity is shown for the production of methanol *or* of higher alcohols by the catalytic systems tested until now. The same selectivity is not shown for the formation of hydrocarbons, where both C1 and higher homologues are obtained simultaneously. On the other hand from the data presented by Noda *et al.* (1989) related to the formation of aldehydes, alcohols, and hydrocarbons with the same number of carbon atoms (mainly 2 and 3), in which the relative ratios of these products are influenced by the cathode potential, it seems possible to deduce that aldehydes and alcohols represent the intermediate steps of the same reduction sequence.

Nonaqueous media The reduction of carbon dioxide in nonaqueous aprotic media is reviewed in its mechanistic aspects, in Chapter 12. From the synthetic point of view the most interesting derivative obtained under these conditions appears to be oxalic acid, which already has quite a large market and could become an interesting starting material for various commodities in a perspective of large scale utilization of carbon dioxide. Some years ago the electrochemical synthesis of oxalic acid was developed to a pilot scale in a process with sacrificial zinc anodes by the Dechema Institut (Fischer *et al.* 1981). In this system zinc is anodically dissolved in the diaphragmless electrocarboxylation unit, and electrolytically recovered in separate cells, to be recycled after hydrolysis of the zinc oxalate. This process, which in principle does not pose problems of waste disposal, gave an estimated production cost comparable to the selling price of oxalic acid (West Germany in 1978). A different approach to this synthesis, in a diaphragm cell with aprotic solvents in the catholyte and aqueous media in the anolyte, gave oxalate in very poor current yields (Goodridge and Presland 1984).

Partially protic systems (Kaiser and Heitz 1973) gave oxalic, glycolic, glyoxylic, tartaric, malic, and succinic acids, as the result of several reduction, hydrogenation, and coupling steps.

High-temperature molten salts represent a different way of having an anhydrous medium in an electrochemical reactor, and these systems have also been tested for the reduction of carbon dioxide on metal cathodes. In various molten alkali chlorides, fluorides or carbonates, the reduction of CO_2 to CO was obtained on various metal cathodes (Halmann and Zuckerman 1987). In molten alkali chlorides the coelectroreduction of molybdate anions and CO_2 affords MoC (Kushkov *et al.* 1987).

Homogeneous or heterogeneous catalysis by metal complexes

Lower overvoltages and in some cases better selectivities than those offered by metal cathodes are presented by catalytic systems based on various metal complexes. The cations used as coordination centres for these complexes belong to the transition metals (Ti, V, Mn, Fe, Co, Ni, Cu, Zn, Ru, Rh, Pd, Ag, Re, Os, Ir, Pt) and to the third (Al, Ga, In) and fourth (Sn, Pb) groups of the periodic table. Various ligands have been proposed: tetrazamacrocycles such as in porphyrins and phtalocyanines (PC), and [14] rings both unsubstituted (cyclam) or bearing various substituents with various degrees of unsaturation (teta), phosphines (phos), polypyridines (poly-py), carbon monoxide. Some of these ligands (cyclam, teta, phos, poly-py) are neutral, and the formal charge of the complex is due to the metal cations, others (Porphyrins, PC) derive change from deprotonation of the corresponding starting molecules and bring to the complex a formal charge of –2. The association of the above listed metal cations and ligands results in precursors of the catalytic systems which have square planar coordination, and often distorted, or octahedral structures. Several of the square planar complexes are able to catalyse the reduction of carbon dioxide in aqueous media, whereas in general octahedral complexes are stable and active in non-aqueous systems.

Square planar complexes Much attention is devoted to the electroreduction of carbon dioxide catalysed by metal–phtalocyanines, which are active, very stable, and unsoluble in aqueous systems, and therefore very promising for applicative developments. The latest investigations on metal–PC complexes embedded in gas diffusion electrodes showed very good selectivities towards the formation of CO with Co, Ni or Pd–PC complexes (Mahmood *et al.* 1987*b*; Furuya and Matsui 1989; Furuya and Koide 1991), of formic acid with Sn or

Pb–PC and of methane with Ti, Cu or Ga–PC (Furuya and Matsui 1989; Furuya and Koide 1991). Cathodic potentials ranging from –1.30 V SHE, at a current density of 54 mA cm^{-2} for Co–PC to –1.98 V SHE at a current density of 55.7 mA cm^{-2} for Ti–PC are reported (Furuya and Koide 1991). The reaction mechanism, in which both the central atom and the ligand are involved, is strongly influenced by the nature of the metal and by the charge distribution in the complex in the various stages of the reaction (Tanabe and Ohno 1987; Masheder and Williams 1987; Christensen *et al.* 1988; Furuya and Koide 1991).

Metal-porphyrin complexes in aqueous media catalyse the reduction of CO_2 to CO with good current yields and overvoltages ranging from 0.7 to 1.2 V (Cao *et al.* 1983, 1986, 1989; Hu *et al.* 1990; Ogura and Yoshida 1988; Atoguchi *et al.* 1991).

In nonaqueous media Ag and Pd octaethylporphyrino complexes gave oxalic acid without carbon monoxide nor carbonate, but the complexes did not prove to be stable for prolonged experiments (Becker *et al.* 1985). The reduction to CO is, on the contrary, catalysed by Fe–porphyrin complexes, both planar and basket-shaped (Hammouche *et al.* 1988). Ni(II) and Co(II) [14]tetrazamacrocycle complexes are interesting electrocatalysts of the reduction of carbon dioxide in mixed water–organic solvent systems, leading to mixtures of hydrogen and carbon monoxide in ratios from 1:1 to 1:2, depending on the nature of ligand and metal (Fisher and Eisenberg 1980; Tinnemans *et al.* 1984). Further investigations (Pearce and Pletcher 1986) showed low turnover numbers, and disappointingly low maximum current densities. An heterogenized Ni–teta complex, $(Ni[Me4Bzo2[14]tetraeneN4])_n$ electropolymerized on Pt, in acetonitrile/methanol, gave formate at a cathodic potential of –1.85 V SCE (Bailey *et al.* 1986).

The Ni-cyclam complex is active in aqueous media with moderate overpotentials, shows high stability and relevant turnover numbers, and is highly selective towards the formation of carbon monoxide (Beley *et al.* 1986; Collin *et al.* 1988). The complex is catalytically active when adsorbed on the mercury cathode surface (Collin and Sauvage 1989). Recent studies (Fujihira *et al.* 1990; Hirata *et al.* 1990) confirmed the role played by the adsorption on the cathode in the reduction process, and explored the product selectivity in a wide range of cathodic potentials. Some members of a new series of Ni(II) complexes of fluorinated cyclams, recently synthesized and characterized (Shionoya *et al.* 1990), show low overpotentials and appreciable selectivities towards the reduction of CO_2 to CO.

Much attention has been devoted to the mechanism by which metal–teta complexes catalyse CO_2 reduction in water (Collin and Sauvage 1989) and in aprotic media (Schmidt *et al.* 1990; Fujita *et al.* 1991). The binding of carbon dioxide to the complexes should take place in an axial position on the reduced central atom and, depending on the mode of linking (on carbon or on oxygen of the CO_2 molecule), different evolutions are to be expected. Most likely formation of a metal–carbon bond is the intermediate step for the formation of carbon monoxide, whereas a bond involving the oxygen atom opens the route to formate anion.

Quasisquare planar metal–phosphino complexes are active for eletroreduction of carbon dioxide only in nonaqueous media. [Rh $diphos_2]^+$ Cl^- catalyses the reduction of CO_2 to formate, at the expense of the solvent acetonitrile (Slater and Wagenknecht 1984). [Pd(triphos)L]$(BF_4)_2$ complexes, where L are various monodentate ligands, such as $P(OMe)_3$ or PPh_3 in acetonitrile catalyse the reduction to CO. A reaction mechanism in which L dissociates from the complex, and a hydrido intermediate is formed before the insertion of carbon dioxide, is proposed (DuBois and Miedaner 1987). IrCl(CO)$(PPh_3)_2$ catalyses the reduction to both CO and formate anion in dimethylformamide–water mixtures (Szymaszek and Pruchnik 1989).

The behaviour of metal–Schiff base complexes in acetonitrile is substantially analogous: both Co-salen (salen = 2,2′-[1,2-ethylene-bis(nitrilomethylidyne)]bis phenol) (Pearce and Pletcher 1986) and Co-salophen (salophen = 2,2′-[1,2-phenylene-bis(nitrilomethylidyne)] bis phenol) (Gennaro *et al.* 1987) catalyse the reduction to CO in acetonitrile. A mechanism analogous to that seen for the metal-teta complexes was proposed for these reactions.

A selective formation of methane, at very high current efficiencies and very moderate overpotentials, on Ru-modified glassy carbon electrodes has been recently published (Arai *et al.* 1989). In this case the metal cations are fixed on poly(hydroquinone/*p*-benzoquinone). Unfortunately the very low current densities and the limited amount of charge make it difficult to evaluate the reliability of this catalyst for long range electrolyses.

Octahedral complexes Polypyridyl and phenantroline complexes of various transition metals are very good catalysts for the reduction of carbon dioxide to CO and formate. Oxalate is generally obtained in minor yields depending on the catalytic system adopted and on the reaction conditions. Two detailed reviews constitute the basis of information for this interesting area (Ziessel 1987; Sullivan *et al.* 1988). Bypy complexes of Re and Ru (Haweker *et al.* 1984, 1986; Ishida *et al.* 1990) and of Rh and Ir (Bolinger *et al.* 1988) have been tested for their electrocatalytic properties. An investigation on the way in which *cis*-[Ru$(bpy)_2$(CO)H^+] catalyses the reduction of CO_2 in acetonitrile with added water disclosed a complex reaction mechanism, going through the formation of a metal-hydride bond and insertion of carbon dioxide, bonded by the oxygen moiety to the metal, thus leading to formate, or through a second parallel pathway leading to CO, involving as associative attack of CO_2 on the twice reduced complexes (Pugh *et al.* 1991). Previous demonstration of the occurrence of an associative activation of carbon dioxide was given for the reduction of CO_2 catalysed by the complex *cis*-[Os$(bpy)_2$(CO)H^+] PF_6^- (Bruce *et al.* 1988). Ni-2,2′-bipyridine systems, which proved to be very effective for many electrocarboxylation reactions, have been also tested for the reduction of CO_2 to CO (Garnier *et al.* 1989). New Rh and Ir complexes with polypyridyl ligands gave current efficiencies up to 70 per cent for the electrocatalysis of the CO_2 reduction to formate in acetonitrile (Rasmussen *et al.* 1990). As it happens in all these systems, the protons needed for the formate synthesis come out of the organic system: in this case, it is suggested that the deprotonation involves the quaternary ammonium salts used as supporting electrolytes. Helical heptacoordinate complexes of cobalt with quinquepyridines gave poor catalytic performances for the CO_2 reduction (Gheysen *et al.* 1990).

The majority of the complexes belonging to this area are poorly stable in water, and this aspect constitutes a severe restriction to the possibility of their large scale application. Interesting possibilities can, however, be seen in the several attempts at heterogeneization of these catalytic systems, realized by electropolymerization on the cathode of vinyl or pyrrole groups bonded to the polypyridyl ligands. The results up to now obtained with polypyridyl complexes of Re (Cabrera and Abruna 1986; Cosnier *et al.* 1986, 1988, 1990; O'Toole *et al.* 1985, 1989), Rh and Ru (Bolinger *et al.* 1985) and Co (Guadalupe *et al.* 1988; Hurrell *et al.* 1989) are very promising, and should be further developed towards the setting up of heterogeneous catalytic systems active in aqueous media.

Polypyridyl complexes of vanadium(II) in various organic solvents are totally inactive for CO_2 activation (Dobson and Taube 1989). Several catalytic systems based on electrochemical activation of two subsequent redox cycles have been tested: the largest investigations have been made into a system in which a first redox couple consisting of $K_2Fe[Fe(CN)_6]/KFe[Fe(CN)_6]$ is fixed on the cathode, and is the reducing agent for a second redox couple dissolved in the reaction medium, for example a suitably complexed Cr(II)/Cr(III) couple, which is the reducing agent of carbon dioxide to methanol (Ogura *et al.* 1990). In several cases good turnover numbers were quoted, although for a limited amount of charge passed. These systems can be active in alcoholic media and the reduction product is methanol, therefore interesting developments could be imagined involving limited separation costs of the product after the reaction step.

Intriguing connections of the CO_2–pyrite interactions hypothesized by Wächtershäuser (see Chapter 16) (and Drobner *et al.* 1990) could possibly be found with the behaviour of macrocyclic Fe_4S_4 clusters, which in aprotic media showed a moderate catalytic activity towards the reduction of CO_2 to formate (Nakazawa *et al.* 1986; Tomohiro *et al.* 1990).

Interesting developments in electrocatalytic activation of carbon dioxide with the $[Ru(bipy)_2(CO)_2]^{2+}$ complex have been reported: depending on the proton availability of the medium, carbon monoxide or formate are obtained (Ishida *et al.* 1987*a*,*b*), and, if dimethylamine is present, dimethylformamide is also formed (Ishida *et al.* 1987*c*). In strictly anhydrous conditions the strongly basic intermediates of the reaction can act as deprotonating agents of suitable organic compounds, such as acetophenone or cyclohexanone, leading to the corresponding anions which are carboxylated by CO_2 present in the reaction medium (Tanaka *et al.* 1988, 1989).

Electrochemically triggered enzymatic systems

Investigations in this area have taken the first steps in the last few years, along a line according to which some enzymatic systems are reactivated in their carbon dioxide fixation capability by reduction with the methylviologen (1,1′-dimethyl-4,4′-dipyridinium) radical cation which acts as homogeneous charge-transfer agent between the electrode and the enzyme. Mild conditions, aqueous-buffered media, high turnover numbers (up to 22 000), and almost quantitative current efficiencies, appear exciting, but the very limited amount of charge passed (1–3 C) and the extremely low current densities adopted hinder a proper evaluation of these results. An interesting feature of these syntheses is that carbon dioxide is selectively fixed in already existing organic molecules, thus posing this CO_2 fixation procedure in between the already seen reductions of carbon dioxide alone and the electrocarboxylation reactions to which the next paragraph is devoted. Thus isocitric acid was obtained from oxoglutaric acid by electroactivation of the isocitrate dehydrogenase enzyme in aqueous media (Sugimura *et al.* 1989); pyruvic acid was obtained from acetyl-coenzyme A (Kuwabata *et al.* 1990) and malic enzyme has been proposed for the carboxylation of pyruvic acid to malic acid (Sugimura *et al.* 1990).

Electrocarboxylations

The term electrocarboxylation was proposed in 1972 (Tyssee *et al.*) to define numerous electrochemical processes by which it is possible to introduce one or more carboxylic groups into suitable organic starting materials.

Electrocarboxylations, making use of carbon dioxide, which is stable, safe, and relatively cheap, offer a valid alternative to the existing carboxylation procedures, many of which make use of dangerous reagents such as phosgene or cyanides, or involve systems very sensitive to contamination, such as those of the Grignard type. The applied final products of this area are quite different from those of the reduction of carbon dioxide alone. Here the transformations have as the main purpose the production of compounds belonging to the field of fine chemistry and parameters such as the product yield, the set up of a synthetic procedure with cheaper and/or more stable starting materials, a lower number of stages in the synthetic sequence, or with easier isolation procedures, have much more relevance in the choice of a process, compared to current yields.

The versatility of this method is demonstrated by the wide choice of organic costarting materials already used in these syntheses (mono- or polyalkenes and alkynes, aromatics, aldehydes, ketones, imines, mono or polyhalides, thioethers) and by the corresponding variety of carboxylic acids obtained from them. The most important synthetic features of electrocarboxylations are shown in the reaction stoichiometries reported in Scheme 13.3.

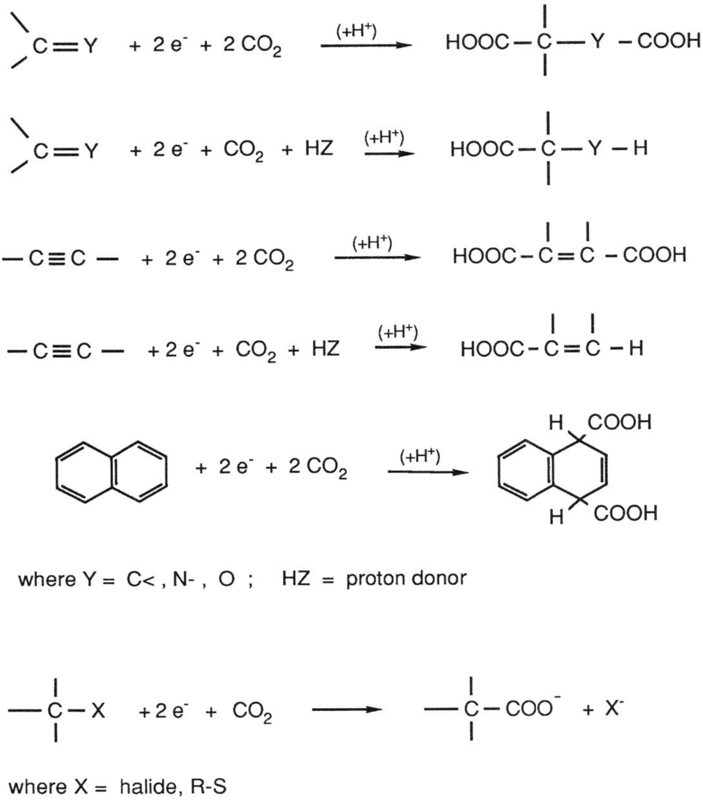

$$>C{=}Y + 2e^- + 2CO_2 \xrightarrow{(+H^+)} HOOC{-}\overset{|}{\underset{|}{C}}{-}Y{-}COOH$$

$$>C{=}Y + 2e^- + CO_2 + HZ \xrightarrow{(+H^+)} HOOC{-}\overset{|}{\underset{|}{C}}{-}Y{-}H$$

$$-C{\equiv}C- + 2e^- + 2CO_2 \xrightarrow{(+H^+)} HOOC{-}\overset{|}{C}{=}\overset{|}{C}{-}COOH$$

$$-C{\equiv}C- + 2e^- + CO_2 + HZ \xrightarrow{(+H^+)} HOOC{-}\overset{|}{C}{=}\overset{|}{C}{-}H$$

$$\text{naphthalene} + 2e^- + 2CO_2 \xrightarrow{(+H^+)} \text{1,4-dihydronaphthalene-1,4-dicarboxylic acid (H, COOH; H, COOH)}$$

where Y = C< , N- , O ; HZ = proton donor

$$-\overset{|}{\underset{|}{C}}{-}X + 2e^- + CO_2 \longrightarrow -\overset{|}{\underset{|}{C}}{-}COO^- + X^-$$

where X = halide, R-S

Scheme 13.3 Reaction stoichiometries of electrocarboxylations

Relationships between structure, redox potential, and reaction mechanism

The reaction mechanisms of electrocarboxylations in aprotic media can be divided in three main groups summarized in Scheme 13.4 depending on the difference between the reduction potential of carbon dioxide and that of the organic compound to be carboxylated (S). Pathway A is followed when S acts as cathodic depolarizer, and its radical anion reacts as a nucleophile with carbon dioxide forming a carboxylated anion radical $[S{-}COO]^{\bullet-}$ which is generally reduced at the potential of formation to $[S{-}COO]^{2-}$; the further reaction of the dianion with CO_2 then leads to the corresponding dicarboxylation derivative.

Pathway B is followed when $S^{\bullet-}$, once formed, reduces via homogeneous electron transfer a neutral carbon dioxide molecule, leading to $CO_2^{\bullet-}$ which will evolve towards carboxylated products through radical couplings, or reactions of $CO_2^{\bullet-}$ with neutral molecules.

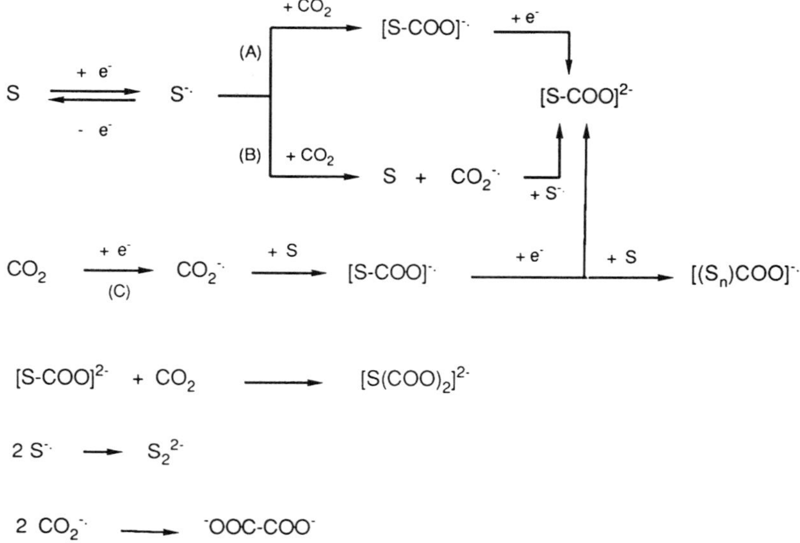

Scheme 13.4 Reaction mechanisms of electrocarboxylation reactions.

Pathway C is followed when carbon dioxide is cathodically reduced to $CO_2^{\bullet -}$ and the reaction involves its radical attack onto S, or more complicated homogeneous electron transfer reactions.

The competition between pathways A and B has been considered in detail in the case of carboxylation of activated olefins (Lamy *et al.* 1979): according to the Marcus theory the rate of the outher sphere charge transfer (case B) is expected to increase with the standard Gibbs energy of the reaction, therefore considering the values of $\Delta E^0 = E^0_{CO_2/CO_2^{\bullet -}} - E^0_{S/S^{\bullet -}}$ the carboxylation should predominantly follow path A for $\Delta E^0 < -400$ mV, and path B for $\Delta E^0 > -400$ mV. From the synthetic point of view the behaviour of systems belonging to case A as the electrocarboxylation of various conjugated olefins (Lamy *et al.* 1979; Gambino *et al.* 1982) and B as the electrocarboxylation of styrene (Filardo *et al.* 1984; Gambino *et al.* 1987) afford good yields of dicarboxylation products in anhydrous conditions. In case B the presence of both $S^{\bullet -}$ and $CO_2^{\bullet -}$ anion radicals leads to the occurrence, as side products, of the corresponding homocoupling derivatives. The electrocarboxylation of ethylene follows case C and various products of carboxylation, of telomerization, and oxalates, were obtained (Gambino and Silvestri 1973; Silvestri *et al.* 1991).

Although not on such a large scale as the latest investigations on carbon dioxide alone, electrocarboxylations have received some attention in the recent years. Mechanistic studies have been developed on the interactions of carbon dioxide with the anion radicals generated by reduction of quinones (Bulhoes and Zara 1988; Simpson and Durand 1990*a*) and of phenazine (Simpson and Durand

1990*b*). From the applicative point of view, some attention has been devoted to the electrocarboxylation of butadiene as an intermediate stage for the production of adipic acid (Chen *et al.* 1985): the regioselectivity of the reaction is influenced by the engineering of the reactor and by the fluodinamic conditions (Pletcher and Girault 1986*a*,*b*).

Catalysis by metal complexes

Uncatalysed electrocarboxylations in conventional diaphragm systems allow moderate product yields in several cases because of the occurrence of side or follow-up reactions. In order to improve yields and selectivities catalytic systems based on transition-metal coordination compounds have been proposed. Among them, nickel complexes appear particularly versatile and have been largely investigated, as it is possible to induce large modifications in the reactivity of several systems, just modifying the type of ligands present in the Ni complex. The latest investigations into the reaction mechanism of the carboxylation of organic halides catalysed by Ni complexes are due to Amatore and Jutand (1990, 1991*a*,*b*) and are described in detail in Chapter 12. The synthetic and applicative aspects of the Ni-catalysed carboxylations of various organic halides have also been explored (Fauvarque *et al.* 1984, 1986, 1988*a*,*b*, 1990), focusing the attention on the carboxylation of some benzylhalides which are precursors of alpha-arylpropionic acids having at present pharmaceutical use as nonsteroidal anti-inflammatory agents (NSAI). Several polypyridine–nickel complexes have shown high and unusual regioselectivities towards the electrocarboxylation of mono- and polyalkynes, in diaphragmless systems in which Mg is anodically dissolved (Duñach and Perichon 1990; Labbe *et al.* 1988; Duñach *et al.* 1989; Derien *et al.* 1990, 1991). Also complexes of Pd (Torii *et al.* 1986), Co (Folest *et al.* 1985), and Fe (Tkatchenko *et al.* 1984) have shown interesting catalytic properties towards the electrocarboxylation of various organic compounds.

The use of sacrificial anodes

Considerable improvements in yields of several electrocarboxylation processes have been observed performing the electrolysis in diaphragmless cells equipped with suitable sacrificial anodes, as already quoted for the electrocarboxylation of alkynes. The anodic reaction

$$\mathrm{M} - n\mathrm{e}^- \rightarrow (\mathrm{M}^{n+})_{\mathrm{solv.}}$$

affords metal cations to the electrolytic medium in stoichiometric balance with the cathodic production of carboxylated species R–COO$^-$, and in the bulk both species react leading to complex salts

$$x\mathrm{R{-}COO}^- + y\,\mathrm{M}^{n+} \rightarrow [(\mathrm{R{-}COO})_x\mathrm{M}_y]^{(yn-x)+}$$

The choice of M has to be oriented towards metals whose cations, in the reaction conditions, should not be reduced at the cathode, nor hinder the desired cathodic process.

As far as the reaction mechanism is considered, the metal cations act as electrophiles, may form complexes with many intermediates of the reaction, and finally stabilize the products as metal salts hindering follow-up reactions (Silvestri *et al.* 1987).

Some examples are presented here to demonstrate how the use of this methodology could be useful for solving problems raised by conventional diaphragm electrolyses.

In the electrocarboxylation of organic halides in conventional systems a chemical follow-up reaction takes place which consumes up to 50 per cent of the halide to be carboxylated leading to the corresponding ester (Baizer and Chruma 1972; Wagenknecht 1974)

$$\text{R–COO}^- + \text{R–X} \rightarrow \text{R–COOR} + \text{X}^-$$

If the synthesis is performed in the presence of metal cations arising from a sacrificial anode, the formation of the complex salts inhibits the nucleophilicity of the carboxylated anions, thus hindering the esterification reaction. The electrocarboxylation of various organic halides has been developed making use of Al (Silvestri *et al.* 1984*a*) or Mg (Sock *et al.* 1985; Heintz *et al.* 1988; Chaussard *et al.* 1989) sacrificial anodes. The electrocarboxylation of benzal chloride in conventional diaphragm systems fails because an internal nucleophilic substitution addresses the reaction towards tars, and only benzylic acid in low yields was obtained (Wawzonek and Shradel 1979). Also in this case the presence of Al cations changed the reaction pattern, allowing the isolation of chlorophenylacetic and phenylmalonic acids in overall yields up to 60 per cent (Silvestri *et al.* 1984*b*). With respect to the conventional carboxylations of imines for the synthesis of substituted alpha amino acids (Hess and Thiele 1982) considerable improvements were reached by the use of Al sacrificial anodes (Silvestri *et al.* 1988*a*). The same effect was observed in the case of aldehydes, from which no electrocarboxylation derivatives were obtained in diaphragm systems (Wawzonek and Gundersen 1960). On the contrary with sacrificial Al anodes the carboxylation of aromatic aldehydes was achieved with yields up to 40 per cent (Silvestri *et al.* 1986). Considerable improvements were obtained in the carboxylation of various ketones with Mg sacrificial anodes (Mcharek *et al.* 1989).

The electrocarboxylation of aromatic ketones, followed by chemical hydrogenation of the hydroxyacid, has received some attention in view of proposing an alternative electrochemical synthesis of some NSAI agents. Considering the electrochemical step the best results in conventional systems were obtained at molar ratios [supporting electrolyte]/[ketone] of 10, with KI as supporting electrolyte (Ikeda and Manda 1985). In this way yields of the hydroxyacids up to 80 per cent were obtained starting from various benzophenones. Syntheses of

NSAI, including the electrocarboxylation of the ketone and the hydrogenation of the hydroxyacid or of the unsaturated acid arising from it, were also claimed (Agency of Industrial Sciences and Technology 1985; Maspero *et al.* 1988; Chan 1990). The sacrificial anodes methodology has been applied to electrocarboxylation of ketones by our group (Filardo *et al.* 1986), by the Enichem laboratories (Maspero *et al.* 1988), and by Monsanto (Wagenknecht 1986*a*,*b*), all devoting particular attention to the synthesis of naproxen (2-hydroxy-2-(6′-methoxy-2′-naphtyl)propionic acid), or of ibuprofen (2[p-(*i*-butyl)phenyl] propionic acid), starting from the corresponding ketones, 2-(6-methoxy-2-naphtyl)acetonaphtone and 2[*p*-(*i*-butyl)phenyl] acetophenone, respectively. In diaphragmless systems equipped with aluminium sacrificial anodes the electrocarboxylation of these ketones to the corresponding acids was performed with yields of up to 85–90 per cent, adopting an isolation procedure consisting of precipitation of the aluminium salts of the carboxylated products by addition of suitable cosolvents such as ethers. In this way the recycle of the solvent/supporting electrolyte system is easy and straightforward. Furthermore, considering the production of NSAI, the direct hydrogenation of the aluminium salts to the corresponding arylpropionic acids has been realized (Maspero *et al.* 1988). The sacrificial anodes technology has been developed using electrochemical cells designed on purpose: the Mg-based electrocarboxylations of halides have been performed and scaled up with tank electrochemical cells in which the soluble anode is fed in bars which are progressively consumed (Chaussard *et al.* 1988), whereas the electrocarboxylations of halides and ketones with Al anodes have been developed using modified plate and frame cells in which the soluble elements are from time to time added in particulate form (Silvestri *et al.* 1988*b*, 1989).

Conclusions

It is hoped that from this review, although compressed into a limited number of pages, the researches involving electrochemical syntheses starting from carbon dioxide can emerge in all their lively diversification and promising perspectives. Of course, at the present level of development, it is rather difficult to foresee whether chemistry will contribute to the solution of the large scale problems posed by the accumulation of carbon dioxide in the atmosphere. In fact the dimensions of this problems are far beyond the technology and the scale of industrial productions of chemicals we are at present used to. On the contrary, it is possible to imagine chemical transformations of carbon dioxide into products useful for energy transportation and storage, or useful as building blocks of more complex structures, utilizing local concentrated carbon dioxide emissions. If chemistry can accomplish this task, electrochemistry, utilizing electric energy coming out of nuclear power stations or of photovoltaic systems, will surely afford valid and competitive solutions.

Acknowledgements

This work has been supported by the Italian Consiglio Nazionale delle Ricerche, Progetto Finalizzato Chimica Fine.

References

Agency of Industrial Sciences and Technology (1985). *Jpn. Pat.* 85 24386 (60 24386). *Chem. Abs.*, **103,** 013560.

Alonso, C., Gonzalez Velasco, J., and Arvia, A.J. (1988). *J. Electroanal. Chem. Interfacial Electrochem.,* **250,** 183–9.

Amatore, C. and Jutand, A. (1990). *Acta Chem. Scand.*, **44,** 755–64.

Amatore, C. and Jutand, A. (1991*a*). *J. Electroanal. Chem. Interfacial Electrochem.*, **306,** 141–56.

Amatore, C. and Jutand, A. (1991*b*). *J. Am. Chem. Soc.*, **113,** 2819–25.

Arai, G., Harashina, T., and Yasumori, I. (1989). *Chem. Lett.*, 1215–18.

Atoguchi, T., Aramata, A., Kazusaka, A., and Enyo, M. (1991). *J. Chem. Soc., Chem. Commun.,* 156–7.

Azuma, M., Hashimoto, K., Hiramoto, M., Watanabe, M., and Sakata, T. (1989). *J. Electroanal. Chem. Interfacial Electrochem.,* **260,** 441–5.

Azuma, M., Hashimoto, K., Hiramoto, M., Watanabe, M., and Sakata, T. (1990*a*). *J. Electrochem. Soc.,* **137,** 1772–8.

Azuma, M., Hashimoto, K., Watanabe, M., and Sakata, T. (1990*b*). *J. Electroanal. Chem. Interfacial Electrochem.,* **294,** 299–303.

Bailey, C.L., Bereman, N., Rillema, D.P., and Nowak, R. (1986). *Inorg. Chim. Acta,* **116,** L45–L47.

Baizer, M.M. and Chruma, J.L. (1972). *J. Org. Chem.,* **37,** 1951–60.

Becker, J.Y., Vainas, B.E.R., and Kaufman, L. (1985). *J. Chem. Soc., Chem. Commun.,* 1471–2.

Beketov, N.N. (1869). *Russ. Phys. Chem. Soc.*, **1,** 33–4.

Beley, M., Collin, J.P., Ruppert, R., and Sauvage, J.P. (1986). *J. Am. Chem. Soc.,* **108,** 7461–7.

Bewick, A. and Greener, G.P. (1969). *Tetrahedron Lett.,* 4623–6.

Bewick, A. and Greener, G.P. (1970). *Tetrahedron Lett.*, 391–2.

Bolinger, C.M., Sullivan, B.P., Conrad, D., Gilbert, J.A., Story, N., and Meyer, T.J. (1985). *J. Chem. Soc., Chem. Commun.,* 796–7.

Bolinger, C.M., Story, N., Sullivan, B.P., and Meyer, T.J. (1988). *Inorg. Chem.*, **27,** 4582–7.

Bruce, M.R.M., Megehee, E., Sullivan, B.P., Thorp, H., O'Toole, T.R. Downard, A., and Meyer, T.J. (1988). *Organometallics,* **7,** 238–40.

Bulhoes, L.O. de Sousa and Zara, A.J. (1988). *J. Electroanal. Chem. Interfacial Electrochem.,* **248,** 159–65.

Butenko, V.A., Grishaenkov, B.G., Arkharov, A.M., Vitkovskii, A.V., and Gastev, A.S. (1987). *Elektrokhimiya,* **23,** 1594–605.

Cabrera, C.R. and Abruna, H.D. (1986). *J. Electroanal. Chem. Interfacial Electrochem.,* **209,** 101–7.

Cao, X., Huang, C., and Wang, M. (1983). *Gaodeng Xuexiao Huaxue Xuebao,* **4,** 549–52. *Chem. Abs.*, **100,** 058659.

Cao, X., Mu, Y., Wang, M., and Luan, L. (1986). *Huaxue Xuebao*, **44,** 220–4. *Chem. Abs.,* **104,** 195348.

Cao, X., Zheng, G., and Teng, Y. (1989). *Huaxue Xuebao*, **47,** 575–82. *Chem. Abs.*, **111,** 122682.

Chan, A.S.C. (1990). *PCT Int. Pat.* 90 15790. *Chem. Abs.*, **115,** 028896.

Chaussard, J., Storck, A., Lapicque, F., and Hornut, J.M. (1988). *Fr. Pat.* 2617197. *Chem. Abs.*, **111,** 086198.

Chaussard, J., Troupel, M., Robin, Y., Jacob, G., and Juhasz, J.P. (1989). *J. Appl. Electrochem.,* **19,** 345–8.

Chaussard, J., Folest, J.C., Nedelec, J.Y., Perichon, J., Sibille, S., and Troupel, M. (1990). *Synthesis,* **5**, 369–81.

Chechel, P.S. and Antropov, L.I. (1958). *Zhur. Priklad. Khim.,* **31**, 1856–61. *Chem. Abs.,* **53,** 6832g.

Chen, H.T., Tien, H.J., Lai, T.T., and Chou, T.C. (1985). *J. Chin. Inst. Chem. Eng.,* **16,** 25–30. *Chem. Abs.,* **103,** 029152.

Christensen, P.A., Hamnett, A., and Muir, A.V.G. (1988). *J. Electroanal. Chem. Interfacial Electrochem.,* **241,** 361–71.

Cohen, A. and Jahn, S. (1904). *Ber. d. D. Chem. Gesellschaft,* **37,** 2836–42.

Collin, J.P. and Sauvage, J.P. (1989). *Coord. Chem. Rev.*, **93,** 245–68.

Collin, J.P., Jouaiti, A., and Sauvage, J.P. (1988). *Inorg. Chem.,* **27,** 1986–90.

Cook, R.L., MacDuff, R.C. and Sammells, A.F. (1987*a*). *J. Electrochem. Soc.,* **134,** 1873–4.

Cook, R.L., MacDuff, R.C., and Sammells, A.F. (1987*b*). *J. Electrochem. Soc.*, **134,** 2375–6.

Cook, R.L., MacDuff, R.C., and Sammells, A.F. (1988*a*). *J. Electrochem. Soc.,* **135,** 1320–6.

Cook, R.L., MacDuff, R.C., and Sammells, A.F. (1988*b*). *J. Electrochem. Soc.,* **135,** 1470–1.

Cook, R.L., MacDuff, R.C., and Sammells, A.F. (1989). *J. Electrochem. Soc.,* **136,** 1982–4.

Cook, R.L., MacDuff, R.C., and Sammells, A.F. (1990*a*). *J. Electrochem. Soc.,* **137,** 607–8.

Cook, R.L., MacDuff, R.C., and Sammells, A.F. (1990*b*). *J. Electrochem. Soc.,* **137,** 187–9.

Cosnier, S., Deronzier, A., and Moutet, J.C. (1986). *J. Electroanal. Chem. Interfacial Electrochem.,* **207,** 315–21.

Cosnier, S., Deronzier, A., and Moutet, J.C. (1988). *J. Mol. Catal.,* **45,** 381–91.

Cosnier, S., Deronzier, A., and Moutet, J.C. (1990). *New J. Chem.,* **14,** 831–9.

Dehn, H., Gutmann, V., Kirch, H., and Shöber, G.. (1962). *Mh. Chem.,* **93,** 1348–52.

Derien, S., Duñach, E., and Perichon, J. (1990). *J. Organomet. Chem.,* **385,** C43–6.

Derien, S., Clinet, J.C., Duñach, E., and Perichon, J. (1991). *J. Chem. Soc., Chem. Commun.,* 549–50.

DeWulf, D.W. and Bard, A.J. (1988). *Catal. Lett.,* **1,** 73–9.

DeWulf, D.W., Jin, T., and Bard, A.J. (1989). *J. Electrochem. Soc.,* **136,** 1686–91.

Dobson, J.C., and Taube, H. (1989). *Inorg. Chem.,* **28,** 1310–15.

Drechsel, E. (1868). *J. Chem. Soc.* (London), **21,** 121.

Drobner, E., Huber, H., Wächtershäuser, G., Rose D., and Stetter, K.U. (1990). *Nature,* **346,** 742–4.

DuBois, D.L. and Miedaner, A. (1987). *J. Am. Chem. Soc.,* **109,** 113–17.

Duñach, E. and Perichon, J. (1990). *Synlett.,* 143–5.

Duñach, E., Derien, S., and Perichon, J. (1989). *J. Organomet. Chem.,* **364,** C33–6.

Eggins, B.R., Brown, E.M., McNeill, E.A., and Grimshaw, J. (1988). *Tetrahedron Lett.,* **29,** 945–8.

Ehrenfeld, R. (1905). *Ber. d. D. Chem. Gesellschaft,* **38,** 4138–43.

Fauvarque, J.F., Chevrot, C., Jutand, A., Francois, M., and Perichon, J. (1984). *J. Organomet. Chem.,* **264,** 273–81.

Fauvarque, J.F., Jutand, A., and Francois, M. (1986). *Nouv. J. Chim.,* **10,** 119–22.

Fauvarque, J.F., Jutand, A., and Francois, M. (1988*a*). *J. Appl. Electrochem.,* **18,** 109–15.

Fauvarque, J.F., Jutand, A., Francois, M., and Petit, M.A. (1988*b*). *J. Appl. Electrochem.,* **18,** 116–19.

Fauvarque, J.F., De Zelicourt, Y., Amatore, C., and Jutand, A. (1990). *J. Appl. Electrochem.,* **20,** 338–40.

Filardo, G., Gambino, S., Silvestri, G., Gennaro, A., and Vianello, E. (1984). *J. Electroanal. Chem. Interfacial Electrochem.,* **177,** 303–9.

Filardo, G., Silvestri, G., and Gambino, S. (1986). Eur. Pat. Appl., (860730) *Chem. Abs.,* **105,** 180522.

Fisher, B. and Eisenberg, R. (1980). *J. Am. Chem. Soc.,* **102,** 7361–3.

Fisher, F. and Prziza, O. (1914). *Ber. d. D. Chem. Gesellschaft,* **47,** 256–60.

Fisher, J., Lehman, Th., and Heitz, E. (1981). *J. Appl. Electrochem.,* **11,** 743–50.

Folest, J.C., Duprilot, J.M., Perichon, J., Robin, Y., and Devynck, J. (1985). *Tetrahedron Lett.,* **26,** 2633–6.

Frese, K.W., Jr., and Leach, S. (1985). *J. Electrochem. Soc.,* **132,** 259–60.

Frese, K.W., Jr., and Summers, D.P. (1988). In *Catalytic activation of carbon dioxide,* ACS Symp. Ser., **363,** 155–70.

Fujihira, M., Hirata, Y., and Suga, K. (1990). *J. Electroanal. Chem. Interfacial Electrochem.,* **292,** 199–215.

Fujita, E., Creutz, C., Sutin, N., and Szalda, D.J. (1991). *J. Am. Chem. Soc.,* **113,** 343–53.

Furuya, N. and Koide, S. (1991). *Electrochim. Acta,* **36,** 1309–13.

Furuya, N. and Matsui, K. (1989). *J. Electroanal. Chem. Interfacial Electrochem.,* **271,** 181–91.

Gambino, S. and Silvestri, G. (1973). *Tetrahedron Lett.,* 3025–8.

Gambino, S., Filardo, G., and Silvestri, G. (1982). *J. Appl. Electrochem.,* **12,** 549–55.

Gambino, S., Gennaro, A., Filardo, G., Silvestri, G., and Vianello, E. (1987). *J. Electrochem. Soc.,* **134,** 2172–5.

Garnier, L., Rollin, Y., and Perichon, J. (1989). *New J. Chem.,* **13,** 53–9.

Gennaro, A., Isse, A.A., and Vianello, E. (1987). In *Recent advances in electroorganic synthesis,* Stud. Org. Chem., **30,** 321–4. Elsevier, Amsterdam, The Netherlands.

Gheysen, K.A., Potts, K.T., Hurrell, H.C., and Abruna, H.D. (1990). *Inorg. Chem.,* **29,** 1589–92.

Goodridge, F. and Presland, G. (1984). *J. Appl. Electrochem.,* **14,** 791–6.

Grishaenkov, B.G., Vassiliev, V.K., Zorina, N.G., and Zhukov, A.K. (1986). *Kosm. Biol. Aviakosm. Med.,* **20,** 76–9. *Chem. Abs.,* **106,** 074855.

Guadalupe, A.R., Usifer, D.A., Potts, K.T., Hurrell, H.C., Mogstad, A.E., and Abruna, H.D. (1988). *J. Am. Chem. Soc.,* **110,** 3462–6.

Halmann, M. and Zuckerman, K. (1987). *J. Electroanal. Chem. Interfacial Electrochem.,* **235,** 369–80.

Hammouche, M., Lexa, D., Saveant, J.M., and Momenteau, M. (1988). *J. Electroanal. Chem. Interfacial Electrochem.,* **249,** 347–51.

Haynes, L.V. and Sawyer, D.T. (1967). *Anal. Chem.,* **39,** 332–8.

Hawecker, J., Lehn, J.M., and Ziessel, R. (1984). *J. Chem. Soc., Chem. Commun.,* 328–30.

Hawecker, J., Lehn, J.M., and Ziessel, R. (1986). *Helv. Chim. Acta,* **69,** 1990–2012.

Heintz, M., Sock, O., Saboureau, C., Perichon, J., and Troupel, M. (1988). *Tetrahedron,* **44,** 1631–6.

Hess, U. and Thiele, R. (1982). *J. Prakt. Chem.,* **324,** 385–99.

Hirata, Y., Suga, K., and Fujihira, M. (1990). *Chem. Lett.,* 1155–8.

Hohn, H., Fitzer, E., Schlager, O., and Nedwed, H. (1953). Austrian Pat. 175237. *Chem. Abs.,* **47,** 8978f.

Hooker, M., Rast, H.E., Rogers, D.K., Borja, L., Clark, K., Fleming, K. *et al.* (1990). *Sci. Tech. Aerosp. Rep.,* **28,** Abstr. No. N90-16886. *Chem. Abs.,* **115,** 059314.

Hori, Y., Kikuchi, K., and Suzuki, S. (1985). *Chem. Lett.,* 1695–8.

Hori, Y., Kikuchi, K., Murata, A., and Suzuki, S. (1986). *Chem. Lett.,* 897–8.

Hori, Y., Murata, A., Kikuchi, K., and Suzuki, S. (1987). *J. Chem. Soc., Chem. Commun.,* 728–9.

Hori, Y., Murata, A., and Takahashi, R. (1989*a*). *J. Chem. Soc., Faraday Trans.,* **85,** 2309–26.

Hori, Y., Murata, A., Ito, S.Y., Yoshinami, Y., and Koga, O. (1989*b*). *Chem. Lett.,* 1567–70.

Hori, Y., Murata, A., and Ito, S.Y. (1990*a*). *Chem. Lett.,* 1231–4.

Hori, Y. and Murata, A. (1990*b*). *Electrochim. Acta,* **35,** 1777–80.

Hori, Y., Murata, A., and Yoshinami, Y. (1991). *J. Chem. Soc., Faraday Trans.,* **87,** 125–8.

Hu, G., Xu, L., and Dong, S. (1990). *Wuli Huaxue Xuebao,* **6,** 710–15. *Chem. Abs.,* **114,** 071056.

Hurrell, H.C., Mogstad, A.L., Usifer, D.A., Potts, K.T., and Abruna, H.D. (1989). *Inorg. Chem.,* **28,** 1080–4.

Ikeda, Y. and Manda, E. (1985). *Bull. Chem. Soc. Jpn,* **58,** 1723–6.

Ikeda, S., Takagi, T., and Ito, K. (1987). *Bull. Chem. Soc. Jpn,* **60,** 2517–22.

Ikeda, S., Amakusa, S., Noda, H., Saito, Y., and Ito, K. (1988). *Proc. Electrochem. Soc.,* **14,** 130–6.

Ishida, H., Tanaka, H., Tanaka, K., and Tanaka, T. (1987*a*). *J. Chem. Soc., Chem. Commun.,* 131–2.

Ishida, H., Tanaka, K., and Tanaka, T. (1987*b*). *Organometallics,* **6,** 181–6.

Ishida, H., Tanaka, H., Tanaka, K., and Tanaka, T. (1987*c*). *Chem. Lett.,* 597–600.

Ishida, H., Fujiki, K., Ohba, T., Ohkubo, K., Tanaka, K., Terada, T. *et al. J. Chem Soc., Dalton Trans.,* 2155–60.

Islam, M.S. and Kunimatsu, K. (1989). *J. Bangladesh Acad. Sci.,* **13,** 165–74. *Chem. Abs.,* **112,** 167704.

Ito, K., Ikeda, S., and Okabe, M. (1980). *Denki Kagaku,* **48,** 247–52.

Ito, K., Ikeda, S., Iida, T., and Niwa, H. (1981). *Denki Kagaku,* **49,** 106–12.

Jordan, J. and Smith, P.T. (1960). *Proceedings Chem. Soc.,* 246–7.

Kaiser, U. and Heitz, E. (1973). *Ber. Bunsenges. Phys. Chem.,* **77,** 818–23.

Kim, J.J., Summers, D.P., and Frese, K.W., Jr (1988). *J. Electroanal. Chem. Interfacial Electrochem.,* **245,** 223–44.

Kitani, A., Yamada, H., and Sasaki, K., (1978). *Denki Kagaku Oyobi Kogyo Butsuri Kagaku.,* **46,** 570–2. *Chem. Abs.,* **90,** 78305f.

Koga, O., Nakama, K., Murata, A., and Hori, Y. (1989). *Denki Kagaku Oyobi Kogyo Butsuri Kagaku.,* **57,** 1137–40. *Chem.. Abs.,* **112,** 065318.

Koike, K., Yamane, Y., and Kishi, T. (1989). *Hyomen Gijutsu,* **40,** 1445–6. *Chem. Abs.,* **112,** 086752.

Kolbe, H. and Schmitt, R. (1861). *Ann. der Chem.,* **119,** 251. (The authors reported they had not found formic acid by electrochemical reduction of aqueous solutions of carbon dioxide.)

Kushkhov, Kh.B., Shapoval, V.I., and Novoselova, I.A. (1987). *Elektrokhimiya,* **23,** 952–6.

Kuwabata, S., Morishita, N., and Yoneyama, H. (1990). *Chem. Lett.,* 1151–4.

Labbe, E., Duñach, E., and Perichon, J. (1988). *J. Organomet. Chem.,* **353,** C51–6.

Lamy, E., Nadjo, L., and Saveant, J.M. (1979). *Nouv. J. Chim.,* **3,** 21–9.

Leitz, F.B. (1967). *US Clearinghouse Fed. Sci. Tech. Inform.,* AD 654146. *Chem. Abs.,* **68,** 26306.

Lieben, A. (1895). *Monatsh.* **16,** 211–47.

Liebig, J. (1847). *Annalen der Chemie und Pharmazie,* **58,** 335–48.

Maeda, M., Kitaguchi, Y., Ikeda, S., and Ito, K. (1987). *J. Electroanal. Chem. Interfacial Electrochem.,* **238,** 247–58.

Mahmood, M.N., Masheder, D., and Harty, C.J. (1987*a*). *J. Appl. Electrochem.,* **17,** 1159–70.

Mahmood, M.N., Masheder, D., and Harty, C.J. (1987*b*). *J. Appl. Electrochem.,* **17,** 1223–7.

Mahoney, M., Howard, M., and Cooney, P. (1980). *Chem. Phys. Lett.,* **71,** 59–63.

Masheder, D. and Williams, K.P.J. (1987). *J. Raman Spectrosc.,* **18,** 391–8.

Maspero, F., Piccolo, O., Romano, U., and Gambino, S. (1988). Eur. Pat. 286944. *Chem. Abs.,* **110,** 075087.

Mcharek, S., Heintz, M., Troupel, M., and Perichon, J. (1989). *Bull. Soc. Chim. Fr.,* 95–7.

McQuillan, A.J., Hendra, P.J., and Fleischmann, M. (1975). *J. Electroanal. Chem.,* **65,** 933–44.

Meller, F.H. (1968). US Clearinghouse Fed. Sci. Tech. Inform., AD 678427. *Chem. Abs.,* **70,** 83579.

Miller, R.M., Knorr, H.V., Eichel, H.J., Meyer, C.M., and Tanner, H.A. (1962). *J. Am. Chem. Soc.,* **84,** 2646–8.

Murata, A. and Hori, Y. (1991). *Chem. Lett.,* 181–4.

Nakagawa, S., Kudo, A., Azuma, M., and Sakata, T. (1991). *J. Electroanal. Chem. Interfacial Electrochem.,* **308,** 339–43.

Nakagawa, M., Mizobe, Y., Matsumoto, Y., Uchida, Y., Tezuka, M., and Hidai, M. (1986). *Bull. Chem. Soc. Jpn,* **59,** 809–14.

Noda, H., Ikeda, S., Oda, Y., and Ito, K. (1989). *Chem. Lett.*, 289–92.

Noda, H., Ikeda, S., Oda, Y., Imai, K., Maeda, M., and Ito, K. (1990). *Bull. Chem. Soc. Jpn,* **63,** 2459–62.

O'Connell, C., Hommeltoft, S.I., and Eisenberg, R. (1987). In *Carbon dioxide as a source of carbon: biochemical and chemical uses,* NATO ASI Series, (Ser. C), **206,** 33–54. Kluwer, Dordrecht, The Netherlands.

O'Toole, T.R., Margerum, L.D., Westmoreland, T.D., Vining, W.J., Murray, R.W., and Meyer, T.J. (1985). *J. Chem. Soc., Chem. Commun.,* 1416–17.

O'Toole, T.R., Sullivan, B.P., Bruce, M.R.M., Margerum, L. D., Murray, R.W., and Meyer, T.J. (1989). *J. Electroanal. Chem. Interfacial Electrochem.,* **259,** 217–39.

Ogura, K. and Yoshida, I. (1988). *J. Mol. Catal.,* **47,** 51–7.

Ogura, K., Migita, C.T., and Imura, H. (1990). *J. Electrochem. Soc.,* **137,** 1730–2.

Osetrova, N.V., Vassiliev, Y.B., Bagotskii, V.S., Sadkova, R.G., Cherashev, A.F., and Khrushch, A.P. (1984). *Elektrokhimiya,* **20,** 286.

Paik, W., Andersen, T.N., and Eyring, H. (1969). *Electrochim. Acta,* **14,** 1217–32.

Pearce, D.J. and Pletcher, D. (1986). *J. Electroanal. Chem. Interfacial Electrochem.,* **197,** 317–30.

Pletcher, D. and Girault, J.T. (1986*a*). *J. Appl. Electrochem.,* **16,** 791–802.

Pletcher, D. and Girault, J.T. (1986*b*). In *Electrochemical engineering,* Inst. Chem. Eng. Symp. Ser., **98,** 13–21, pp. 321–2. Institution of Chemical Engineers, Rugby, UK.

Pugh, J.R., Bruce, M.R.M., Sullivan, B.P., and Meyer, T.J. (1991). *Inorg. Chem.*, **30,** 86–91.

Rabinowitsch, M. and Maschowetz, A. (1930). *Z. Electrochem.,* **36,** 846–51.

Rasmussen, S.C., Richter, M.M., Yi, E., Place, H., and Brewer, K.J. (1990). *Inorg. Chem.,* **29,** 3926–32.

Roberts, J.L., Jr and Sawyer, D.T. (1965). *J. Electroanal. Chem.,* **9,** 1–7.

Royer, M.E. (1870). *C.R. Hebd. S. Acad. Sci. Fr. (Sect. Chimie),* **70,** 731–3.

Russel, P.G., Kovac, N., Srinivasan, S., and Steinberg, M. (1977). *J. Electrochem. Soc.,* **124,** 1329–38.

Ryu, J., Andersen, T.N., and Eyring, H. (1972). *J. Phys. Chem.,* **76,** 3278–86.

Sawada, T., Kajima, K., Otsuji, K., Miyamoto, T., and Takenaka, H. (1990). *Soda to Enso,* **41,** 11–22. *Chem. Abs.,* **113,** 227470.

Schmidt, M.H., Miskelly, G.M., and Lewis, N.S. (1990). *J. Am. Chem. Soc.,* **112,** 3420–6.

Shionoya, M., Kimura, E., and Iitaka, Y. (1990). *J. Am. Chem. Soc.,* **112,** 9237–45.

Shöber, G. (1964). *Abh. Deutsch. Akad. Wissensch. Berlin, Klasse für Chem. Geol. Biol.,* **1,** 496–7.

Silvestri, G. (1987). In *Carbon dioxide as a source of carbon, biochemical chemical uses,* NATO ASI Series, (Ser. C), **206,** 339–69. Kluwer, Dordrecht, The Netherlands.

Silvestri, G., Gambino, S., Filardo, G., and Gulotta, A. (1984*a*). *Angew. Chem.,* **96,** 978–9.

Silvestri, G., Gambino, S., Filardo, G., Greco, G., and Gulotta, A. (1984*b*). *Tetrahedron Lett.,* **25,** 4307–8.

Silvestri, G., Gambino, S., and Filardo, G. (1986). *Tetrahedron Lett.,* **27,** 3429–30.

Silvestri, G., Gambino, S., and Filardo, G. (1987). In *Recent advances in electroorganic synthesis,* Stud. Org. Chem., **30,** 287–94. Elsevier, Amsterdam, The Netherlands.

Silvestri, G., Gambino, S., and Filardo, G. (1988*a*). *Gazz. Chim. Ital.,* **118,** 643–8.

Silvestri, G., Filardo, G., and Gambino, S. (1988*b*). Eur. Pat. 283796. *Chem. Abs.,* **110,** 065730.

Silvestri, G., Gambino, S., Filardo, G., and Tedeschi, F. (1989). *J. Appl. Electrochem.,* **19,** 946–8.

Silvestri, G., Gambino, S., and Filardo, G. (1990). In *Enzymatic model carboxylation reduction reactions and carbon dioxide utilization,* NATO ASI Series, (Ser. C), **314,** 101–27. Kluwer, Dordrecht, The Netherlands.

Silvestri, G., Gambino, S., and Filardo, G. (1991). *Acta Chem. Scand.,* 987–92.

Simpson, T.C. and Durand, R.R., Jr (1990*a*). *Electrochim. Acta,* **35,** 1399–403.

Simpson, T.C. and Durand, R.R., Jr (1990*b*). *Electrochim. Acta,* **35,** 1405–10.

Slater, S. and Wagenknecht, J.H. (1984). *J. Am. Chem. Soc.,* **106,** 5367–8.

Sock, O., Troupel, M., and Perichon, J. (1985). *Tetrahedron Lett.,* **26,** 1509–12.

Srivastava, S.C. and Shukla, S.N. (1970). *Electrochim. Acta,* **15,** 2021–2.

Sugimura, K., Kuwabata, S., and Yoneyama, H. (1989). *J. Am. Chem. Soc.,* **111,** 2361–2.

Sugimura, K., Kuwabata, S., and Yoneyama, H. (1990). *Bioelectrochem. Bioenerg.,* **24,** 241–7.

Sullivan, B.P., Bruce, M.R.M., O'Toole, T.R., Bolinger, C.M., Megehee, E., Thorp, H. *et al.* (1988). In *Catalytic activation of carbon dioxide* ACS Symp. Ser., **363,** 52–90. American Chemical Society, Washington DC, USA.

Summers, D.P., Leach, S., and Frese, K.W., Jr (1986). *J. Electroanal. Chem. Interfacial Electrochem.,* **205,** 219–32.

Szymaszek, A. and Pruchnik, F.P. (1989). *J. Organomet. Chem.,* **376,** 133–40.

Tanabe, H. And Ohno, K. (1987). *Electrochim. Acta,* **32,** 1121–4.

Tanaka, K., Matsui, T., and Tanaka, T. (1989). *J. Am. Chem. Soc.,* **111,** 3765–7.

Tanaka, K., Miyamoto, H., and Tanaka, T. (1988). *Chem. Lett.,* 2033–6.

Taniguchi, I. (1989). *Mod. Aspects Electrochem.,* **20,** 327–400.

Teeter, T.E. and Van Rysselberghe P. (1954). *J. Chem. Phys.,* **22,** 759–60.

Tinnemans, A.H.A., Koster, T.P.M., Thewissen, D.H.M.W., and Mackor, A. (1984). *Recl.: J.R. Neth. Chem. Soc.,* **103,** 288–95.

Tkatchenko, I.B.M., Ballivet-Tkatchenko, D.A., Murr, N.E., Tanji, J., and Payne, J.D. (1984). Fr. Pat. 2542764. *Chem. Abs.,* **102,** 069341.

Tomohiro, T., Uoto, K., and Okuno, H. (1990). *J. Chem. Soc., Chem. Commun.,* 194–5.

Torii, S., Tanaka, H., Hamatani, T., Morisaki, K., Jutand, A., Pfluger, F. *et al.* (1986). *Chem. Lett.,* 169–72.

Tyssee, D.A., Wagenknecht, J.H., Baizer, M.M., and Chruma, J.L. (1972). *Tetrahedron Lett.,* 4809–12.

Udupa, K.S., Subramanian, G.S., and Udupa, H.V.K. (1971). *Electrochim. Acta.,* **16,** 1593–8.

Van Rysselberghe, P. (1946). *J. Am. Chem. Soc.,* **68,** 2047–9.

Van Rysselberghe, P., and Alkyre, G.J. (1944). *J. Am. Chem. Soc.,* **66,** 1801.

Van Rysselberghe, P., Alkyre, G.J., and McGee, J.M. (1946). *J. Am. Chem. Soc.,* **68,** 2050–5.

Vassiliev, Y. B., Bagotskii, V.S., Osetrova, N.V., Khazova, O.A., and Mayorova, N.A. (1985). *J. Electroanal. Chem. Interfacial Electrochem.,* **189,** 271–94.

Vlcek, A.A. (1953). *Nature,* 861–2.

von Stackelberg, M. and Stracke, W. (1949). *Z. Electrochem.,* **53,** 118–25.

Wagenknecht, J.H. (1974). *J. Electroanal. Chem.,* **52,** 489–92.

Wagenknecht, J.H. (1986*a*). U.S. Pat. 4601797. *Chem. Abs.,* **106,** 024982.

Wagenknecht, J.H. (1986*b*). U.S. Pat. 4582577. *Chem. Abs.,* **106,** 050861.

Watanabe, M., Shibata, M., Katoh, A., Sakata, T., and Azuma, M. (1991). *J. Electroanal. Chem. Interfacial Electrochem.,* **305,** 319–28.

Wawzonek, S. and Gundersen, A. (1960). *J. Electrochem. Soc.,* **107,** 537–40.

Wawzonek, S. and Shradel, J.M. (1979). *J. Electrochem. Soc.,* **126,** 401–3.

Wolf, F. and Rollin, J. (1977). *Z. Chem.,* **17,** 337–8.

Ziessel, R. (1987). In *Carbon dioxide as a source of carbon: biochemical chemical uses,* NATO ASI Series, (Ser.C), **206,** 113–38. Kluwer, Dordrecht, The Netherlands.

Note added in proof: in between the submission of this manuscript and its printing, a book on electrochemistry of carbon dioxide has been published:

Sullivan, B.P., Krist, K., and Guard, H.E. (eds) (1993) *Electrochemical and electrocatalytic reactions of carbon dioxide.* Elsevier, Amsterdam, The Netherlands.

14

Photochemical electron transfer applied to the reduction of carbon dioxide

Thomas J. Meyer

Introduction

Biological systems which utilize carbon dioxide are well represented in this book. Progress in this area is impressive, our understanding of the mechanism is growing, and there are interesting prospects for tinkering with the photosynthetic apparatus to improve its efficiency. I will take the opposite tack. What happens if we turn our backs on natural photosynthesis and start over again? Do we have sufficient understanding of chemistry and chemical systems to design total artificial systems which use visible light to reduce carbon dioxide?

In thinking about this problem it is useful to consider model systems and reaction schemes. One scheme which captures the essence of molecular artificial photosynthesis is shown in Scheme 14.1. The first step is light absorption which gives an excited state (ES). It is followed by conversion of the excited state energy into the stored chemical energy of the products. This occurs at a reaction interface by the transfer of electrons from water to carbon dioxide.

An artificial system has the advantage of relative simplicity compared to natural photosynthesis. There is no need for building the complex skeletal framework of a plant nor for coexisting with a variety of enzymic functions. Only one thing is required, a relatively complex chemical transformation.

The example shown in Scheme 14.1 is illustrative, not encyclopaedic. There are other reasonable targets. In these schemes the real difficulty comes at the reaction interface where excited state energy is converted into the chemical energy of the products. Difficulties arise because of the requirement to couple the *single* photon events associated with creating excited states with the *multiple*

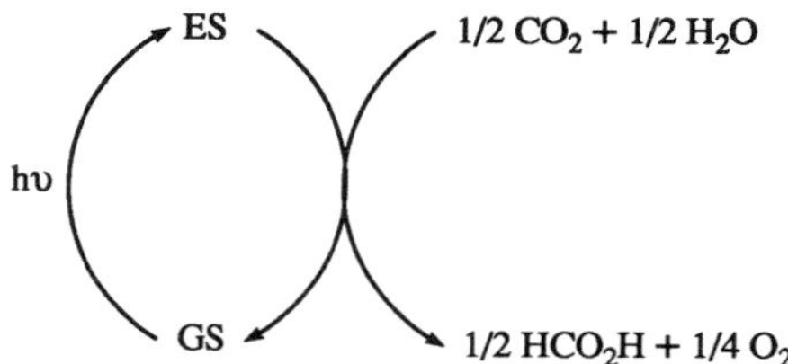

Scheme 14.1 A scheme for artificial photosynthesis.

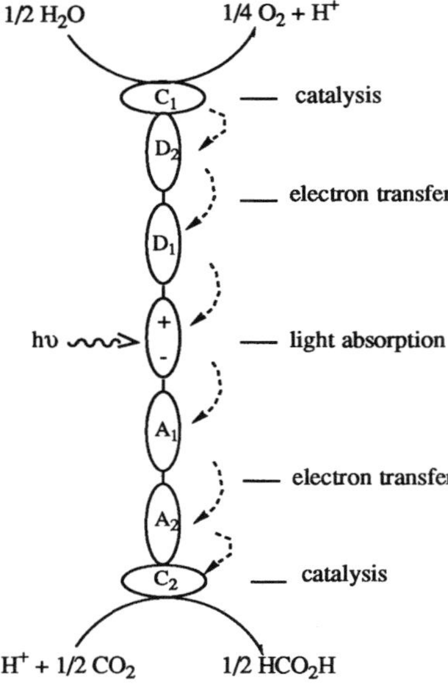

Scheme 14.2 A molecular device for artificial photosynthesis.

electron changes required to reach the products, for example $CO_2 + 2H^+ + 2e^- \rightarrow HCO_2H$.

A possible design for the reaction interface is shown in Scheme 14.2 (Meyer 1989, 1991). It consists of a multicomponent array in which a series of single electron steps add up to give multielectron chemistry. It is driven by the absorption of a photon. It excites an electron to a higher level and creates a molecular electron-hole pair. The energy stored in the resulting excited state provides the driving force for transferring electrons across the array. Electrons are transferred from the excited state to an acceptor or from a donor by taking advantage of free energy gradients so that $\Delta G^0 < 0$ for each electron transfer step. The donor and acceptor sites are separated spatially to avoid back electron transfer. The net effect is to create an electron transfer chain for delivering electrons (as reducing equivalents) to a catalyst where carbon dioxide is reduced and electron holes (as oxidative equivalents) to a second catalyst where water is oxidized.

The array in Scheme 14.2 provides a means for transferring electrons from water to carbon dioxide after photoexcitation. It consists of several interconnected parts each of which contributes a function that is well understood and well documented in the chemical literature. The problem is to assemble the parts in an appropriate way.

This chapter will survey the progress that has been made in designing such arrays by first introducing components and then possible architectures. The emphasis will be on carbon dioxide reduction. For this account, it will be sufficient to note that progress has been made on water oxidation as well, and although difficulties exist, there is promise for achieving catalytic reduction of CO_2. (Meshitsuk *et al.* 1974; Hiratsuka *et al.* 1977; Takahashi *et al.* 1979; Ishida *et al.* 1984; Kapusta and Hackerman 1984; Lieber and Lewis 1984; Bolinger *et al.* 1983; Andre and Wrighton 1986; Hawecker *et al.* 1986; Sullivan and Meyer 1986; Bruce *et al.* 1988; Eisenberg and Fisher 1990).

Catalytic reduction of carbon dioxide

Before considering possible mechanisms, it is useful to recall the thermodynamics for CO_2 reduction and H_2O oxidation which are shown in Table 14.1. Both reactions require multiple electron transfers to reach stable products and therefore more than one photon if they are to be made by the mechanism in Scheme 14.1. The two-electron reduction of carbon dioxide to formic acid requires a moderate reducing agent (E^0<~0.61 V), but if the mechanism is to involve one-electron transfer and $CO_2^{\bullet-}$, a very powerful reducing agent is required (E^0<~2.0 V) in the first step. This is beyond the reach of any practical photochemical assembly which uses visible light. Rather, catalysts must be found which can be reduced by single electron transfer at potentials as near to –0.82 V as possible and which can reduce CO_2 in complex, multiple electron pathways.

The two reactions, electron transfer and CO_2 reduction, can be studied separately by a technique called electrocatalysis. The principles behind this technique are illustrated in Fig. 14.1.

With some notable exceptions, most electrodes can transfer electrons at the electrode–solution interface but are not capable of more complex pathways. In the first panel in Fig. 14.1 the direct reduction of carbon dioxide is illustrated. It necessarily requires an applied potential far more negative than –0.82 V since $CO_2^{\bullet-}$ is formed as an intermediate.

Electrocatalysis is illustrated in the second panel. Here the catalyst is reduced by simple electron transfer at a potential as close to –0.82 V as possible and utilizes the stored redox equivalents in the catalyst to reduce CO_2. Such molecules are immediate candidates for the role of the catalyst in Scheme 14.2 for CO_2 reduction. In this case electrons would come from the excited state through the chemical bridge rather than from the electrode.

Table 14.1 Reduction potentials

Reaction	E^0V*
$2H_2O \longrightarrow O_2 + 4H^+ + 4e^-$	–0.82
$H_2O \longrightarrow OH + OH^- + e^-$	–2.0
$CO_2 + 2H^+ + 2e^- \longrightarrow HCO_2H$	–0.61
$CO_2 + e^- \longrightarrow CO_2^-$	–1.90

* Driving force relative to the normal hydrogen electrode, NHE, at pH = 7.

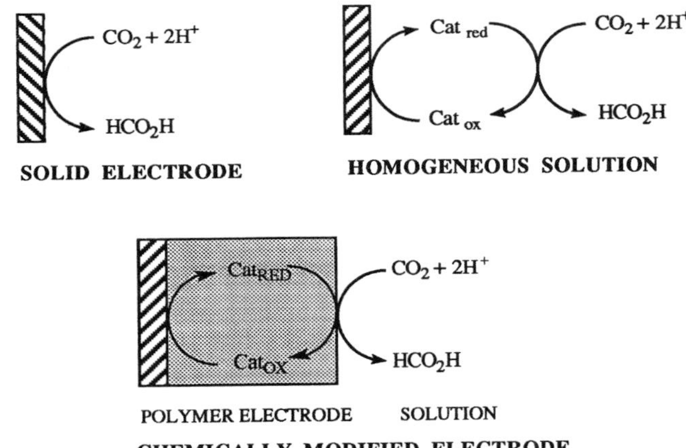

Fig. 14.1 Electrocatalysis.

A second approach to electrocatalysis is illustrated in the third panel in Fig. 14.1. Here the catalyst is bound at the electrode–solution interface by incorporation into a polymeric film. This configuration opens new possibilities for the design of hybrid electrode-molecular assemblies for the photochemical reduction of CO_2.

Several transition metal complexes have been found to act as electrocatalysts for CO_2 reduction. (Meshitsuk *et al.* 1974; Hiratsuka *et al.* 1977; Takahashi *et al.* 1979; Ishida *et al.* 1984; Kapusta and Hackerman 1984; Lieber and Lewis 1984; Bolinger *et al.* 1983; Andre and Wrighton 1986; Hawecker *et al.* 1986;

Table 14.2 Representative catalysts for the electrocatalysed reduction of CO_2*

Catalyst†	Medium	E^0 (V versus NHE)‡	Products
$Rh(bpy)_2^{2\dagger}$	CH_3CN/H_2O	−1.36	HCO_2H
Co(pc)	H_2O	−0.95	CO
$Ni(cyclam)^{2\dagger}$	H_2O	−1.05	CO
poly-[$Re(bpy)(CO)_3Cl$]	CH_3CN	−1.26	CO, oxalate

* Meshitsuka *et al.* 1974; Hiratsuka *et al.* 1977; Takahashi *et al.* 1979; Ishida *et al.* 1984; Kapusta and Hackerman 1984; Lieker and Lewis 1984; Bolinger *et al.* 1985, Davensbaurg *et al.* 1985*a,b*; Keene *et al.* 1985; Andre and Wrighton 1986; Hawecker *et al.* 1986; Sullivan *et al.* 1986; Behr 1988; Bruce *et al.* 1988; Sullivan *et al.* 1988; Alvarez *et al.* 1989; Hurrell *et al.* 1989; O'Toole *et al.* 1989; Tanaka *et al.* 1989; Eisenberg and Fisher 1990; Schmidt *et al.* 1990; Amatore and Jutand 1991; Pugh *et al.* 1991.

† Pc is phthalocyanine dianion; bpy is 2,2′-bipyridine; cyclam is 1,4,8,11′-tetrazacyclotetradecane; poly refers to a polymeric film on a Pt electrode.

‡ *E* is the applied potential in V versus the normal hydrogen electrode.

Sullivan and Meyer 1986; Bruce *et al.* 1988; Eisenberg and Fisher 1990). Representative examples are listed in Table 14.2. Their reactivity toward CO_2 arises from two common features: (1) an ability to undergo rapid electron transfer with an electrode by using orbitals at the metal or its ligands, and (2) possession of an orbital basis for bonding to CO_2 or for its incorporation into a metal hydride bond. The last entry in Table 14.2 gives an example of a catalyst in a polymeric film.

Detailed mechanistic information about how CO_2 is reduced is available in some cases. An example is shown in Scheme 14.3 where *cis*-$[Os^{II}(bpy)_2(CO)H]^+$(bpy is 2,2′-bipyridine) in acetonitrile undergoes sequential one-electron reductions at the bpy ligands. The reduced complex binds and then reduces CO_2 to give formate or CO depending on the water content of the sol-

CO Pathway

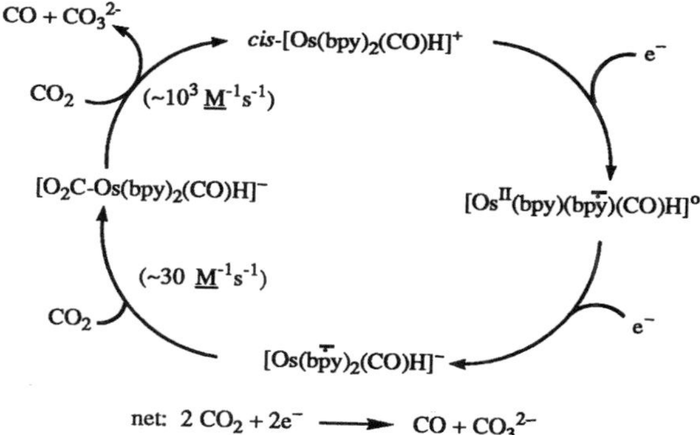

HCO_2^- Pathway

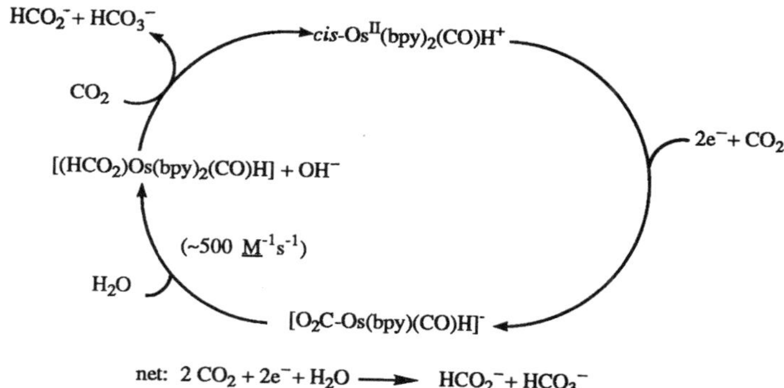

Scheme 14.3 Reduction of CO_2 by *cis*-$[Os(bpy)_2(CO)H]^+$.

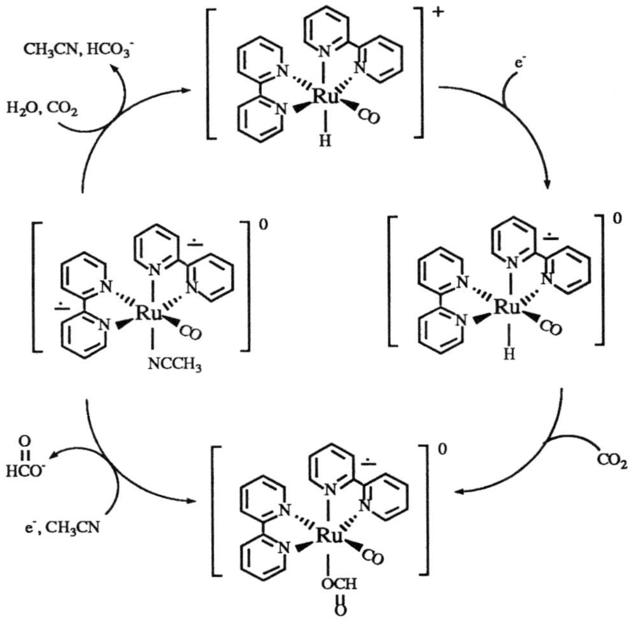

Scheme 14.4 Reduction of CO_2 by *cis*-$[Ru(bpy)_2(CO)H]^+$.

vent. This mechanism was established by a combination of kinetic, isotopic labelling, and product yield studies (Bruce *et al.* 1988).

The analogous complex of Ru^{II} takes a different route to CO_2 reduction. As shown by the mechanism in Scheme 14.4, one-electron reduction is followed by insertion into the Ru–H bond. This is followed by further one-electron reduction, labilization of formate, and finally, re-reduction to return to the initial hydride (Pugh *et al.* 1991).

Considerable progress has been made in finding complexes which act as electrocatalysts for CO_2 reduction, holding out the promise that viable candidates for this role in molecular assemblies can be found, and that the catalyst can be incorporated into the assemblies by chemical synthesis. There are difficulties to overcome in developing practical systems. The catalysts that have been investigated tend to have limited stabilities, it is difficult to control product selectivities, and there are often competing reactions with water (to give hydrogen) or with oxygen.

Photochemical electron transfer

Progress has also been made in designing the photochemical apparatus in Scheme 14.2 (Balzani and Scandola 1991). The most popular molecules for

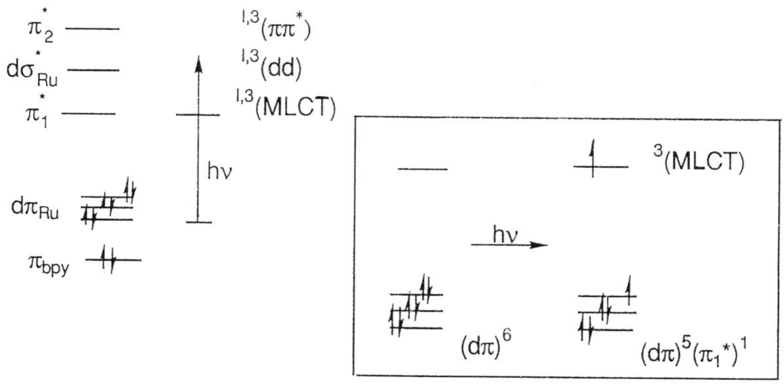

Scheme 14.5 MLCT excited states.

these studies have been porphyrin derivatives and derivatives of polypyridyl complexes of d^6 transition metal ions such as Ru^{II}, Os^{II}, or Re^{I} (Schanze *et al.* 1986; Hofstra *et al.* 1988; Schmidt *et al.* 1988; Gust *et al.* 1989, 1991; Wasielewski *et al.* 1990; Balzani and Scandola 1991; Chen *et al.* 1991; Meyer 1989, 1991).

The electronic structures of the polypyridyl complexes are illustrated in Scheme 14.5. The light absorptivity arises from metal-to-ligand charge transfer (MLCT) transitions in which an electron is promoted from a $d\pi$ orbital at the

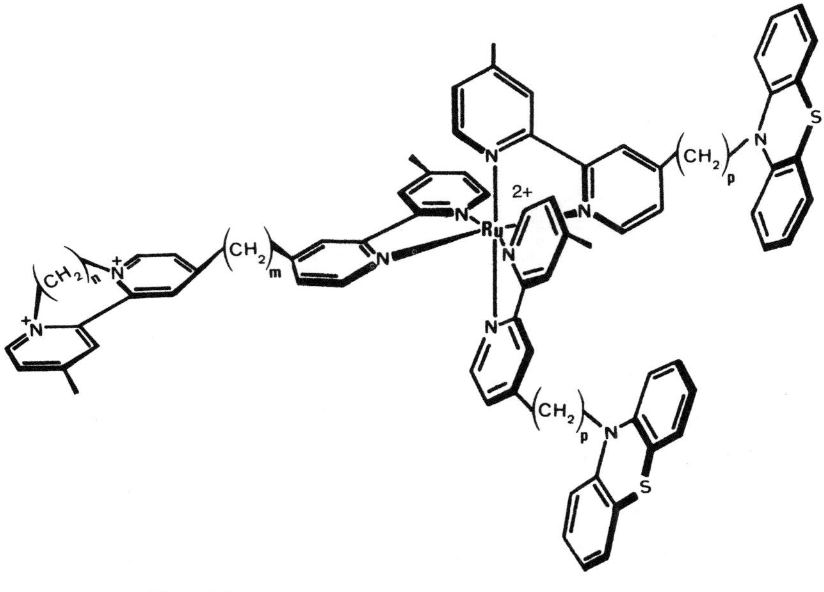

Fig. 14.2 A chromophore-donor-acceptor assembly.

metal to a π^* acceptor level on the ligand. The resulting excited states have been shown to undergo rapid electron transfer to appropriate acceptors or from appropriate donors.

The electron transfer chemistry has been extended to 'chromophore-quencher' complexes where electron donors and/or acceptors are covalently attached to the ligands of the polypyridyl complexes (Schanze *et al.* 1986; Chen *et al.* 1991; Meyer 1989, 1991). An example is shown in Fig. 14.2 where following Ru → bpy excitation, electron transfer occurs to the attached bipyridinium ion and from phenothiazine (Danielson *et al.* 1987). These reactions are conveniently followed by observing the absorption charges that occur after laser flash photolysis.

The changes in electron distribution cause appreciable colour changes which reveal what species have formed, and when time resolved, the kinetics of formation and decay.

These experiments demonstrate the feasibility of constructing supramolecular assemblies which incorporate the initial excitation/electron transfer sequence in Scheme 14.2. There are other approaches to preparing such assemblies which have special advantages of their own. One is the use of amide coupling to assemble the required components. An example of an amino acid derivatized in this way is shown in Scheme 14.6. This molecule was prepared by sequential amide couplings starting with the Boc-protected amino acid L-lysine. Following laser flash excitation in CH_3CN, a series of electron transfers occurs to reach the final 'redox-separated' state shown in the scheme. This state is reached with an efficiency of ~30 per cent and lasts for 146 ns in acetonitrile at room temperature before back electron transfer returns it to the ground state (Mecklenberg *et al.* 1991).

The importance of the lysine derivative and amide coupling is that they open the possibility of using the Merrifield technique for solid-phase peptide synthesis

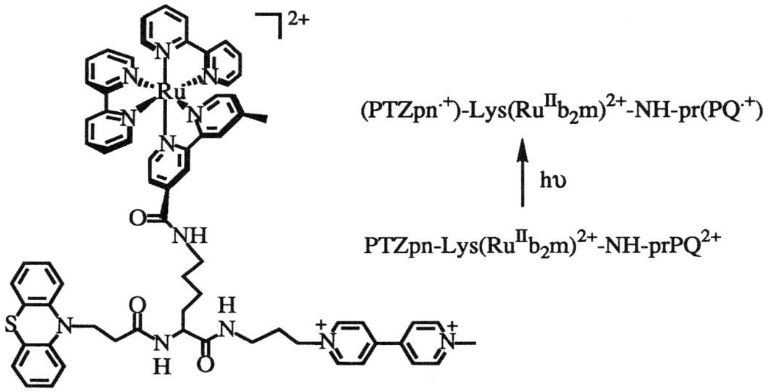

$(PTZpn^{\cdot+})\text{-}Lys(Ru^{II}b_2m)^{2+}\text{-}NH\text{-}pr(PQ^{\cdot+})$

↑ hυ

$PTZpn\text{-}Lys(Ru^{II}b_2m)^{2+}\text{-}NH\text{-}prPQ^{2+}$

$(PTZpn\text{-}Lys(Ru^{II}b_2m)^{2+}\text{-}NH\text{-}prPQ^{2+})$

Scheme 14.6 An amino acid-based molecular assembly for photo-induced electron transfer.

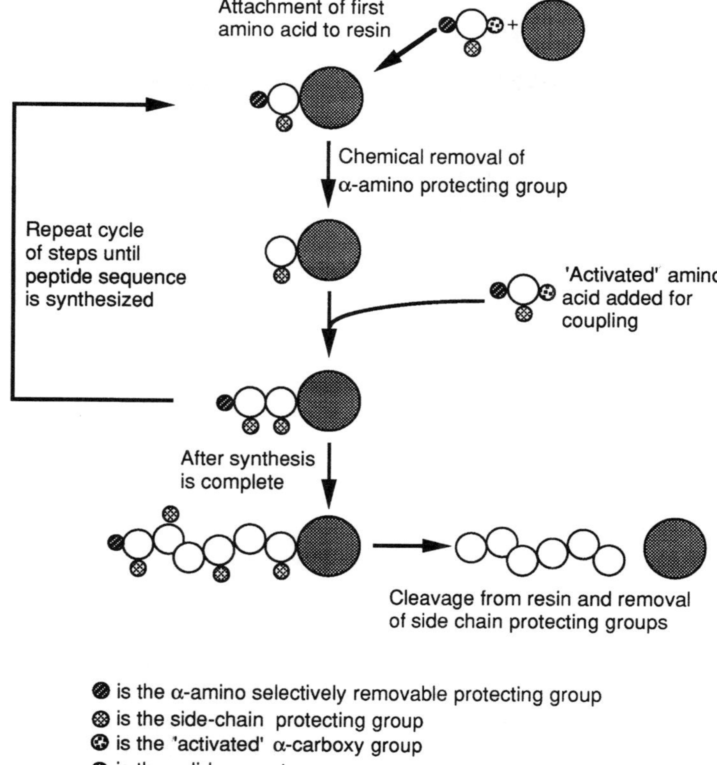
Attachment of first amino acid to resin
Chemical removal of α-amino protecting group
Repeat cycle of steps until peptide sequence is synthesized
'Activated' amino acid added for coupling
After synthesis is complete
Cleavage from resin and removal of side chain protecting groups
is the α-amino selectively removable protecting group
is the side-chain protecting group
is the 'activated' α-carboxy group
is the solid support

Scheme 14.7 Solid state synthesis of peptides by the Merrifield technique.

for preparing molecular assemblies (Erickson and Merrifield 1976). The principle behind the Merrifield technique is outlined in Scheme 14.7. A protected amino acid is attached to a support, the protecting group removed, and a dipeptide formed by amide coupling. By repeated cycles it is possible to build up assemblies in a stepwise manner and, therefore, to control both composition and spatial order.

We have prepared a number of lysine derivatives containing light absorbers and quenchers and used them to prepare lysine-based assemblies. A generalized example, shown in Fig. 14.3, contains electron transfer donors and acceptors and a derivatized polypyridyl complex. This chemistry is being extended to assemblies that directly mimic the model in Scheme 14.2.

In the future, this work will be extended to assemblies which contain both a photochemical electron transfer array and an appropriate catalyst for CO_2 reduction which will function as shown in Scheme 14.8. Sequential excitation-electron transfer steps will be used to deliver two reductive equivalents to the

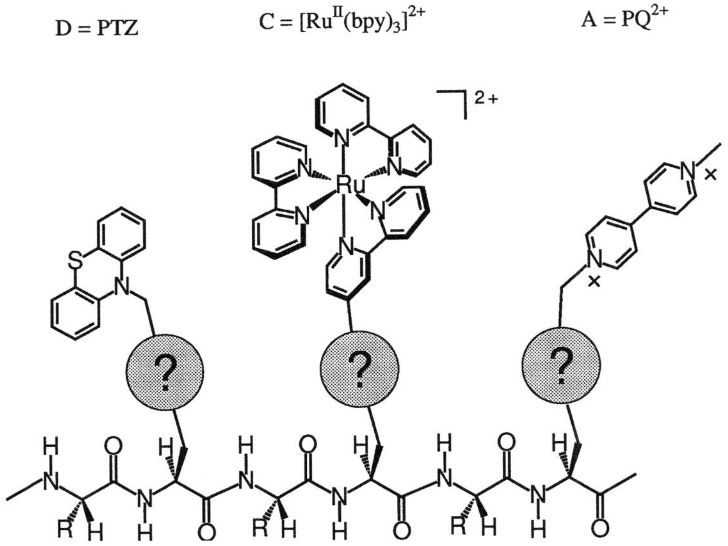

D = PTZ $C = [Ru^{II}(bpy)_3]^{2+}$ $A = PQ^{2+}$

Fig. 14.3 Designing a light-harvesting peptide.

catalyst where CO_2 is to be reduced. Although not a complete cycle, this would be an important step in showing that molecular systems can be used to drive chemical reactions by coupling excitation and electron transfer with catalysis.

We have taken another approach to preparing molecular assemblies based on soluble polymers. These are derivatives of a 1:1 copolymer of styrene/*p*-chloromethyl styrene prepared by copolymerization of the monomers under free radical conditions (AIBN) (Arshady *et al.* 1984). The derivatization chemistry is

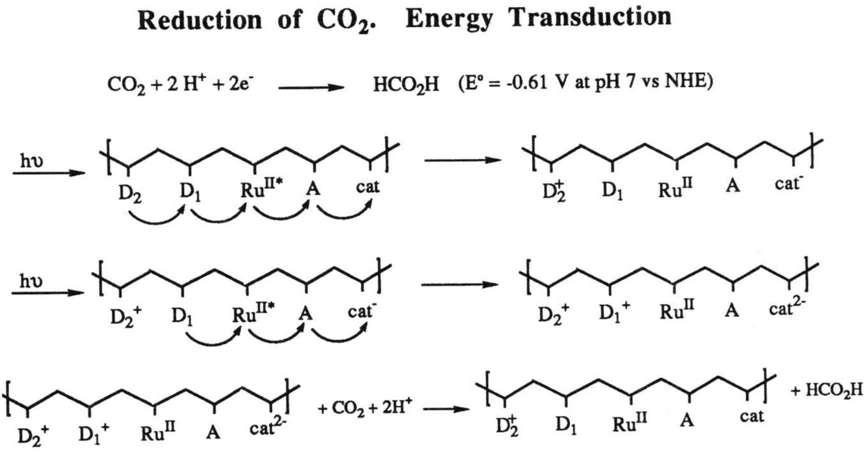

Reduction of CO_2. Energy Transduction

$$CO_2 + 2\,H^+ + 2e^- \longrightarrow HCO_2H \quad (E^\circ = -0.61\ \text{V at pH 7 vs NHE})$$

$$\xrightarrow{h\nu} [D_2\ D_1\ Ru^{II*}\ A\ cat] \longrightarrow [D_2^+\ D_1\ Ru^{II}\ A\ cat^-]$$

$$\xrightarrow{h\nu} [D_2^+\ D_1\ Ru^{II*}\ A\ cat^-] \longrightarrow [D_2^+\ D_1^+\ Ru^{II}\ A\ cat^{2-}]$$

$$[D_2^+\ D_1^+\ Ru^{II}\ A\ cat^{2-}] + CO_2 + 2H^+ \longrightarrow [D_2^+\ D_1\ Ru^{II}\ A\ cat] + HCO_2H$$

Scheme 14.8 Photo-induced reduction of CO_2.

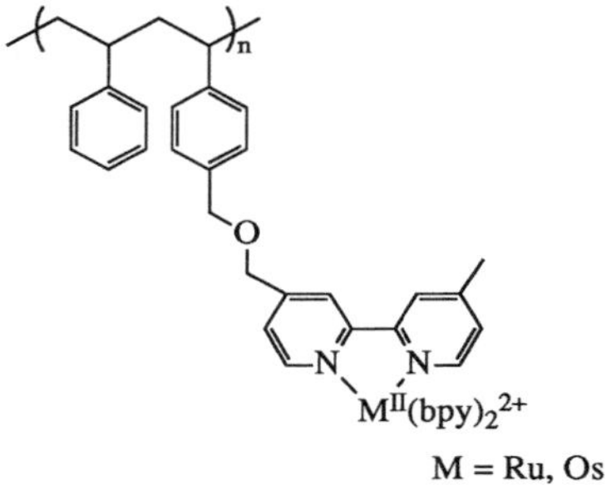

$M^{II}(bpy)_2^{2+}$

M = Ru, Os

Fig. 14.4 A chromophore-attached polymeric unit.

based on nucleophilic displacement at the C–Cl bond, and it has proven useful for attaching chromophores and quenchers by ester, ether or amine links. For example, under basic conditions the alcoholic groups in the complexes $[M(bpy)_2(bpyCH_2OH]^{2+}$ (M = Ru, Os) exist as alkoxides and undergo quantitative reactions with the polymer to form ether links. With an excess of complex, total derivatization occurs to give polymers whose repeat units are illustrated in Fig. 14.4 (Olmsted *et al.* 1987; Worl *et al.* 1990).

The diameter of the complex (~14 Å) is considerably greater than the repeat distance along the polymeric backbone (~5–6 Å). Because of the large excluded volume of the complexes, the polymers can not coil and are constrained to form extended structures in solution. From the results of a molecular modelling study, which included the dicationic charges on the complexes, structures of minimum energy exist at internuclear separations between complexes of ~21 Å which translates into a separation distance between their peripheries of ~7 Å (Danielson *et al.*, unpublished). From the results of photophysical studies on these polymers, it has been possible to identify multiphoton effects, to measure rate constants for intramolecular electron and energy transfer (Danielson *et al.*, unpublished), and to demonstrate that sequential photoexcitation and quenching can lead to the buildup of multiple redox equivalents on individual polymeric strands (Worl *et al.* 1990).

One disadvantage of soluble polymers is that, although composition can be varied systematically, there is no control over the spatial distribution of the components. Nonetheless, a number of interesting possibilities exist in the context of the multicomponent demands of the model in Scheme 14.2. For example, we have been able to show that during a single, intense laser pulse in the presence of the electron donor phenothiazine (PTZ), as many as 8–9 reductive equivalents can be transferred to an average, individual polymeric strand, Scheme 14.9 (Jones, unpublished).

The next step in this work in the context of carbon dioxide reduction will be to prepare mixed polymers that contain both Ru^{II} or Os^{II} chromophores and a molecular catalyst for the reduction of CO_2. This would also represent only part

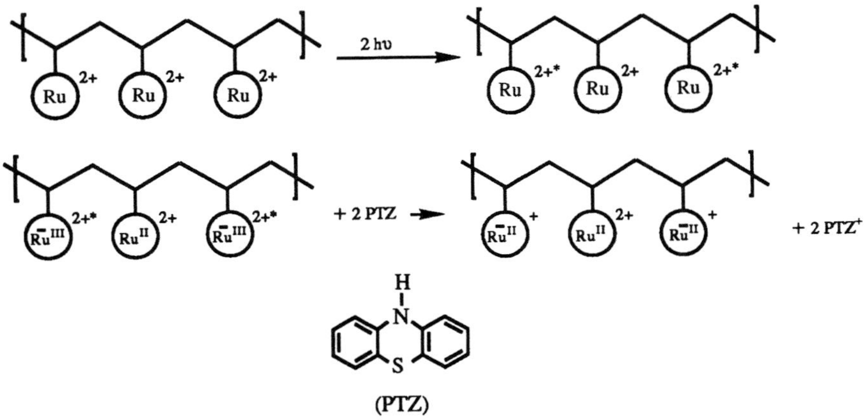

Scheme 14.9 Multi-excitation, multi-electron transfer in a polymeric assembly.

of the overall strategy implied by Scheme 14.2 but it would be an important demonstration of the possibility of combining multiple functions in a single assembly based on a soluble polymer.

Some final comments

The chemistry described here has come from a number of areas of fundamental science ranging from molecular physics to organic synthesis. It is, in the end, a problem in chemical synthesis and will require the continued development of synthetic procedures for the preparation of complex structures. It is for this reason that recent progress made with peptides and soluble polymers provide especially promising leads for future studies. Neither the products nor the model in Scheme 14.2 are unique. There are a number of alternatives both for the products and for the assembly in the latter case, based on hybrid chemical–electrode arrays. All of this work is of relatively recent vintage, most of it from the past 15 years. At this point it is only a study in fundamental science. The prospects for continuing advances are high and it seems clear that interesting developments will continue to appear in the future.

Acknowledgements

The author would like to acknowledge the support of the Gas Research Institute (Grant No. 5087-260-1455, the Office of Naval Research (Grant No. N00014-87-K-0430), the Department of Energy (Grant No. DE-FG05-86ER13633), and the National Science Foundation (Grant No. CHE-90222493) for support of the research described in this account.

References

Alvarez, R., Carmona, E., Galindo, A., Gutiérrez, E., Marin, J., Monge, A., *et al.* (1989). *Organometallics*, **8**, 2430.

Amatore, C. and Jutand, A. (1991). *J. Am. Chem. Soc.*, **113**, 2819.

Andre, J.-F. and Wrighton, M.S. (1986). *Inorg. Chem.*, **34**, 67.

Arshady, R., Reddy, B.S.R., and George, M.H. (1984). *Polymer*, **25**, 716.

Balzani, V. and Scandola, F. (1991). *Supramolecular photochemistry* (ed. E. Harwood) New York.

Behr, A. (1988). *Carbon dioxide activation by metal complexes*. VCH.

Bolinger, C.M., Sullivan, B.P., Conrad, D., Gilbert, J.A., Story, N., and Meyer, T.J. (1985). *J. Chem. Soc., Chem. Commun.*, 796.

Bruce, R.M.M., Megehee, E., Sullivan, B.P., Thorp, H., O'Toole, T.R., Downard, A., and Meyer, T.J. (1988). *Organometallics*, **7**, 238.

Chen, P., Duesing, R., Graff, D.K., and Meyer, T.J. (1991). *J. Phys. Chem.*, **95**, 5850.

Danielson, E., Elliott, C.M., Merkert, and J.W., Meyer, T.J. (1987). *J. Am. Chem. Soc.*, **109**, 2519.

Darensbourg, D.J., Hanckel, R.K., Bauch, C.G., Pala, M., Simmons, D., and White, J.N. (1985). *Coord. Chem. Rev.*, **107**, 7463.

Eisenberg, R. and Fisher, B. (1990). *J. Am. Chem. Soc.*,**102**, 7363.

Erickson, B.W. and Merrifield, R.B. (1976). *Proteins*, **2**, 255.

Gust, D., Moore, T.A., Moore, A.L., Lee, S.-J., Bittersmann, E., Luttrull, D.K., *et al.* (1989). *Science*, **244**, 35.

Gust, D., Moore, T.A., Moore, A.L., *et al.* (1991). *J. Am. Chem. Soc.*, **113**, 3638.

Hawecker, J., Lehn, J.M., and Ziessel, R. (1986). *Helv. Chem. Acta*, **69**, 1990.

Hiratsuka, K., Takahashi, K., Sasaki, and H., Toshima, S. (1977). *Chem. Lett.*, 1137.

Hofstra, U., Schaafsma, T.J., Sanders, G.M., Van dijk, M., Van Der Plas, Johnson, D.G., *et al.* (1988). *Chem. Phys. Lett.*, **151**, 169.

Hurrell, H.C., Mogstad, A.-L., Usifer, D.A., Potts, K.T., and Abruña, H.D. (1989). *Inorg. Chem.*, **28**, 1080.

Ishida, H., Tanaka, K., and Tanaka, T. (1984). *Chem. Lett.*, 405.

Jones, Jr, W.E. (1993). Unpublished results.

Kapusta, S. and Hackerman, N.J. (1984). *Electrochem. Soc.*, **131**, 1511.

Keene, F.R., Creutz, C., and Sutin, N. (1985). *Coord. Chem. Rev.*, **64**, 247.

Lieber, C.M. and Lewis, N.S. (1984). *J. Am. Chem. Soc.*,**106**, 5033.

Marcus, R.A. (1965). *J. Chem. Phys.*, **454**, 1261.

Mecklenburg, S.L., Peek, B.M., Erickson, B.W., and Meyer, T.J. (1991). *J. Am. Chem. Soc.*, **113**, 8540.

Meshitsuka, S., Ichikawa, M., and Tamaru, K. (1974). *J. Chem. Soc., Chem. Commun.*, 158.

Meyer, T.J. (1989). *Accounts of Chemical Research*, **22**, 163.

Meyer, T.J. (1991). In *Photochemical processes in organized molecular systems* (ed. K. Honda), pp. 133. Elsevier, Yokohama, Japan. Weinheim, Germany; see also references therein.

O'Toole, T.R., Sullivan, B.P., Bruce, M.R.M., Margerum, L.D., Murray, R.W., and Meyer, T.J. (1989). *J. Electroanal. Chem.*, **259**, 217.

Olmsted III, J., McClanahan, S.F., Danielson, E., Younathan, J.N., and Meyer, T.J. (1987). *J. Am. Chem. Soc.*, **109**, 3297.

Pugh, J.R., Bruce, M.R.M., Sullivan, B.P., and Meyer, T.J., (1991). *Inorg. Chem.,* **30,** 86.

Sakaki, S. (1990). *J. Am. Chem. Soc.*, **112**, 7813.

Schanze, K.S., Neyhart, G.A., and Meyer, T.J. (1986). *J. Phys. Chem.*, **90**, 2182.

Schmidt, J.A., McIntosh, A.R., Weedon, A.C., Bolton, J.R., Connolly, J.S., Hurley, J.K., *et al.* (1988). *J. Am. Chem. Soc.*, **110**, 1733.

Schmidt, M.H., Miskelly, G.M., and Lewis, N.S. (1990). *J. Am. Chem. Soc.*, **112**, 3420.

Sullivan, B.P. and Meyer, T.J. (1986). *Organometallics*, **5**, 1500.

Sullivan, B.P., Bruce, M.R.M., O'Toole, T.R., Bolinger, C.M., Megehee, E., Thorp, H. *et al.* (1988). *Adv. Chem. Ser.*, **363**, 52; see also references therein.

Sullivan, B.P., Meyer, T.J., Stershic, M.T., and Keefer, L.K. (1991). In *Relevance to human cancer of N-nitroso compounds, tobacco smoke and mycotoxins* (ed. I.K. O'Neill, J. Chen., H. Bartsch). International Agency for Research on Cancer, Lyon.

Takahashi, K., Hiratsuka, K., Sasaki, H., and Toshima, S. (1979). *Chem. Lett.*, 305.

Tanaka, K., Wakita, R., and Tanaka, T. (1989). *J. Am. Chem. Soc.*, **111**, 2428.

Wasielewski, M.R., Gaines III, G.L., O'Neil, M.P., Svec, W.A., and Niemaczyk, M.P. (1990). *J. Am. Chem. Soc.*, **112**, 4559.

Worl, L.A., Strouse, G.F., Younathan, J.N., Baxter, S.M., and Meyer, T.J. (1990). *J. Am. Chem. Soc.*, **112**, 7571.

15

Room-temperature catalytic and photocatalytic fixation of carbon dioxide

Michael Grätzel

Introduction

At the start of this chapter we recall the salient features of natural photosynthesis which serves as a model in the development of artificial systems that accomplish the fixation of carbon dioxide. Nature has built a fascinating device to make use of sunlight in order to drive a thermodynamically uphill reaction to generate carbon-containing compounds. The reaction is the reduction of carbon dioxide to carbohydrates by water. In green plants, algae, and cyanobacteria, the overall photosynthesis reaction is

$$CO_2 + H_2O \rightarrow \tfrac{1}{6}\, C_6H_{12}O_6 + O_2 \qquad (1)$$

The Gibbs free energy change, associated with this reaction is +125 kcal mol^{-1} for the conditions under which most photosynthesis occur. The amount of energy trapped and stored by photosynthesis is enormous. More than 10^{17} kcal of free energy from the sun is harvested annually by plants. This is equivalent to the continuous generation of 13 000 GW of electrical power. This process is associated with the assimilation of more than 10^{10} tons of carbon dioxide into carbohydrates.

Photosynthetic energy conversion comprises two parts, as shown in Fig. 15.1. The first, photophosphorylation, involves the two-electron reduction of nicotinamide adenine dinucleotide phosphate ($NADP^+$) by water to produce NADPH and oxygen. This redox reaction is coupled to the generation of adenosine triphosphate (ATP) from adenosine diphosphate (ADP)

$$2H_2O + 2NADP^+ + 3ADP + 3P \rightarrow 2NADPH + H^+ + 3ATP + O_2 \qquad (2)$$

where P stands for phosphate anions. This sunlight-driven light reaction takes place in the thylakoid membranes located in the interior part of the chloroplasts of plant cells. The photosynthetic unit assembled in these membranes is composed of antenna pigments for light energy harvesting (that is, they absorb the light energy required), and a reaction centre consisting of two photosystems. The absorption of light causes electrons to be ejected from chlorophyll pigments and then passed between various electron-transferring components of the photosystems. The judicious spatial arrangement of these components allows the electron transfer to proceed in a vectorial fashion

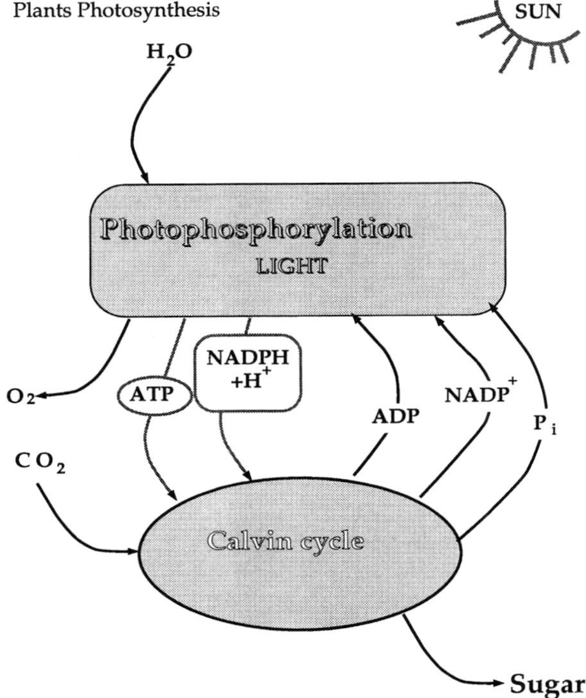

Fig. 15.1 The light reaction and dark fixation of carbon dioxide in natural photosynthesis.

across the thylakoid membrane. Positive charges are accumulated on the inside of the thylakoid membrane while the negative countercharges are transferred to the outer surface. The resulting electrochemical gradient is used to drive the reduction of $NADP^+$ by water and the phosphorylation of ADP. The overall reaction, despite its complexities in detail corresponds to the simple equation above.

The second part of the process, known as the Calvin cycle, uses the NADPH as well as the free energy stored in the ATP to assimilate carbon dioxide in the form of carbohydrates. The way by which nature achieves this is via the reaction of CO_2 with ribulose biphosphate (RuBP) to give two molecules of 3-phosphoglycerate, a process which is catalysed by RuBP carboxylase. The elucidation of the very complex mode of operation of this enzyme has been one of the focal points of this Symposium. Our colleagues have demonstrated the tremendous progress that has been achieved in the understanding of this key process and deserve to be congratulated for their superb scientific achievements in this important field.

Mimicking photosynthesis

Scientists have succeeded in unravelling in remarkable detail the complex processes underlying photosynthetic light energy conversion. There are still important points left to be elucidated, such as the structure and function of the oxygen-evolving complex. Another challenge in the coming years will be to develop artificial systems that can harvest solar energy to drive the fixation of carbon dioxide. The worldwide quest for clean and renewable energy sources has already stimulated, over the last few years, a significant research effort in this domain, triggered off by the oil crisis in the early 1970s. The sharp increase in the price of petroleum in 1973 together with an increased awareness that fossil reserves would be exhausted, at the current rate of consumption, within less than a century, initiated a large scientific activity to provide for renewable sources of energy. More recently, growing concerns about the so-called 'greenhouse effect' have heightened this research effort. The 'greenhouse effect' is a term used to describe the global warming due to the accumulation of gases in the atmosphere that absorb IR radiation in the 10–20 μm region, the most prominent such gas being carbon dioxide. As a result of the accelerated combustion of fossil fuel reserves, the amount of CO_2 in the atmosphere is currently increasing at an annual rate of several billion tons. This has already increased the CO_2 content of the atmosphere from 280 to 350 ppm resulting in a 0.5 °C increase in the temperature of the earth's atmosphere. Further warming could entail catastrophic climatic consequences. Therefore, it is important to develop artificial routes of CO_2 fixation that will complement the natural assimilation process.

Our artificial systems should not attempt blindly to imitate all the intricacies of nature's photosynthetic apparatus. It is not our goal to build a very complex device that would generate sugars from carbon dioxide and water, since complex carbohydrates are not the most ideal fuels we could create. The photosynthetic conversion of carbon dioxide and water into simpler fuels, such as methane or methanol, is a much more attractive concept. These components contain only one carbon atom per molecule and are distinguished by a high chemical potential. Moreover, they are widely used in present-day technology and there already exists the infrastructure necessary for large-scale distribution and employment. A system accomplishing the photosynthetis of methane from carbon dioxide and water is presented schematically in Fig. 15.2. In analogy to nature's assimilation there are two cycles operating in series. The light reaction involves the splitting of water into hydrogen and oxygen

$$4H_2O \rightarrow 4H_2 + 2O_2 \quad (3)$$

and this is coupled to a dark process in which carbon dioxide is reduced to methane by combination with the hydrogen (known as the 'Sabatier' reaction)

$$CO_2 + 4H_2 \rightarrow CH_4 + 2H_2O \quad (4)$$

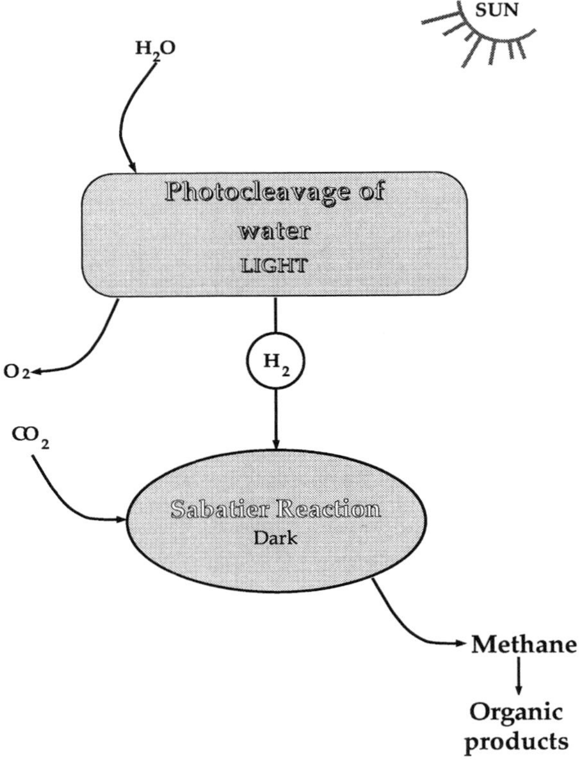

Fig. 15.2 Photocleavage of water and methanation of carbon dioxide in artificial photosynthesis.

The overall reaction is the fixation of carbon dioxide in the form of methane, using water as an electron source

$$CO_2 + 2H_2O \rightarrow CH_4 + 2O_2 \tag{5}$$

Under standard conditions (a pressure of 1 atm and temperature of 298 K) this process is associated with a free energy increase of +198.3 kcal mol^{-1}.

The methane produced is an attractive energy vector. It is safer to handle than hydrogen and could be distributed through pipelines that are already available for the transport of natural gas (which is largely methane). However, care will have to be taken to avoid leakage in the atmosphere since methane itself is a greenhouse gas. In addition, methane provides a feedstock for value-added chemicals. For example, catalysts have recently been developed that allow for the selective dimerization of methane to ethylene and ethane (Tomoayasi and Lundsford 1985). This opens up a route for the permanent fixation of carbon dioxide in the form of stable organic compounds, such as a polymers. It should be noted that if methane is used as a fuel, CO_2 is released again into the atmosphere during its combustion. Thus, carbon is continuously moved back and forth between the atmosphere and the fuel user. The problem with such an approach is that each such cycle generates toxic pollutants as byproducts of methane combustion. Hydrogen burns much more cleanly than organic compounds and is therefore the preferred fuel in the future. Thus, rather than generating organic fuels, the primary task of future artificial photosynthetic systems should be to claim CO_2 from the atmosphere and convert it into useful lasting products. Methane would be only an intermediate in this process. The fact that systems are available now allowing for the surprisingly facile conversion of CO_2 into methane renders this proposition realistic.

The practical implementation of the scheme proposed in Fig. 15.2 requires that a process will become available that allows the generation of hydrogen from sunlight in an economic fashion. The combination of a photovoltaic device with a water electrolyser offers at this time the best perspective. While the cost of conventional silicon-based photocells is still prohibitively high, a new, efficient, and very low cost photovoltaic system based on transparent nanocrystalline oxide semiconductor films has recently been discovered (O'Regan and Grätzel 1991). Water photoelectrolysis with such systems has already been demonstrated opening up the possibility for an economic source of solar hydrogen. In the present chapter hydrogen generation will not be dwelled on further. Rather, attention will be focused on the artificial analogue of the Calvin cycle, i.e. the fixation of carbon dioxide via the methanation reaction (eqn (5)). This is a thermodynamically downhill reaction for which the standard Gibbs free energy is 31.3 kcal mol^{-1}. Because the conversion of CO_2 to methane is an eight-electron process involving high energy intermediates one might expect it to be difficult to achieve under mild conditions. Indeed, previous studies have shown that with a conventional catalyst high temperatures and pressures are usually required to obtain reasonable rates of methane generation (Weatherbee and Bartholomew 1981). However, it was recently discovered in our laboratory that a ruthenium oxide catalyst dispersed onto titanium oxide afforded methane production from mixtures of hydrogen and CO_2 even at room temperature and atmospheric pressure (Thampi *et al.* 1987). This observation will now be discussed in more detail.

Preparing of an ambient-temperature methanation catalyst

The route to hydrocarbon production from carbon dioxide begins with the Sabatier-type process (eqn (4)). The product methane can be used as feedstock for further reactions yielding methanol as a storable liquid fuel or other petrochemicals. Despite its exothermic nature ($\Delta G_{298\,K} = -27$ kcal mol^{-1}), this reaction is difficult to activate, and previously high temperatures and pressures were thought necessary (Weatherbee and Bartholomew 1981). Based on the experience of our laboratory with redox catalysts of highly dispersed substoichiometric ruthenium oxide supported on ceramic semiconductor powders for catalysis and photocatalysis at the liquid–solid interface a similar system was successfully evaluated for the heterogeneous catalysis of the above gas-phase reaction.

The standard substrate powder was a titanium dioxide grade P25 (Degussa, Germany), which consists of approximately 80 per cent anatase phase, the remainder being rutile. It is finely divided, with a BET specific surface of 55 $m^2\ g^{-1}$ and a density of 3.8 g ml^{-1}. The dispersion of the ruthenium oxide on the support was carried out by deposition/precipitation.

A hydrated ruthenium chloride, $RuCl_3.3H_2O$ was dissolved in 0.1 M HCl, to a concentration of 1 mg ml^{-1}. The TiO_2 powder was then added (10 mg ml^{-1}) and the ruthenium species was hydrolytically precipitated on this substrate at 70 °C by adding 0.1 M NaOH as necessary over 5 h (to maintain a pH of 4–4.5. This pH during hydrolysis is a critical controlling parameter for the dispersion of the catalyst on the semiconductor substrate. The suspension was then evaporated to dryness, followed by two calcination steps, at 170 °C and 370 °C, each for 18 h. To remove the sodium chloride formed in the reaction, the product was dialysed and redried. The loading of RuO_2 on TiO_2 consequent on this process, is 5 per cent wt (3.8 per cent Ru), in the form, typically, of 20 Å clusters. Comparison with pure anatase or rutile substrates shows that on the former, high dispersion is not attained, while on the latter a very thin continuous surface film is produced (Rutheranat *et al.* 1990). Immediately prior to use in the methanation reaction, the catalyst was partially reduced at 200 °C for 1 h in a H_2/Ar mixture. The shape of the Ru line in the XPS spectrum establishes that after this conditioning, about 20 per cent of the ruthenium is in a zero oxidation state. Systematic studies have established that this reductive conditioning to a substoichiometric RuO_x gives optimum catalytic properties: both stoichiometric RuO_2, and fully-reduced metallic Ru prepared under H_2 at 500 °C are less effective, even at the same high dispersion on TiO_2.

Room-temperature methanation and photomethanation reactions

Initial investigations of catalysis of reaction (4) took place in a batch-type reactor fabricated in Pyrex glass, of volume 20 ml and charged with 1 ml CO_2, 12 ml H_2, and the remainder Ar at atmospheric pressure. 0.1 g of catalyst was used for each experiment. The gas mixture was sampled at regular intervals and analysed by gas chromatrography, confirmed with mass spectroscopy measurements. Even at ambient temperature (25 °C) a slow production of methane was observed (Fig. 15.3). The selectivity of the reaction for methane was remarkable: rates of CO_2 absorption and CH_4 evolution were equal to within 1 per cent, and no trace of other reaction products such as CO or CH_3OH could be detected, either in the gas samples or by liquid extraction and HPLC from the catalyst at

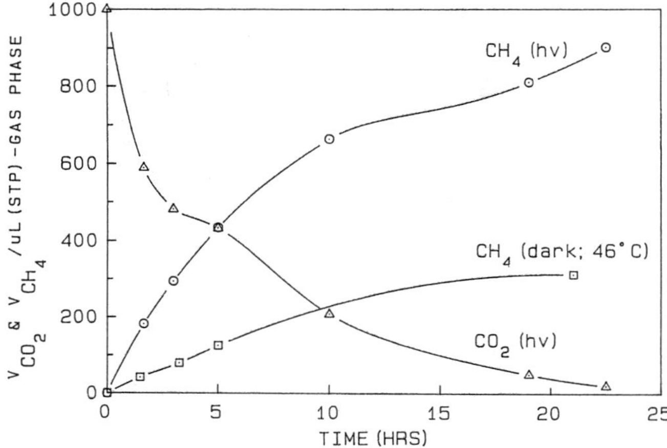

Fig. 15.3 Methanation of CO_2 in a batch-type reactor: under photocatalytic conditions the reaction goes to completion at near-ambient temperature.

the conclusion of the reaction. The reaction rate increased rapidly with temperature, from an initial value of 0.17 μmol h^{-1} at 25 °C to 10.5 μmol h^{-1} at 90 °C, the latter value representing a turnover frequency of 1.6×10^{-4} s^{-1} based on the 50 per cent dispersion of the ruthenium component deduced from H_2-adsorption studies.

The light source for the initial photocatalytic investigations was a Suntest (Hanau, Germany) solar simulator giving an intensity of 80 mW cm^{-2} on the cat-

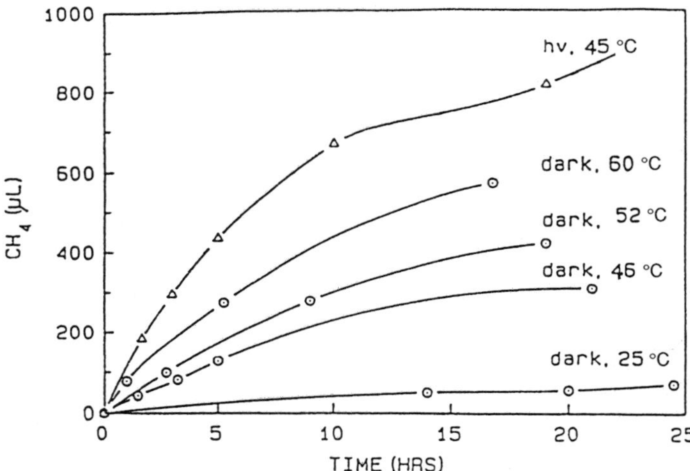

Fig. 15.4 Thermal and photocatalysis of the methanation reaction in a batchtype reactor.

alytic layer in the batch reactor, an intensity comparable to terrestrial sunlight-conditions. When the catalyst surface temperature was 46 °C, an initial methanation rate of five times higher than that in darkness was observed. Of equal importance is the fact that whereas some CO_2 remains after one day of contact of the gas mixture with the catalyst in darkness, under light methanation proceeds practically to completion in the same time (Fig. 15.4).

Catalytic and photocatalytic behaviour of a series of variants of the optimised catalyst were investigated. With RuO_x supported on conventional insulating ceramic substrates, such as alumina or silica no methanation is recorded. On both crystalline phases of titania and on a titanate ($SrTiO_3$), some catalytic effect and its photoenhancement are noted. However, it is clear that the overall efficiency of the catalyst is sensitive to the dispersion attained on the substrate, which is maximum in P25.

Low temperature methanation of CO_2 was also observed with the catalyst being mounted in a fixed-bed flow reactor. Results obtained with 150 mg RuO_x/TiO_2 spread out in a flat thermostated quartz cell are shown in Fig. 15.5. The conversion of CO_2 into methane is again augmented by illumination under

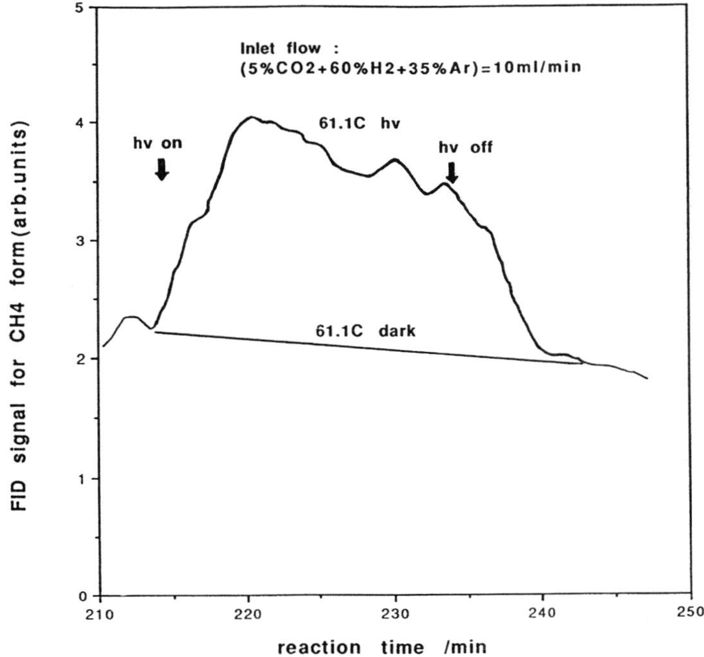

Fig. 15.5 Carbon dioxide conversion to methane in a fixed-bed flow reactor thermostated at 61 °C. Light enhances the catalytic conversion.

these conditions although the increase is smaller than that observed in a batch reactor.

The pathway of the low temperature dark methanation was investigated by *in situ* FTIR. Hydrogenation of CO_2 produces surface-adsorbed formate which subsequently is dehydrated to yield adsorbed CO. The latter exists in three different forms which exhibit greatly different reactivity towards further hydrogenation to produce absorbed polymethylene. This explains the important finding that CO reacts less rapidly with hydrogen at room temperature than CO_2. Apparently the reactive form of surface adsorbed CO is generated by CO_2 hydrogenation on our catalyst while the adsorption of CO from the gas phase produces the less active intermediate. The surface absorbed hydrocarbon chains generated during hydrogenation are very reactive towards hydrogen and oxygen. Thus in the presence of excess hydrogen methane is formed while oxygen reacts to yield C1 oxygenates.

The photoenhancement of methanation was initially suggested to involve band gap excitation of the TiO_2 support which generates electron-hole pairs. These charge carriers were though to participate in interfacial redox reactions facilitating CO_2 fixation. However, a close examination of the action spectrum of the photoeffect showed its onset to be located around 700 nm which is at much lower energy than the 3 eV fundamental absorption edge of TiO_2. On the basis of these results alternative models have recently been proposed and these include the photo-induced detrapping of trapped electrons and the light excitation of the RuO_x catalyst. While RuO_2 is a metallic conductor as a bulk material, it is present here in highly dispersed form, i.e. as 2 nm sized clusters. Such small particles exhibit quantum size effect leading to discrete electronic energy states. Photoexcitation of electronic transitions within the clusters may affect its catalytic activity explaining the observed photoeffect.

Photothermochemistry of methanation

Continuous methanation of CO_2 under concentrated solar irradiation could be a significant process in the scenario of a future solar-hydrogen economy, and it has therefore been modelled experimentally on a laboratory scale using the reactor shown in Fig. 15.7. Machined in stainless steel, the conical section of the reactor allowed the incidence of a convergent beam of high-intensity light from a 150 W xenon lamp with an ellipsoidal mirror (AMKO-LTI, Germany) on the surface of the catalyst bed. A simulated intensity of 300 suns was achieved. With the geometry and dimensions given – the catalyst surface has a diameter of 7 mm – a flow rate of reagent gas of 150 Nml min^{-1} fluidized the catalyst bed. With the reactor in a recycling loop and with a recycling ratio (V_{cycle}/V_{total}) > 7, conditions close to those of an ideally mixed reactor were maintained for kinetics measurements. The loop consisted of a viton membrane pump, a mass-flow meter and a condenser containing Raschig rings cooled to 10 °C. This condenser

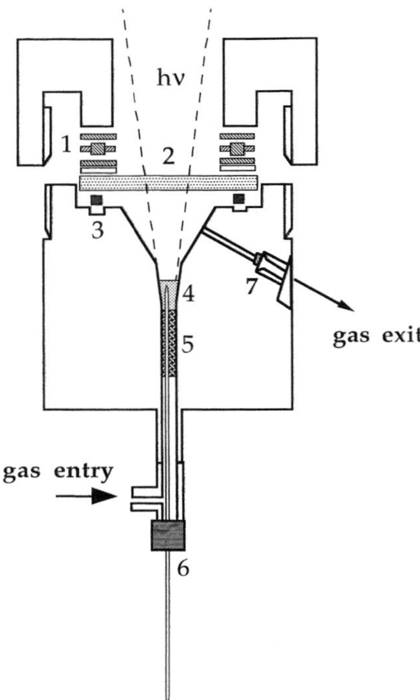

Fig. 15.6 Fluidized bed reactor for photothermochemical investigations. 1: flange. 2: quartz window. 3: gaskets. 4: catalyst bed. 5: glass frit porous plug. 6: thermocouple.

abstracted water from the gas flow in the loop, thereby inhibiting the reverse reaction and extending the life of the catalyst which was adversely affected by excess water vapour. The reactor was fitted in a purpose-built oven, so that the comparison with thermally-activated reactions could be made, and to supplement the thermal effect of the incident radiation when photoassisted measurements were to be carried out at higher temperatures. Product analysis, as previously, was by gas chromatography.

The analysis of the kinetics of the reaction was complicated by the fact that under illumination a temperature gradient exists in the fluidized catalyst bed, so that the rate of the thermally activated reaction is not constant with depth. This temperature profile was measured with a thermocouple and taken into account during data analysis. It is, of course, evident that at any given temperature the photocatalytically generated component cannot be measured separately from the product of thermally activated processes. However, it is assumed that there is a linear superposition of both effects

$$V_{ph+th} = V_{ph} + V_{th}, \qquad \text{(ph = photo, th = thermal)} \quad (5)$$

In practice, therefore, the total rate, $V_{ph}{}^{+}{}_{th}$, was measured. Thereafter, V_{th} was calculated on the basis of isothermal calibrations in darkness and the temperature

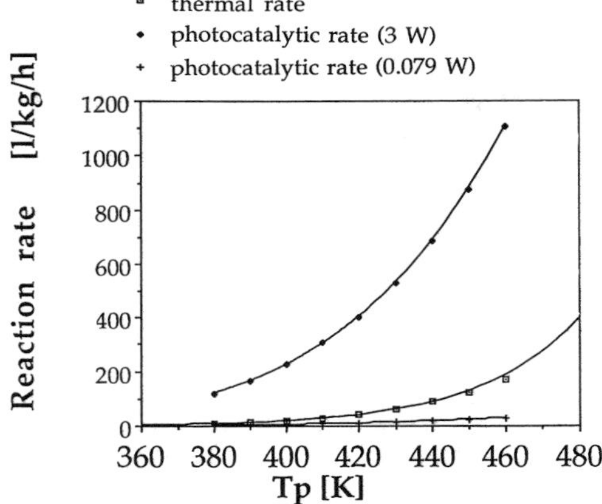

Fig. 15.7 Photocatalytic methanation rate in a fluidized bed reactor under approximately 100 suns equivalent irradiation, as a function of catalyst particle temperature. The thermal activity of the catalyst and photoenhancement effects at low light intensity are also shown.

profile in the catalyst bed. The photocatalysed reaction rate is specified by subtraction. The photocatalytic reaction rate thus derived under approximately 100 suns equivalent irradiation is given as a function of catalyst particle temperature in Fig. 15.7. To confirm that a photocatalytic effect was being observed, and not a thermal artefact during irradiation, two control experiments were carried out; (1) hydroxylation of the ruthenium catalyst with steam at 300 °C poisoned specifically the photoactivity while leaving the thermal activation behaviour unchanged; (2) a substrate which is not itself a photoactive semiconductor, zeolite SK40, was evaluated. In the first case, the photocatalytic component of the reaction rate was suppressed. In the latter case, only a very small residual photoactivity was observed, Fig. 15.8. This clearly establishes that both the high dark methanation activity and the pronounced photoenhancement is contingent on the use of the TiO_2/P25 as catalyst support.

Meanwhile a detailed kinetic analysis of the methanation process has been performed (Revilliod 1991). A significant result is that the activation enthalpy for methanation is lower for the photoinduced pathway ($\Delta H^{\neq}$ = 15.4 kcal mol^{-1}) compared to the thermal reaction ($\Delta H^{\neq}$ = 18.5 kcal mol^{-1}. In the latter case the reaction orders for CO_2 and H_2 are 0.16 and 0.36, respectively.

Conclusion

In the context of a commitment to control the carbon dioxide emissions in a solar-hydrogen technology and economy, the use of carbon as a vector for

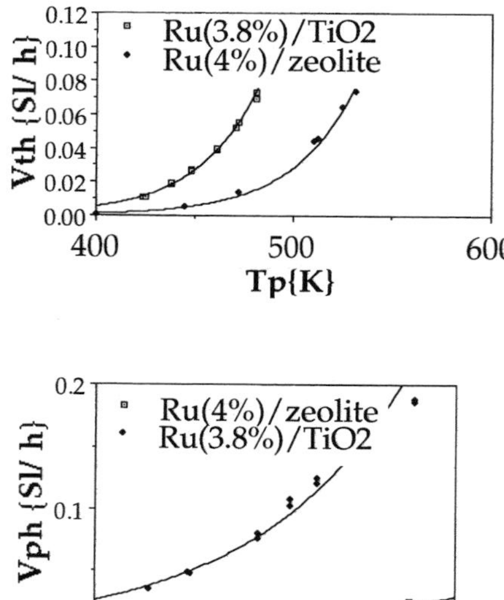

Fig. 15.8 Effect of the support material on the methanation of the activity of the RuO_x catalyst in the dark and under photoexcitation.

hydrogen, as a feedstock for liquid fuel and commodity, chemical synthesis will find a place. One technology which will then be required is the catalytic or photocatalytic reduction of gaseous carbon dioxide. In particular, the exploitation of quantum energy conversion for the production of electricity, fuels and chemicals using solar radiation will be of fundamental importance. Cheap, efficient, and durable photovoltaic devices are required to produce the hydrogen vector which serves as a reductant in the subsequent fixation of carbon dioxide much in the same way as NADPH, an organic hydride acts in natural photosynthesis. Apart from being the primary energy source in photovoltaic conversion, solar radiation may also assist the catalytic hydrogenation of carbon dioxide. Using solar radiation concentrated by parabolic mirrors of heliostats has been simulated under laboratory conditions, with promising results.

Acknowledgement

Work on photocatalysis and photothermochemistry in EPFL has been supported by the Gas Research Institute (USA) and by the *CORE* programme of the Swiss Federal Office of Energy. In particular, we appreciate the constant interest and encouragement of Dr Paul Kesselring (PSI, Switzerland), and Dr C. Courvoisier (Geneva).

References

O'Regan, B. and Grätzel, M. (1991). *Nature,* 353, 737.

Revilliod, C. (1991). Ph.D thesis. Ecole Polytechnique de Lausanne, Lausanne, Switzerland.

Rutheranat, P., Buffat, P.-A., Thampi, K.R., and Grätzel, M. (1990). *Ultramicroscopy,* **34,** 66.

Thampi, K.R., Kiwi, J., and Grätzel, M. (1987). *Nature*, **327,** 506.

Tomoyasi, I. and Lundsford, J.H. (1985). *Nature*, **314,** 721.

Weatherbee, G.D. and Bartholomew, C.H. (1981). *J. Catal.* **63,** 67–70.

16

Retrodiction of carbon dioxide fixation towards a chemoautotrophic origin of life

Günter Wächtershäuser

The problem

Scientific theories aim at explaining the world. To explain means to reduce known facts to unknown assumptions. This is Popper's crisp formulation. And it is true. When we explain in chemistry, we reduce known reaction products to unknown intermediates or unknown reaction mechanisms. When we explain in biology, we reduce known organisms to unknown missing links or an unknown mechanism of evolution. And when we explain in biochemistry, we try to reduce the known biochemical pathways to an unknown primordial metabolism, the origin of life. Alternative theories compete for the best explanation: for explaining more facts with fewer assumptions. They compete for explanatory power (Popper 1963).

There is a most fundamental alternative in the biosphere. It is the alternative between two ways of life: autotrophy and heterotrophy. Autotrophs are producers, capable of synthesizing all their constituents from inorganic scratch. Heterotrophs are consumers or parasites, dependent on taking up organic compounds as food. This gives biology its primary problem: was the first organism an autotrophic producer (auto-origin) or a heterotrophic consumer (hetero-origin). If put in this way the answer seems obvious: of course producers must come first. Yet for more than 60 years the field of biology has been dominated by Oparin's (1924) theory of a heterotrophic origin in a prebiotic broth. If we think of the widespread occurrence of autotrophy within the domains of life, it makes us wonder why the auto-origin alternative has remained neglected for so long. Two reasons come to mind. The first reason is the problem of chemical consistency. In an auto-origin all chemical reactions must occur within the same tiny locale and they must all be mutually compatible. This condition is difficult to satisfy.

The second reason is concerned with the difficulty of conceiving a plausible energy source for a primordial carbon fixation. Such an energy source must satisfy six conditions (Wächtershäuser 1990*b*)

1. It must be a source of reducing power.
2. The reducing potential must be sufficient for all reductions in the metabolism.
3. The electron flow must proceed directly and linearily from the reducing agent to the organic compounds.

4. The energy flow must be somewhat inhibited in the absence of a metabolism. Otherwise a high chemical potential could not build up and be tapped by a metabolism.
5. The energy source must be operative within the organism, because the reducing agent is required at several steps along the metabolic pathways. Therefore, it must be mild and selective. This excludes for example UV light.
6. The energy source must be geochemically plausible.

The central hypothesis

If we go through all the geochemically plausible sources of reducing power only one possibility seems to come to mind which satisfies the above conditions. This brings us to the central thesis of this proposal (Wächtershäuser 1988*a*).

The first organism is a chemoautotroph. It uses as its energy source for carbon fixation the oxidative formation of pyrite from hydrogen sulphide and ferrous ions.

$$FeS + H_2S \rightarrow FeS_2 + 2e^- + 2H^+$$

This energy source is a powerful battery. It has a standard potential of –620 mV, more than enough for all biochemical reductions. It does not require geochemically obscure assumptions. Ferrous ions and hydrogen sulphide have already been abundant. Pyrite is the most stable iron mineral under anaerobic conditions.

All biological redox energy sources show kinetic inhibition. Think of a mixture of hydrogen and oxygen. It does not explode without ignition. This inhibition is the precondition for the geochemical build up of a high chemical potential. The pyrite-forming energy source has precisely this characteristic. It is an inhibited energy source. The reaction of depletion would be the formation of hydrogen

$$FeS + H_2S \rightarrow FeS_2 + H_2$$

It has been proven that this reaction occurs with a sufficiently slow rate (Drobner *et al.* 1990). It is not a reaction of rapid depletion.

The proposed primordial energy source can explain many features of the central pathways. They are all seen as rooted in a reductive primordial metabolism based on pyrite and sulphur chemistry. Table 16.1 gives explanatory precursor–successor relationships between features of an archaic metabolism and features of extant metabolisms.

Hydrogen sulphide is seen as the precursor of all biomolecules with catalytic sulphydryl groups. A majority of enzymes is dependent on such functional sulphydryl groups. Enzymes are in fact largely sulphydryl carriers. Therefore, if we need a name for this world of life, we should say that it is today what it always has been: the iron–sulphur world (Wächtershäuser 1990*b*).

Table 16.1 Precursor–successor relationships between an archaic and ezu extant metabolism

Evolutionary precursors	Evolutionary successors
H_2S	Cys, CoA, CoM, CoB, lipoate
Thioacids (–COSH)	Thioesters (–COSR)
H_2S/FeS reducing power	Ferredoxins, NAD (P) H, $FADH_2$ F_{420}
$^-S–S^-$ in pyrite	R–S–S–R
Pyrite surface	Iron–sulphur clusters, iron coenzymes
H_2S + FeS as energy source	extant sulphur-dependent energy sources $H_2 + S \rightarrow H_2S$ sulphate reducers H_2S + light (purple bacteria)

The main consequences

Some interesting consequences of the proposed primordial energy source are new considered. A first consequence can be derived by appealing to two facts:

1. Pyrite crystals have positive surface charges, capable of binding organic molecules with anionic groups such as $–COO^-$, $–S^-$, $–O–PO_3^{2-}$.
2. CO_2 fixation leads immediately to anionic groups ($CO_2 \rightarrow –CO_2^-$).

This means that the products of carbon dioxide fixation become bonded onto the pyrite surface in their nascent state. Their residence time on the pyrite surface corresponds to their surface-bonding strength. Weak bonding molecules become detached. Strong-bonding products have time to undergo subsequent reactions. They form a two-dimensional reaction system which is self-selective for strong bonding polyanionic constituents. Its rate is largely limited by the rate of two-dimensional diffusion.

The strong bonding between anionic constituents and a cationic pyrite surface provides a degree of thermodynamic isolation. It establishes a flow-through system: with an input of inorganic nutrients, an output of detached organic products of decay, and an 'internal' surface reaction system, a surface metabolism (Wächtershäuser 1988*b*). Such a reaction system has several peculiar thermodynamic and kinetic properties, which are in sharp contrast to the solution metabolism listed in Table 16.2.

Table 16.2 Thermodynamic properties of surface and solution metabolism

Hetero-origin	Auto-origin
$\lvert\Delta S_r\rvert$ large Depolymerization Hydrolysis	$\lvert\Delta S_r\rvert$ small Polymers tend to be thermodynamically stable on a surface
Temperature as low as possible (psychrophilic origin)	Elevated temperature required (thermophilic origin)
Three-dimensional kinetics (inherently chaotic)	Two-dimensional kinetics (inherently orderly)
Evolution Liquid ⇒ solid Chaos ⇒ order Chance ⇒ chance	Evolution Solid ⇒ liquid Order ⇒ chaos Necessity ⇒ chance
Racemic solution → biochirality	Chiral pyrite → biochirality

1. In a solution, cleavage reactions are thermodynamically favoured by virtue of the reaction entropy (ΔS_r). Proteins and nucleic acids are thermodynami cally unstable and subject to hydrolysis. This is one of the main problems of all soup theories. In a surface metabolism surface-bonded polymers tend to be thermodynamically stable. This has an important consequence for the overall process of evolution. According to the soup theories life begins in solution and evolves toward the solid state. According to the auto-origin theory, life begins in a next to solid state and it evolves toward the liquid state.
2. All problems of degradation of the soup theories are aggravated by temperature increase. This is the reason why it has been said that the soup must be chilled or frozen, the colder the better. A surface metabolism has a small entropy burden. Therefore, it can tolerate higher temperatures. It even requires higher temperatures for kinetic reasons. An auto-origin is thermophilic. This result is in agreement with the facts of phylogeny. The deepest branches in the universal tree of life are occupied by extreme thermophiles.
3. Solution chemistry is characterised by a large number of paths of approach and of reaction possibilities. It is inherently chaotic. Therefore, a hetero-origin in a prebiotic broth means an evolution from chaos to order. A surface metabolism is inherently orderly. Its reactions are quasi-intramolecular

rearrangements. They proceed from a simple order with highly restricted possibilities toward a complex order with an unfolding of possibilities. This means an evolution from order to order or, more precisely, from necessity to change.

4. Perhaps the most unexpected consequence concerns the problem of the origin of biochirality. In contrast to high-temperature pyrite (i.e. metamorphic or magmatic pyrite) the low temperature pyrite of sedimentary or hydrothermal origin seems to have a noncubic crystal structure. If the triclinic structure proposed by Bayliss (1977) is correct, low-temperature pyrite must be optically active. This would mean that biochirality can arise by a dual chiral feedback between the pyrite surface and its organic coating: by a transference of the chirality of a pyrite crystal to its organic coating and by a retransferrence of the chirality of the organics in the pyrite crystals that arise by secondary nucleation (Wächtershäuser 1991).

Cellularization

The auto-origin theory yields a straightforward model for cellularization. A reductive carbon dioxide fixation means essentially a conversion of CO_2 molecules into $-CH_2-$units. This amounts to a self-lipophilization of the pyrite surface. Moreover, all surface-bonded constituents, say dicarboxylic acids, will undergo recursive reactions. Take as an example an essential step in the reductive citric acid cycle

$$\mathrm{HOOC{-}CH_2)_{\mathit{n}}{-}COOH \rightarrow HOOC{-}(CH_2)_{\mathit{n}}{-}CO{-}COOH \rightarrow} \\ \mathrm{HOOC{-}(CH_2)_{\mathit{n}+1}{-}COOH}$$

We have here a recursive reaction from n to n+1. This means that the surface-bonded organic molecules tend to grow longer and longer. They turn into lipids. At a certain chain length a one-sided detachment by protonation or reduction produces a surface-bonded membrane. This membrane is a monolayer membrane. Now, with a growing pyrite crystal the surface-to-volume ratio decreases. This means that the lipids become more and more crowded. Finally, we have isolated hydrophilic domains in an otherwise lipophilic membrane coating. Such holes in a membrane are thermodynamically unfavourable. By closure of these holes with detached membrane portions we obtain semicellular structures. These have a closed membrane with a membrane metabolism; a cytosol with a cytosol metabolism; and still an internal pyrite surface with a surface metabolism. This means an increase of complexity, with metabolic continuity throughout the process of cellularisation.

The restricted isolated hydrophilic domains are places of growth. This means that in these restricted spots there may be secondary nucleation. Every such secondary pyrite nucleus starts again with a high surface-to-volume ratio. This means the secondary nucleation may well be at the origin of cell division. It will lead to a cluster of pyrite crystals which looks like a raspberry. And raspberry pyrite, so-called framboidal pyrite, is precisely what we find abundantly in nature. Raspberry structures are typical for pyrite. They seem to be unique for pyrite and they are unexplained to this day.

The mechanism of evolution

The origin of life coincides with the origin of evolution. The RNA world variety of the soup theory claims that the origin of evolution coincides with the origin of something like autocatalytic RNA replication. Living RNA molecules are supposed to feed on a soup of activated nucleotides. Just in time before the depletion of this broth they invent intermediary metabolism and finally they learn how to work by inventing autotrophy. In short: parasitic intellectuals turn into workers.

The auto-origin theory suggests a completely different course of events. It preserves the idea that life originates with the ignition of a synthetic autocatalysis. Everything else, however, is different. The first autocatalytic cycle is not a template cycle. It is a surface-bound autocatalytic carbon fixation cycle. To understand how such a cycle can constitute reproduction and inheritance, we begin with one molecule of a surface-bonded CO_2 acceptor. It grows by CO_2 uptake and reduction. Growth by CO_2 uptake leads to an additional surface bonding by additional formation of $-CO_2^-$ in *statu nascendi*. By assuming further that such molecular growth leads to an increasing molecular instability we may postulate a final cleavage into two molecules of the surface-bonded CO_2 acceptor. The CO_2 acceptor comes to occupy a previously vacant lot which inherits the CO_2 fixation capability. This reproduces by spreading in two dimensions.

It is easy to see how such an autocatalytic carbon fixation process can evolve by inheritable variations. All chemical reactions have side reactions. From the point of view of an autocatalytic cycle a side reaction is a reaction of decay. From the point of view of evolution the side reactions produce the material for variation by the expansion of the autocatalytic cycle into a more and more complex network. Most side reactions are not autocatalytic. But occasionally a low-propensity branch reaction will lead to a new kind of molecule K which is catalytic for the production cycle. This turns the branch reaction from a burden into a benefit. However, this expansion of the autocatalytic cycle will disappear again if the conditions become unfavourable for the branch reaction. It is not inheritable. Occasionally, a low-propensity catalytic branch product K′ will appear, which is not only catalytic for the production cycle but also catalytic for its own branch pathway. Now it will persist even if the conditions become unfavourable for the *de novo* formation of K′. It is truly inheritable. This situation is typical for the extant coenzymes. They are catalytic for a large class of reactions including reactions in their own biosynthesis. Such autocatalytic constituents K′ are called 'vitalysts'. Organisms, which loose the capability of producing a vitalyst are dependent of taking it in as a vitamin. Some branch products K″ may be catalytic for their own branch pathway but not for the production cycle. This means that they are destructive. They are called 'virulysts'. Viruses are evolvable virulysts which can leave and re-enter an organism.

Finally, we need a mechanism for selection. In a surface metabolism a chemical selection results most easily. It occurs by selective detachment. Weak-bonding constituents disappear from the surface metabolist into the water phase. The surface metabolist is self-selective for strong surface bonding. This explains why all constituents of the central metabolism, including the coenzymes, are polyanionic.

Nucleic acids are late extensions of this process of evolution: they are glorified coenzymes. They are catalytic for the production cycle and autocatalytic for their own branch pathway. Their evolution begins with catalytic imidazole bases in the purine pathway. It gives rise to thiamin, pterins, and flavins. As a byproduct purine base-pairing appears and now replicating nucleic acids (DNA, RNA) come into the picture. Some of them are vitalysts and others virulysts. Finally, very late heterotrophs appear. They are nothing but opportunistic shortcuts, benefiting from the work of others like all parasites. With this we have arrived at a complete reversal of the soup theory. The auto-origin begins with honest-to-goodness workers and these turn into intellectuals, and from there finally into parasites.

The evolution of carbon fixation

The best candidate for the retrodiction of the primordial carbon fixation cycle is the extant reductive citric acid cycle (RCC) (Wächtershäuser 1990*a*)

succinate → citrate	(growth)
citrate → oxaloacetate + acetyl–CoA	(cleavage)
acetyl–SCoA → oxaloacetate	(growth)
2 oxaloacetate → 2 succinate	

It is truly autocatalytic. It doubles with every turn. For the retrodiction of the primordial autocatalytic CO_2–fixation cycle the following rules of retrodiction are required.

1. Replace all reducing agents by FeS/H_2S.
2. Replace thioester activations by thioacid equilibria.
   $-COOH + H_2S \rightleftharpoons -COSH + H_2O$
3. Replace carbonyl groups by thioenol equilibria.
   $-CH_2-CO- + H_2S \rightleftharpoons -CH{=}C_1-SH + H_2O$

With these rules we come to postulate two reactions of primordial CO_2 fixation

a. $-CH{=}C(SH) - COOH + CO_2 \rightarrow HOOC{-}C{=}C(SH) - COOH$
b. $R{-}COOH + CO_2 + FeS/H_2S \rightarrow R{-}CO{-}COOH + FeS_2 + H_2O$

Both proposals are speculative. Proposal b is supported by an analogous nonenzymatic CO_2 fixation of a thioester in a nonaqueous solution with $Na_2S_2O_4$ as reducing agent and an iron–sulphur complex as catalyst (Nakajima 1978).

In an archaic nonenzymatic metabolism higher homologues of the RCC would also occur. The next higher homo-RCC is partially preserved in the extant amino-adipate pathway to lysine. Almost all extant biosynthetic pathways are seen as radiating from the RCC (and the homo-RCC). Thiamin appears as an important vitalyst for the RCC and it catalyses the formation of phospho-ribose from the phospho-trioses, which in turn results by a reductive pathway from pyruvate (constituent in the RCC). Glycine arises by ammonia and CO_2 fixation via methylene-THF. The purines are formed from phosphoribose, glycine, NH_3, CO_2, and methenyl-THF (obtainable by the reduction of CO_2). THF arises from the purines, which means that it is a vitalyst. THF (or the methanopterins) give rise to the reductive acetyl-CoA pathway, which is an anaplerotic pathway for the RCC. Biotin arises by a process of carbon fixation. It comes to catalyse extant versions of the reaction type of proposal b. It seems also to be a vitalyst since this reaction is expected to occur within the biosynthesis of the dicarboxylate precursor for biotin. Finally, Rubisco appears in some bacteria which closes the sugar interconversion pathways to the Calvin cycle. *Chloroflexus* seems to have a variant of the RCC which proceeds from acetyl-CoA via malonyl-CoA (CO_2 fixation), propionyl-CoA (reduction), methyl-malonyl-CoA (CO_2 fixation) to succinate (B_{12} catalysis) (Holo 1989; Strauß *et al.* 1991). This unfolding of the potentialities of CO_2 fixation has been accompanied by diverse processes of streamlining with the abandonment of some CO_2 fixation pathways and the interruption of the RCC. In aerobic organisms a portion of the RCC is reversed to the oxidative Krebs cycle.

Conclusion

The battle cry of the soup theorists is 'Order out of order'. And they have to say this. For the prebiotic broth, as it exits in their imagination, is truly chaotic: a chaos of compounds, a chaos of movements, and a chaos of reaction possibilities. That this chaotic situation could have ever turned into the wonderful order of biochemistry has often been claimed but never explained. The theory of a chemoautotrophic origin of life leads to a totally different slogan. It is the slogan 'Order from order from order'. At the beginning there is a simple order consisting of a pyrite crystal with a surface coating. From here evolution goes in the direction of self-emancipation; an emancipation from surface bonding and from the narrow chemical confines of pyrite formation; a conquering of the third dimension and of ever new chemical spaces. This process of self-liberation has now been going on for some four billion years. But at what price: at the price of unfathomable complications and ever more sophisticated controls.

It may be complained that throughout this chapter nothing but speculations have been presented. Therefore, as a compensation of sorts, the chapter concludes with a prophesy. Pyrite is frequently encountered by gold diggers. It glitters like gold. This is the reason why it is called fool's gold. But the author would like to predict that once the secrets of pyrite are known, only a fool will mistake this mineral with something as base as gold.

References

Bayliss, P. (1977). *Am. Min.*, **62,** 1168–72.

Drobner, E., Huber, H., Wächtershäuser, G., Rose, D., and Stetter, K.O. (1990). *Nature*, **346,** 742–4.

Holo, H. (1989). *Arch. Microbiol.*, **151,** 252–6.

Oparin, A.I. (1924). *Proiskhozhdenie zhizny*. Izd. Moskovshii Rabochii, Moscow.

Popper, K.R. (1963). *Conjectures and refutations: the growth of scientific knowledge*. Routledge & Kegan Paul, London.

Strauß, G., Eisenreich, W., Bacher, A., and Fuchs, G. (1992). *Eur. J. Biochem.* (In press.)

Wächtershäuser, G. (1988*a*). *Syst. Appl. Microbiol.*, **10,** 207–10.

Wächtershäuser, G. (1988*b*). *Microbiol. Rev.*, **52,** 452–84.

Wächtershäuser, G. (1990*a*). *Proc. Nat. Acad. Sci.* USA, **87,** 200–4.

Wächtershäuser, G. (1990*b*). *Origins Life*, **21.**

Wächtershäuser, G. (1991). *Medical Hypotheses*. (In press.)

17

Structure–activity correlations in carbamoyl phosphate synthetases

Vicente Rubio

General aspects

Carbamoyl phosphate synthetases, CPSs, catalyse the first step of the routes of urea, pyrimidine, and arginine synthesis (reaction 1).

$$2\ \mathrm{ATP} + \mathrm{HCO_3^-} + \mathrm{NH_3} \xrightarrow{\mathrm{Mg^{2+}},\ \mathrm{K^+}} 2\ \mathrm{ADP} + P_i + \text{carbamoyl phosphate} \quad (1)$$

Although all CPSs catalyse reaction (1), many have a low affinity for ammonia and use glutamine as an endogenous source of ammonia. However, CPS I, which is involved in urea synthesis and thus in the detoxification of ammonia, exhibits considerable affinity for ammonia and cannot use glutamine (Marshall 1976).

CPSs are composed of a region of about 120 kDa which is responsible for catalysing reaction (1) and for the binding of allosteric effectors, and a region of about 40 kDa which includes the domain responsible for binding and cleaving glutamine (Meister 1989; Simmer *et al.* 1990) (Fig. 17.1). The 40- and 120-kDa regions exist either as separate subunits (bacterial and arginine-specific yeast and *Neurospora* CPSs) or they are integrated into a single polypeptide in which a short sequence (< 40 residues) bridges the COOH terminus of the 40-kDa region and the NH_2 terminus of the 120-kDa region (summarized by Simmer *et al.* 1990). In CPS II, which is involved in pyrimidine synthesis in eukaryotes, the 120-kDa region is also fused by its COOH terminus to dihydroorotase (Simmer *et al.* 1990) (mammalian enzyme) or a dihydroorotase-like region (Souciet *et al.* 1989) (yeast enzyme), and the dihydroorotase component is fused to aspartate transcarbamylase. The resulting multienzymatic polypeptide (Mw about 240 kDa) is designated CAD in vertebrates (the initials for its three enzyme activities) (Simmer *et al.* 1990).

CPS is under allosteric control in many organisms. CPS I is strongly activated (> 50-fold) by acetylglutamate (Rubio *et al.* 1983). The CPSs involved in arginine biosynthesis are generally activated by ornithine, and the CPSs involved in pyrimidine biosynthesis are activated by IMP or phosphoribosyl pyrophosphate and are inhibited by UMP or UTP (Makoff and Radford 1978). The *E. coli* enzyme, which is involved in both pyrimidine and arginine biosynthesis, is inhibited by UMP and is activated by IMP and ornithine (Meister 1989). Another potential control mechanism in CPSs is phosphorylation, since CPS II

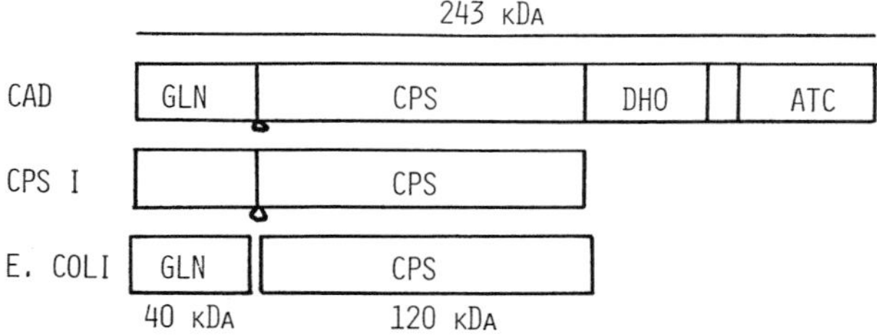

Fig. 17.1 Structural types of carbamoyl phosphate synthetase, CPS. GLN, small subunit or equivalent region in CAD; the COOH-terminal half of this region exhibits glutaminase activity; the homologous region in CPS I is also indicated, but it is not labelled because it has no glutaminase activity. CPS, large subunit or homologous region in monomeric CPSs; it catalyses carbamoyl phosphate synthesis from ammonia. DHO, dihydroorotase component; an homologous region exists in the URA2 gene product in yeast, but it has lost the ability to catalyse the dihydroorotase reaction. ATC, aspartate transcarbamylase. The small loop shown in monomeric CPSs symbolizes the bridge (10–40 residues) linking the GLN and CPS components. The zone shown in CAD between the DHO and ATC components is a linking region of unknown function; it can be phosphorylated by cAMP-dependent protein kinase (Carrey and Hardie 1988).

from hamster (but not CPS I); (Itarte and Rubio, unpublished observations) is phosphorylated *in vitro* by cAMP-dependent protein kinase, and the phosphorylated enzyme exhibits decreased susceptibility to inhibition by UTP (Carrey *et al.* 1985; Carrey and Hardie 1988).

We are only at the beginning of the process of locating physically and characterizing the sites in CPSs for the substrates, effectors, and point(s) of phosphorylation, the regions of interaction between subunits and between different enzymatic components, and the conformational changes associated with the binding of substrates and effectors. The large size of the polypeptide, the multiplicity of sites to be mapped and the lack of X-ray diffraction data are obvious difficulties. Nevertheless, substantial data have accumulated already, and a review of this information will be attempted.

Mechanistic aspects

Carbamoyl phosphate synthesis by CPSs (reaction (1)) is essentially irreversible (Jones 1976). Another enzyme, carbamate kinase, synthesizes carbamoyl phosphate reversibly (reaction (2)), although the equilibrium favours ATP synthesis ($K = 0.027$) (Marshall and Cohen 1966).

$$\text{ATP} + \text{carbamate} \underset{}{\overset{Mg^{2+}}{\rightleftharpoons}} \text{ADP} + \text{carbamoyl phosphate} \qquad (2)$$

Carbamate is formed non-enzymatically from HCO_3^- and NH_3 (reaction (3)), and the equilibrium favours slightly its formation (K = 1.89 M^{-1}; calculated from Marshall and Cohen 1966)

$$HCO_3^- + NH_3 \rightleftharpoons CO_2NH_2^- + H_2O \tag{3}$$

Adding up reactions (2) and (3) carbamoyl phosphate is formed from HCO_3^- and NH_3 (reaction (4)).

$$HCO_3^- + NH_3 + ATP \rightleftharpoons \text{carbamoyl phosphate} + ADP + H_2O \tag{4}$$

The equilibrium in this reaction does not favour carbamoyl phosphate synthesis ($\Delta G^0(4) = \Delta G^0(2) + \Delta G^0(3)$ = 7.4 kJ mol^{-1}, yielding $K = 0.051^{-1}$ M^{-1}; ΔG^0 for reactions (2) and (3) was estimated from their equilibrium constants). If reaction (4) is coupled with the hydrolysis of an extra ATP molecule (ΔG^0 for ATP hydrolysis, –30.5 kJ mol^{-1}, see Lehninger 1971) reaction (1) is obtained and carbamoyl phosphate formation becomes virtually irreversible ($\Delta G^0(1) = \Delta G^0(4)$–30.5 kJ mol^{-1} = –23.1 kJ mol^{-1}; therefore, K = 11273). Thus, on the basis of chemical plausibility it may be proposed that (a) carbamate is an intermediate and carbamate phosphorylation by ATP is a step in the reaction catalysed by CPSs; (b) the ATP molecule that is cleaved to ADP and P_i (called from here on ATP_A; the ATP molecule that phosphorylates carbamate will be designated ATP_B) is used to push the synthesis of carbamate and to generate a high local concentration of carbamate at the catalytic centre; and (c) CPS may be related evolutionarily to carbamate kinase.

A number of approaches (summarized by Rubio 1986) have demonstrated that the first step of the reaction catalysed by CPSs is the phosphorylation of bicarbonate by ATP_A, forming enzyme-bound carbonic-phosphoric anhydride (also called carboxyphosphate; Fig. 17.2). This intermediate appears to be formed in all enzyme carboxylations in which HCO_3^- rather than CO_2 is the carboxylating agent (CPSs, biotin-dependent carboxylases, phosphoenolpyruvate carboxylase) (Rubio 1986). In the absence of NH_3 (and glutamine) CPS catalyses a bicarbonate-dependent ATPase partial reaction (reaction (5), Fig. 17.2) as a consequence of the slow decomposition of carboxyphosphate (see Rubio 1986). In the absence of carbamate, carbamate kinase also catalyses HCO_3^--dependent ATP hydrolysis (Marshall and Cohen 1970), indicating that this enzyme also phosphorylates bicarbonate. Thus, carbamate kinase catalyses the two phosphorylations that are catalysed by CPS, further supporting the relation between these two enzymes.

CPS also catalyses a partial reaction of ATP synthesis (reaction (6), Fig. 17.2) with the same stoichiometry as carbamate kinase (Metzenberg *et al.* 1958). There were doubts about the true significance of this reaction, which might represent the reversal of the step of carbamate phosphorylation, or, since carbamoyl phosphate is an analogue of carboxyphosphate, might reflect the reversal of carboxyphosphate synthesis (Jones 1976). These doubts are now dissipated, for the two partial reactions of CPS were inactivated independently in site-directed mutagenesis (Post *et al.* 1990) and oxidative inactivation experiments (Alonso

GLOBAL REACTION - MECHANISM

$$ATP + HCO_3^- \longrightarrow ADP + {}^-O-\overset{O}{\overset{\parallel}{P}}(O^-)-O-\overset{O}{\overset{\parallel}{C}}-OH \xrightarrow{NH_3} P_i + H_2N-\overset{O}{\overset{\parallel}{C}}-OH \xrightarrow{ATP} ADP + H_2N-\overset{O}{\overset{\parallel}{C}}-O-\overset{O}{\overset{\parallel}{P}}(O^-)-O^-$$

(Carboxyphosphate) (Carbamate) (Carbamoylphosphate)

PARTIAL REACTIONS

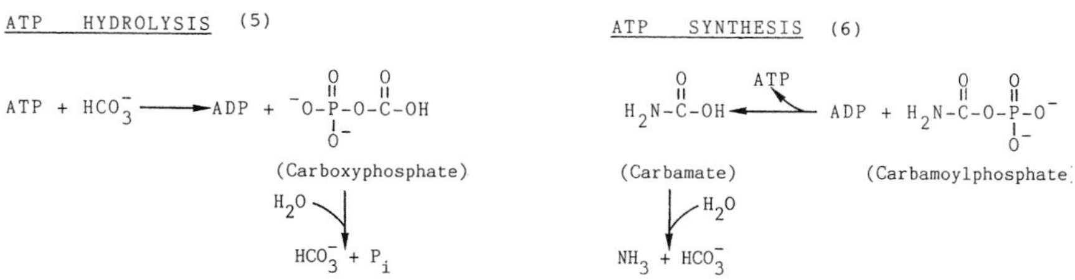

Fig. 17.2 Schematic reaction mechanism and partial reactions of CPS.

et al. 1992). Therefore, it is now clear that the partial reaction of ATP synthesis is the reversal of the step of carbamate phosphorylation.

In summary, carbamoyl phosphate synthesis by CPS takes place in three steps (Fig. 17.2): bicarbonate phosphorylation, carbamate synthesis, and carbamate phosphorylation. To associate the two phosphorylation steps in a catalytic cycle, either a single ATP site and catalytic centre has to be used twice (one time per phosphorylation step) in each catalytic cycle, or two ATP sites and catalytic centres, possibly derived from duplication of carbamate kinase, have to be associated, one for each phosphorylation. The demonstration in CPS of a separate binding site for each ATP molecule (Rubio *et al.* 1979; Britton *et al.* 1979), and more recent information on the primary structure of CPSs (see below) supports the latter possibility. Much effort is being invested in localizing the ATP sites and the regions involved in the catalysis of the different steps of the reaction.

Mapping of functions

The amino acid sequences of a number of CPSs have been deduced from their cDNAs (see Simmer *et al.* 1990, for a review). All sequences share considerable homology over the entire length of the molecule, irrespective of the structural type of the enzyme. We will discuss separately the findings for the small and large subunits (or the homologous regions in monomeric CPSs). These finding are summarized in graphic form (Figs. 17.3 and 17.4). The numbering of the residues follows that for the subunits of CPS from *E. coli* (Nyunoya and Lusty 1983; Nyunoya *et al.* 1985). The estimation of the degree of conservation includes identities and conservative replacements as defined by Baur *et al.* (1987). Total conservation means identity.

Fig. 17.3 Small subunit in heterodimeric CPSs and equivalent region in other CPSs. The scale corresponds to the numbering of residues in CPS from *E. coli*. The long bar represents the entire subunit. The black stripes within correspond to residues that are conserved (identities and conservative replacements, as defined by Baur *et al.* 1987) in all of the following CPSs: hamster CAD, rat liver CPS I, *Drosophila melanogaster* CAD, *Dictyostelium discoideum pyr*1–3 gene product, *Saccharomyces cerevisiae* pyrimidine-specific CPS (URA2 gene product), *Saccharomyces cerevisiae* arginine-specific CPS (CPA1 gene product), *E. coli car*A gene product, and *Salmonella typhimurium car*A gene product. The alignement of the sequences is that given by Simmer *et al.* (1990) and Bein *et al.* (1991). For simplicity, gaps are not represented. The region of subunit interaction and the glutaminase region are represented at the bottom as the dotted and open bar, respectively. The horizontal bars above the glutaminase portion define the regions identified by Simmer *et al.* (1990) as of particularly high homology within the glutaminase component; regions 1, 2, and 4 are characteristic of trpG-like amidotransferases, whereas region 3 appears only in CPSs. The arrows at the bottom denote the residues that have been replaced in site-directed mutagenesis experiments.

Small subunit

About 35 per cent of the amino acids of the small subunit (length of the subunit in *E. coli*, 382 residues) are conserved in all CPSs studied (Simmer *et al.* 1990; Bein *et al.* 1991). The COOH-terminal half of this region (residues 190–382) is identified as the glutaminase domain, for it is homologous to the glutaminase component of glutamine amidotransferases of the trpG-type (Nyunoya and Lusty 1984). These glutaminases form a γ-glutamyl thioester intermediate between a cysteine in the catalytic centre and the glutamate moiety of glutamine. A histidine is also essential for catalysis, probably by assisting the deprotonation of the thiol group (Chaparian and Evans 1991). The formation of the thioester was demonstrated long ago in *E. coli* CPS (Wellner *et al.* 1973) and more recently in CAD (Chaparian and Evans 1991). Site-directed mutagenesis experiments (Rubino *et al.* 1986; Miran *et al.* 1991) have demonstrated the involvement of cysteine 269 [which is conserved in all CPSs except CPS I, which cannot use glutamine (Nyunoya *et al.* 1985)] and histidine 353 in the catalysis of glutamine cleavage, and have shown that histidine 312 is critical for glutamine binding, but not for catalysis. In contrast, replacement of histidines 272 or 341 by asparagine had little influence on the enzyme activity or on the binding of glutamine, excluding a key role of these residues.

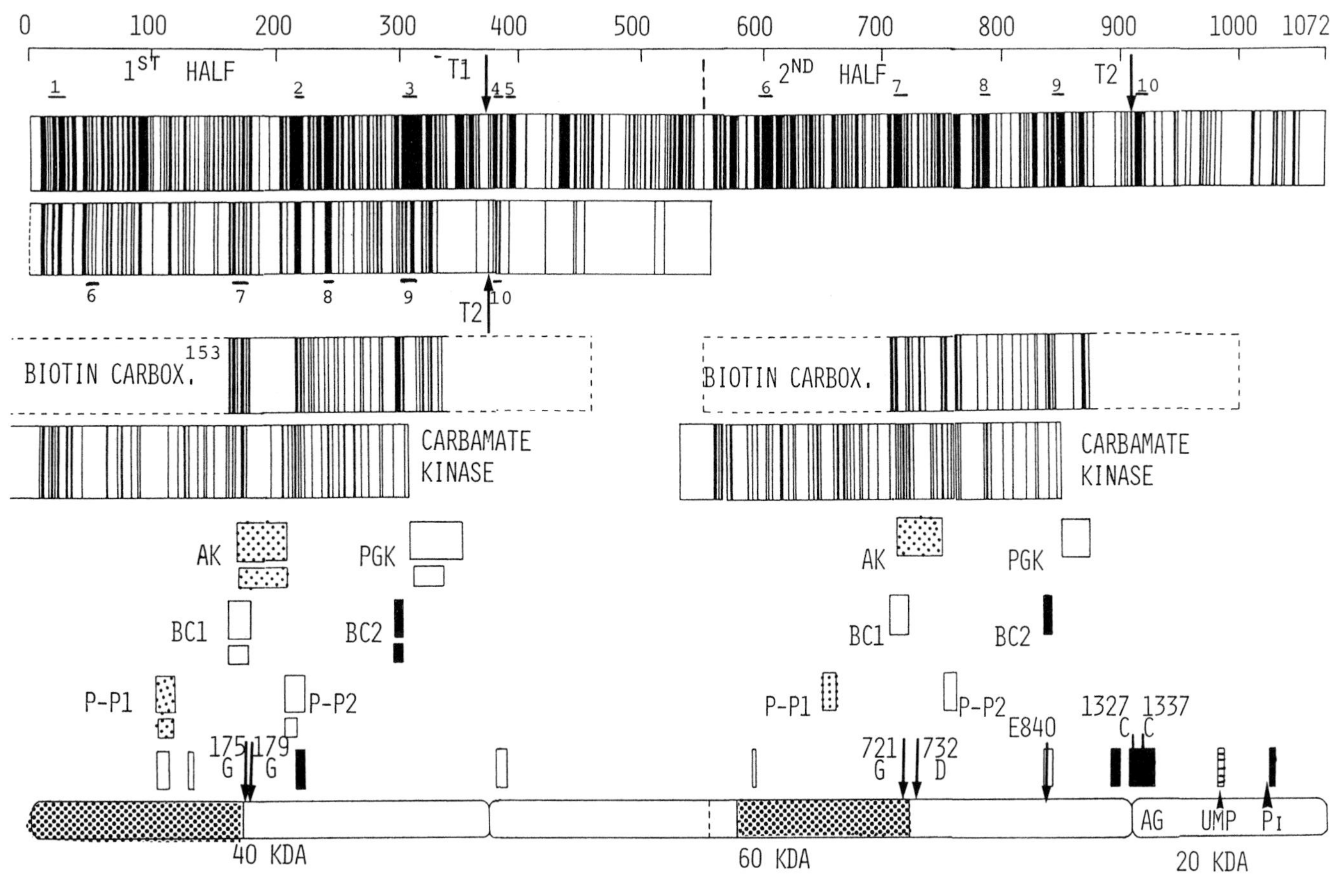
0
100
200
300
400
500
600
700
800
900
1000
1072
1ST HALF
2ND HALF
T1
T2
1
2
3
4
5
6
7
8
9
10
153
BIOTIN CARBOX.
CARBAMATE KINASE
AK
PGK
BC1
BC2
P-P1
P-P2
175 G
179 G
721 G
732 D
E840
1327 C
1337 C
AG
UMP
PI
40 KDA
60 KDA
20 KDA

Fig. 17.4 Large subunit of heterodimeric CPSs and equivalent region in other CPSs. The scale corresponds to the numbering of residues in CPS from *E. coli*. *The long bar at the top* represents the entire subunit. The black stripes within correspond to residues that are conserved (identities and conservative replacements, as defined by Baur *et al.* (1987)) in all of the following CPSs: hamster CAD, rat liver CPS I, *Drosophila melanogaster* CAD, *Saccharomyces cerevisiae* pyrimidine-specific CPS (URA2 gene product), *Saccharomyces cerevisiae* arginine-specific CPS (CPA1 gene product), and *E. coli carB* gene product. The alignment of the sequences is that given by Simmer *et al.* (1990). For simplicity, gaps are not represented. The two homologous halves are identified, and the point where the second half begins is indicated by the broken vertical line. The shorter horizontal bar immediately below the long bar at the top represents the alignement of the two halves (based on the alignment of Nyunoya and Lusty 1983). The stripes identify conserved residues in both halves of all CPSs compared. The small horizontal lines numbered 1–10 above the entire subunit are the regions of strict conservation identified by Simmer *et al.* (1990); regions 6–10 are found in the second half, and are represented also below the bar in which the two subunits are aligned, to compare their position with that of regions 1–5. The arrows above the subunit are the two points of preferential tryptic cleavage (Simmer *et al.* 1990; Potter and Powers-Lee 1992; Rubio *et al.*, unpublished data with *E. coli* CPS); point T2 has been represented also below the comparison of the two halves, to demonstrate its equivalence with T1 in the first half. *The bar located next below* illustrates the residues that are conserved in each half of all CPSs and in the following biotin enzymes (Kondo *et al.* 1991): yeast pyruvate carboxylase, chicken acetyl CoA carboxylase, α subunit of rat propionyl CoA carboxylase, and *E. coli* biotin carboxylase (a subcomponent of acetyl CoA carboxylase). Although only the region identified as homologous to CPS (see Kondo *et al.* 1991) has been compared, the entire length of biotin carboxylase is represented (broken line). The following downwards bar illustrates the residues that are conserved in each half of all CPSs and in carbamate kinase (from *Pseudomonas aeruginosa*; Baur *et al.* 1989). *Below the latter bar*, blocks labelled AK localize in the sequence of CPS the putative nucleotide sites that are homologous to adenylate kinase; blocks labelled PGK identify those that are homologous to phosphoglycerate kinase (Lusty *et al.* 1983); blocks labelled BC1 and BC2 are the two regions of highest conservation between biotin enzymes and CPSs (Kondo *et al.* 1991); BC1 was identified as a glycine-rich putative nucleotide binding sequence. Blocks labelled P–P1 and P–P2 are the two types of consensus sequences separated by about 100 residues that were identified by Powers-Lee and Corina (1987) as parts of a nucleotide binding site. The thinner blocks represent these sites from the second half, placed in the alignment of the two halves, to show that the topography of these sequences is conserved in both halves. The horizontal bar at the bottom of the figure illustrates the domain structure of the subunit (and of the corresponding region in monomeric CPSs) as demonstrated using trypsin and other proteases. The masses of the three domains (40, 60, and 20 kDa) are only approximate. The point where the COOH-terminal half begins is represented by a broken line. The binding of acetylglutamate (AG; in CPS I) and UMP, and the point of phosphorylation (P_I; in CAD) are mapped to the COOH-terminal 20-kDa domain: the arrowtips denote the exact points of phosphorylation (Carrey and Hardie 1988) and of photolabelling with UMP (peptide striped horizontally; Cervera and Rubio, unpublished results). Ornithine and IMP are also likely to bind in the COOH-terminal domain (see text). The residues that were replaced by site-directed mutagenesis are indicated by arrows. The sequences that were labelled with FSBA in CAD (Kim *et al.* 1991) are shown by open blocks, and the sequences labelled with FSBA in CPS I (Potter and Powers-Lee 1992) are shown by filled blocks, immediately above the bar representing the domain structure. Cysteines 1327 and 1337 of CPS I are shown in the filled block that is at the junction between the 60-kDa and 20-kDa domains. These cysteines are highly exposed in the presence of acetyl-glutamate (Geschwill and Lumper 1989), and their oxidation results in inactivation (Alonso *et al.* 1992).

Replacement in the CPS from *E. coli* of the essential cysteine 269 by serine or glycine increases the ATPase partial activity, which is catalysed by the large subunit (Rubino *et al.* 1986). The increased ATPase is the result of a decrease in the stability of the carboxyphosphate intermediate (Mullins *et al.* 1991), indicating a less effective seclusion of this intermediate in the catalytic centre. These changes in large subunit functions as a consequence of a mutation in the glutaminase catalytic centre stress the importance of the interactions between the two subunits. Such interactions are reflected also in the change in the kinetic properties of the reactions catalysed by the large subunit when the two subunits are separated or when noncleavable glutamine analogues bind to the small subunit (Meister 1989).

The importance of these interaction was in evidence directly in scanning microcalorimetry experiments (Cervera *et al.* 1991), in which we observed for the entire *E. coli* enzyme a single denaturational endotherm rather than the sum of the endotherms observed with the isolated large and small subunits. In addition, the T_m value was much larger for the complete enzyme than for any of the subunits. These results indicate that both subunits are strongly stabilized by their interaction, and that the denaturation of the entire enzyme occurs in a concerted way, with simultaneous rather than independent unfolding of the two subunits. In addition, occupation of the glutamine site increases the T_m value for the entire enzyme, but not for the isolated small or large subunit. Thus, the conformational changes induced by the occupation of the glutamine site require the association of the two subunits.

Of the portion of the small subunit that is outside the glutamine domain (residues 1–190), residues 1–114 were shown, using deletion mutants (Guillou *et al.* 1989), to be essential for interaction of the small and large subunits. No known function has been ascribed thus far to the region between residues 114 and 190. Only 16 per cent of the amino acids in this region are conserved in all CPSs (compare with 37 and 41 per cent conservation in the 1–114 and 190–382 regions, respectively), and the sequence from residue 165 to residue 186 is not represented in CAD (Bein *et al.* 1991).

Since CPS I cannot use glutamine, the function of the region which is homologous to the small subunit is unclear. Elastase nicks the hinge between this region and the remainder of the enzyme (which is homologous to the large subunit), but the two moieties remain associated and the enzymatic activity is retained (Guadalajara 1987; Guadalajara *et al.* 1987; Marshall and Fahien 1988). ATP prevents and acetylglutamate promotes further cleavage by elastase of the moiety corresponding to the large subunit. Upon separation of the two moieties by chromatography on DEAE cellulose (Marshall and Fahien 1988), the moiety corresponding to the large subunit becomes unable to catalyse carbamoyl phosphate synthesis, and ATP and acetylglutamate do not protect or promote, respectively, the cleavage of this moiety by elastase. Mixing the two moieties partially restores the activity and the effects of ATP and acetylglutamate on the susceptibility to proteolytic attack. Thus, the moiety of CPS I corresponding to the small subunit is essential to confer the other moiety an appropriate conformation for proper function. (See Fig. 17.3.)

Large subunit

Analysis of the sequence About half (47 per cent) of the 1072 residues that constitute this subunit are conserved in all CPSs of known sequence. However, the conservation decreases to only 20 per cent in the final 150 residues (Fig. 17.4) (Simmer *et al.* 1990). The sequence exhibits considerable internal homology between the amino- and carboxy-terminal halves, indicating that the subunit arose by a single duplication and tandem fusion of a gene coding for half of the subunit. The best homology is found between residues 1–400 and 553–933 (Nyunoya and Lusty 1983).

The internal homology suggests that each half is involved in a phosphorylation step and contains an ATP site. On the basis of sequence homology to known nucleotide-binding sites, Lusty *et al.* (1983) identified adenylate kinase-like (AK, Fig. 17.4) and phosphoglycerate kinase-like (PGK, Fig. 17.4) putative nucleotide-binding sequences separated by about 100 residues in the initial 400 residues of each half of the subunit. More recently, on the basis of consensus sequences for nucleotide sites, Powers-Lee and Corina (1987) also identified in the same regions two additional putative ATP biding sequences (P-P1 and P-P2, Fig. 17.4), separated by about 100 residues. One is located about 70 residues upstream and the other follows the adenylate kinase-like sequence.

The region between residues 160 and 335 of the large subunit and the corresponding region in the carboxy-terminal half (residues 707–873) are homologous to biotin-dependent carboxylases (Simmer *et al.* 1990; Kondo *et al.* 1991), with about 30 per cent of the residues conserved in all CPSs and biotin-dependent carboxylases sequenced thus far. The sequences 161–179 and 295–302, and the homologous sequences in the COOH-terminal half (residues 707–723 and 837–844) are particularly conserved in CPSs and biotin enzymes (Kondo *et al.* 1991). The more proximal of these conserved sequences (BC1, Fig. 17.4) overlaps with the beginning of the adenylate kinase-like putative nucleotide-binding sequence identified by Lusty *et al.* (1983), and the more distal (BC2, Fig. 17.4) ends 5–8 amino residues upstream of the phosphoglycerate kinase-like putative nucleotide-binding sequences.

Carbamate kinase, a 310-residue enzyme, is also homologous to the proximal 305 residues of each half of the synthetase (Baur *et al.* 1989). The region of homology ends immediately before the phosphoglycerate kinase-like region of CPS. Although the number of conserved residues is lower than in biotin enzymes (only 22–24 per cent of the residues in the region of homology is conserved in all CPSs and in carbamate kinase), the homology extends over the entire length of carbamate kinase (except for its initial 28 residues). The homology is somewhat greater for the COOH-terminal half than for the NH_2-terminal half of CPS and is stronger in the glycine-rich regions located at the beginning and at the end of the adenylate kinase-like region of CPS.

Mutational analysis

The study of mutants with deletions in the large subunit of the *E. coli* enzyme demonstrated that the NH_2-terminal 173 residues of the subunit and the equivalent region in the COOH half of the subunit are essential for association with the small subunit (Guillou *et al.* 1989). In contrast, deletions of up to 350 residues from the COOH terminus of the enzyme had no detrimental effect on the association with the small subunit. These deletions and a deletion of the COOH-terminal 171 residues of the large subunit resulted in the inability of the cells containing the mutant enzymes to grow on medium containing either limiting or high concentrations of NH_4^+. Therefore, these deletions impair the catalytic activity of the enzyme.

Post *et al.* (1990) replaced glycines 175, 179, and 721 by isoleucine, and aspartate 732 by alanine in the large subunit of the enzyme from *E. coli* (the numbering is that given by Nyunoya and Lusty (1983) for the mature large subunit, and does not take into consideration the initial methionine, which is cleaved off). Glycines 175 and 721 are homologous residues, and are located in the adenylate kinase-like region of each half (Fig. 17.4). They are totally conserved in all CPSs sequenced thus far, in biotin-dependent carboxylases and in carbamate kinase. Glycine 179 is located in the adenylate kinase-like region of the NH_2-terminal half, and is also conserved completely in CPSs and in biotin enzymes (Fig. 17.4), but not in carbamate kinase or in the COOH-terminal half of CPS. Post *et al.* (1990) showed that these three glycines are essential for carbamoyl phosphate synthesis. Replacement of glycines 175 and 179 drastically decreased the ATPase partial activity but not the partial reaction of ATP synthesis, and the reverse was observed for the replacement of glycine 727. Replacement of aspartate 732 by alanine had little effect. It was concluded that glycines 175 and 179, and thus the NH_2-terminal half of the subunit, are involved in the synthesis of carboxyphosphate, and that glycine 721, and therefore the COOH-terminal half of the subunit, is involved in the phosphorylation of carbamate.

Guillou *et al.* (1992) have replaced in the *E. coli* enzyme glutamate 840 by lysine. This residue is located in the COOH-terminal half of the large subunit, in the more distal of the two regions of highest homology with biotin-dependent enzymes (Fig. 17.4). It is conserved in all CPSs and in biotin enzymes. Its presence in carbamate kinase is less certain, for it is in a region of low homology where different alignments are possible. A glutamate is also present in the equivalent position in the amino-terminal half of CPSs. The results of Guillou *et al.* (1992) show that glutamate 840 is not essential for the binding of substrate nucleotide, but it is essential for the phosphorylation of carbamate, thus confirming the involvement of the COOH-terminal half of the subunit in this phosphorylation step. However, they found that this mutation also substantially affects the kinetic properties of the ATPase partial reaction, suggesting that this residue is involved in the coupling between the two phosphorylation steps of the reaction.

Mapping of functions in structural domains Limited proteolysis experiments with CPS I (Powers-Lee and Corina 1986; Guadalajara 1987; Evans and Balon 1988; Marshall and Fahien 1988) revealed that the enzyme is composed, from the NH_2-terminus, of four domains of about 40, 40, 60, and 20 kDa. We will designate these domains A, B, C, and D. Domain A is homologous to the entire small subunit of the *E. coli* enzyme. Domains B, C, and D correspond to the large subunit; the two points of cleavage that separate these three domains are located in an approximately equivalent place in each homologous half, at the end of the region of high homology between the two halves (Figs 17.4 and 17.5). Thus, the two halves appear to share a similar structural organization revealed by limited proteolysis. Studies with the *E. coli* (Rubio *et al.* 1991) and CAD CPSs (Kim *et al.* 1991) have demonstrated similar preferential points of cleavage and thus a similar domain structure, which is probably the same in all CPSs.

Photoaffinity labelling of CPS I with the analogue of acetylglutamate (the allosteric activator of the enzyme) chloroacetyl-[^{14}C]-L-glutamate, demonstrated (Rodriguez-Aparicio *et al.* 1989) the exclusive labelling of the D domain. Similarly, photolabelling of the *E. coli* enzyme with its inhibitor, UMP, labelled only domain D (Rubio *et al.* 1991) (Fig. 17.4). The binding sites for UMP and for the activator IMP have been reported to overlap (Boettcher and Meister 1982), and thus IMP is likely to bind in domain D. Scanning microcalorimetry experiments (Cervera *et al.* 1991) also support the binding of the activator ornithine in domain D of the *E. coli* enzyme. This effector, as well as IMP and UMP, influence the denaturational isotherm of the intact enzyme but not that of a mutant enzyme in which domain D is deleted. We have proposed that domain D is the allosteric or regulatory domain (Rodríguez-Aparicio *et al.* 1989; Rubio *et al.* 1991). Further evidence for this proposal was provided by the finding that the *in vitro* phosphorylation of CAD by cAMP-dependent protein kinase results in decreased inhibition by UTP (Carrey *et al.* 1985), and that the site of phosphorylation is within domain D (Carrey and Hardie 1988) (Fig. 17.4).

Domain D of CPS from *E. coli* exhibits some similarities to the regulatory chain of *E. coli* aspartate transcarbamylase: similar mass, binding of a purine nucleotide activator and a pyrimidine nucleotide inhibitor at overlapping sites (Kantrowitz and Lipscomb 1988; Allewell 1989), inhibition by UMP or UTP (Wild *et al.* 1989), and some degree of sequence homology (for residues 900–1031 of CPS, 24.8 per cent identities, 15.6 per cent conservative replacements, and 5 gaps/100 residues), which is comparable to the degree of homology between the D domain of different CPSs or between this domain and the corresponding region in the amino-half of the large subunit (Rubio *et al.* 1991). It is tempting to speculate that the D domain and the aspartate transcarbamylase regulatory chain derive from a common ancestor. Since a number of enzymes are activated or inhibited by nucleotides or nucleotide derivatives, it would be worth investigating the structural basis and evolution of this type of regulatory function, because a common origin and mechanism of regulation may exist.

From the data reviewed thus far a model might be proposed in which domain B is responsible for the binding of ATP_A and for carboxyphosphate synthesis, domain C is responsible for ATP_B binding and for the phosphorylation of carbamate, and domain D is responsible for binding of the effectors and for regulation. However, several facts indicate a greater complexity. For example, proteolytic cleavage of domain D inactivates simultaneously the overall reaction and the two partial reactions (Guadalajara 1987; Guadalajara *et al.* 1987), as expected if domain D is involved in the catalysis of the two phosphorylation steps. The parallelism between the effects of acetylglutamate on the binding of ATP_A (Rubio *et al.* 1983) and on the accessibility of the site of cleavage between domains C and D in CPS I, the protection from this cleavage by the binding of MgATP, and the similarity of the requirements for this protection and for the binding of ATP_A (Guadalajara 1987; Marshall and Fahien 1988) are highly suggestive of the binding of ATP_A at the hinge between domains C and D. This is supported also by the results of oxidative inactivation experiments with CPS I (Alonso *et al.* 1992) which demonstrated the preferential oxidation of cysteines 1327 and 1337 (highly reactive cysteines (Geschwill and Lumper 1989) located in the hinge between domains C and D) and of residues in domain B, in the presence of acetylglutamate, and the protection from oxidation by the removal of acetylglutamate or by the addition of ATP. In addition, since the oxidation in the presence of acetylglutamate results in the simultaneous loss of both partial reactions, it appears that the oxidation of these regions interferes with the binding of bicarbonate, for this is the only substrate that participates (as such or as a moiety of carboxyphosphate, carbamate, and carbamoyl phosphate) in all the steps of the reaction. Oxidation in the absence of acetylglutamate and in the presence of ATP resulted in selective inactivation of the partial reaction of ATP synthesis, and was associated with the preferential oxidation of residues in domain C, confirming (Post *et al.* 1990) the involvement of this domain in the reactions in which ATP_B is involved.

To account for the interactions in CPS I between acetylglutamate binding and ATP_A binding (Alonso and Rubio 1983; Rubio *et al.* 1983), it was postulated (Rodríguez-Aparicio *et al.* 1989; Rubio *et al.* 1990) that domain D (the regulatory domain) is involved in facilitating the binding of ATP_A and in closure of the binding site to allow formation of carboxyphosphate in the absence of water. It was proposed, to render possible the binding of ATP_A at the interface between domains B and D and to account for the influence of the effectors on ATP_A binding, that the two homologous halves of the large subunit fold as a homodimer in complementary isologous association (Fig. 17.5), thus placing domain D close to domain B (Rubio *et al.* 1991). This arrangement is also consistent with the regions of interaction between the two subunits mapped recently by Guillou *et al.* (1989) and, in addition, allows the interaction between domains C and D. Such interaction may be demanded by the effects of the allosteric modulators on the partial reaction of ATP synthesis (Meister 1989) and by the results of labelling CPS I with the ATP analog FBSA (see below).

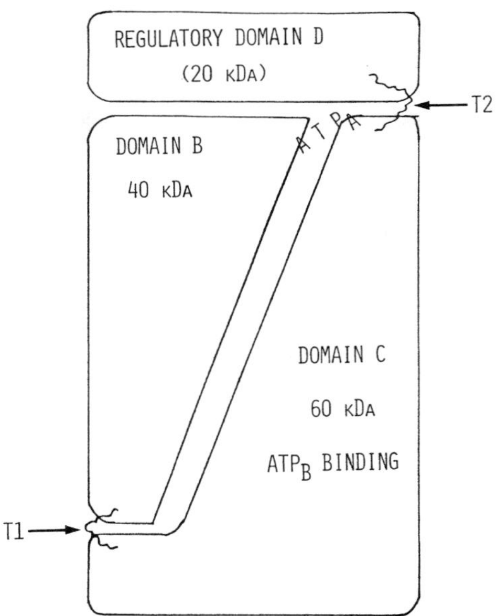

Fig. 17.5 Schematic representation of how isologous pseudohomodimerization of the large subunit of CPSs (or the equivalent part of the chain of those CPSs that do not have a separate large subunit) brings the nucleotide site involved in the reactions of ATP_A close to the regulatory domain. ATP_A is shown binding at the interface between domains B and D. The model also allows the interaction of the regulatory domain with the domain that is involved in the binding of ATP_B. In fact, the reaction is likely to take place at the interface between the domains. The regions that are preferentially cleaved by proteases are shown as flexible hinges labelled T1 and T2. The allosteric effectors bind to the regulatory domain and influence the relative positions of the three domains. In particular, in CPS I acetylglutamate binding relaxes the structure so that the site for ATP_A becomes accessible, and the binding of ATP_A at the interface between the regulatory domain and the 40-kDa domain (domain B) results in a much tighter structure (Rubio *et al.* 1990); Rodríguez-Aparicio *et al.* 1989).

The ATP affinity labels 8-azido-ATP and 5′-*p*-fluorosulphonyl-benzoyl-adenosine (FSBA) have been used to try to locate ATP binding sites on the enzyme. The incorporation of two moles of 8-azido-ATP per mole of CPS I was required for complete inactivation, and all enzyme domains were labelled, but a more precise localization of the label was not attempted (Powers-Lee and Corina 1987). When FSBA was used with CPS I (Potter and Powers-Lee 1992) four sequences were found to be labelled (Fig. 17.4): (1) the more distal of the two ATP-binding sequences proposed by Powers-Lee and Corina (1987) in domain B; (2) a sequence at the end of domain C; (3) the highly conserved sequence at the hinge between domains C and D; and (4) a sequence 50 residues from the COOH-terminus, in domain D. Since acetylglutamate strongly influences ATP_A

binding in CPS I (Rubio *et al.* 1983) and also influences the labelling of sequences 1 and 3, these two peptides were proposed to participate in the steps in which ATP_A is involved, whereas sequences 2 and 4 were proposed to be involved in the steps involving ATP_B (Potter and Powers-Lee 1992). Thus, oxidation experiments and FSBA-labelling agree in implicating not only domain B but also the hinge between domains C and D in the reaction with ATP_A. The involvement of domain D in ATP_B binding is not unreasonable. Domains C and D are contiguous and ATP_B may bind at points of contact between the two domains. Furthermore, as already discussed, the effectors influence the two partial reactions (Meister 1989). Overall, the evidence is accumulating to support the idea that the reaction takes place at the interface between domains B, C, and D and that the effectors bind to domain D and change the geometry of the relations between these three domains.

Surprisingly, when CPS II from hamster CAD was labelled with FSBA (Kim *et al.* 1991) none of the five sequences found to be labelled overlapped with the labelled sequences in CPS I (Potter and Powers 1992). Two of the sequences labelled in CAD (Fig. 17.4) are located in domain B, one in the more proximal ATP-binding sequence proposed by Powers-Lee and Corina (1987) and the other in the region 10 residues downstream. The remaining three labelled sequences are found in domain C: (a) one sequence is a part of the NH_2-terminal half, and begins 10 residues downstream from the NH_2-terminus of domain C; it is highly conserved in the two halves of all CPSs and it is not represented in biotin enzymes or in carbamate kinase; (b) another labelled sequence begins approximately in residue 40 of the COOH-terminal half; and (c) the third labelled sequence encompasses the glutamate that was replaced by Guillou *et al.* (1992) in the *E. coli* enzyme (glutamate 840), within a highly conserved region shared with biotin enzymes.

In view of the multiplicity of sequences labelled with FSBA, additional site-directed mutagenesis studies involving these sequences will be required to assess their actual role in the reaction.

Concluding remarks

The gene coding for the large subunit of CPS may have arisen from fusion or recombination of an ancestral gene for carbamate kinase, which is homologous to the initial 300 residues of the COOH-terminal half, and the gene(s) encoding the phosphoglycerate kinase-like putative nucleotide binding site and the regulatory domain D. Part of the phosphoglycerate kinase-like region is shared also by biotin enzymes. The reaction of phosphoglycerate kinase involves an acylphosphate, and in this respect resembles the synthesis of carboxyphosphate. Therefore, the more conserved residues in the phosphoglycerate kinase-like region of the COOH-terminal half should be mutated to assess the importance of this component of the enzyme sequence in the reaction mechanism.

The COOH-terminal domain of CPSs and the regulatory chain of aspartate transcarbamylase may have evolved from an ancestral nucleotide-binding, regulatory domain. The sequences of the regulatory domains of as many proteins as possible, particularly of those in which nucleotides are effectors, should be compared. Information on the evolution of the regulatory domain of CPSs could shed light on the origins of allosteric regulation.

A proper understanding of CPS will require solving its three-dimensional structure. Nevertheless, mapping of functions in the sequence by the use of site-directed mutagenesis, affinity labelling or by other means should be of great help in solving structure–function relationships in CPSs, even after a complete three-dimensional image of the enzyme becomes available. Therefore efforts to map more functions should continue even if appropriate crystals for diffraction do not become available soon.

Note added in proof. Since this review was written, we have demonstrated by photo cross-linking the binding of IMP in domain D of *E. coli* CPS (Bueso, Lusty, and Rubio, unpublished). Guillou and Lusty (1993, *J. Biol. Chem.*, **32**, 1278–84) have shown that the replacement Glu840 by Lys in the large subunit of *E. coli* CPS alters the catalytic properties of the glutaminase subunit; thus, Glu840 is involved in the coupling of the function of the two subunits. Milles *et al.* (1993, *Biochemistry*, **32**, 232–40) have replaced by Asn the His242, 780, and 787 of the large subunit of *E. coli* CPS, showing that His242 is involved in the reaction of carboxyphosphate with ammonia. The results of the other two mutations, although less clear, support the implication of the COOH-terminal half of the large subunit in the phosphorylation of carbamate.

Acknowledgements

Supported by grants PB87-0189 and PB90-0665 from the Dirección General de Investigación Científica y Técnica.

References

Allewell, N.M. (1989). *Ann. Rev. Biophysics Biophys. Chem.,* **18,** 71–92.

Alonso, E. and Rubio, V. (1983). *Eur. J. Biochem.,* **135,** 331–7.

Alonso, E., Cervera, J., García-España, A., Bendala, E., and Rubio, V. (1992). *J. Biol. Chem.,* **267,** 4524–32.

Baur, H., Stalon, V., Falmagne, P., Luethi, E., and Haas, D. (1987). *Eur. J. Biochem.,* **166,** 11–17.

Baur, H., Luethi, E., Stalon, V., Mercenier, A., and Haas, D. (1989). *Eur. J. Biochem.,* **179,** 53–60.

Bein, K., Simmer, J.P., and Evans, D.R. (1991). *J. Biol. Chem.,* **266,** 3791–9.

Boettcher, B. and Meister, A. (1982). *J. Biol. Chem.,* **257,** 13 971–6.

Britton, H.G., Rubio, V., and Grisolia, S. (1979). *Eur. J. Biochem.,* **102,** 521–30.

Carrey, E.A. and Hardie, D.G. (1988). *Eur. J. Biochem.,* **171,** 583–8.

Carrey, E.A., Campbell, D.G., and Hardie, D.G. (1985). *EMBO J.,* **4,** 3735–42.

Cervera, J., Conejero-Lara, F., Ruiz-Sanz, J., Galisteo, M.L., Mateo, P.L., Lusty, C.J., and Rubio, V. (1993). *J. Biol. Chem.*, **268**, 12 504–11.

Chaparian, M.G. and Evans, D.R. (1991). *J. Biol. Chem.,* **266,** 3387–95.

Evans D.R. and Balon, M.A. (1988). *Biochim. Biophys. Acta,* **953,** 185–96.

Geschwill, K. and Lumper, L. (1989). *Biochem. J.,* **260,** 573–6.

Guadalajara, A.M. (1987). Unpublished D. Phil. thesis. University of Valencia.

Guadalajara, A., Grisolía, S. and Rubio, V. (1987). *Eur. J. Biochem.,* **165,** 163–9.

Guillou, F., Rubino, S.D., Markovitz, R.S., Kinney, D.M, and Lusty, C.J. (1989). *Proc.*

Nat. Acad. Sci. USA, **86,** 8304–8.
Guillou, F., Liao, M., García-España, A., and Lusty, C.J. (1992). *Biochemistry,* **31,** 1656–64.
Jones, M.E. (1976). In *The urea cycle* (ed. S. Grisolía, R. Báguena, and F. Mayor), pp. 107–22. Wiley, New York.
Kantrowitz, E.R. and Lipscomb, W.N. (1988). *Science,* **241,** 669–74.
Kim, H., Lee, L., and Evans, D.R. (1991). *Biochemistry,* **30,** 10 322–9.
Kondo, H., Shiratsuchi, K., Yoshimoto, T., Masuda, T., Kitazono, A., Tsuru, D. *et al.* (1991). *Proc. Nat. Acad. Sci. USA,* **88,** 9730–3.
Lehninger, A.L. (1971). *Bioenergetics* (2nd end). W.A. Benjamin, Menlo Park, CA.
Lusty, C.J., Widgren, E.E., Broglie, K.E., and Nyunoya, H. (1983). *J. Biol. Chem.,* **258,** 14 466–72.
Makoff, A.J. and Radford, A. (1978). *Microbiol. Rev.*, **42,** 307–28.
Marshall, M. (1976). In *The urea cycle* (ed. S. Grisolía, R. Báguena, and F. Mayor). pp. 133–42. Wiley, New York.
Marshall, M. and Cohen, P.P. (1966). *J. Biol. Chem.,* **241,** 4197–208.
Marshall, M. and Cohen, P.P. (1970). *Meth. Enzymol.,* **17,** 229–34.
Marshall, M. and Fahien, L.A. (1988). *Arch. Biochem. Biophys.,* **262,** 455–70.
Meister, A. (1989). *Adv. Enzymol.,* **62,** 315–74.
Metzenberg, R.L., Marshall, M., and Cohen, P.P. (1958). *J. Biol. Chem.,* **233,** 1560–4.
Miran, S.G., Chang, S.H., and Raushel, F.M. (1991). *Biochemistry,* **30,** 7901–7.
Mullins, L.S., Lusty, C.J., and Raushel, F.M. (1991). *J. Biol. Chem.,* **266,** 8236–40.
Nyunoya, H. and Lusty, C.J. (1983). *Proc. Nat. Acad. Sci. USA,* **80,** 4629–33.
Nyunoya, H. and Lusty, C.J. (1984). *J. Biol. Chem.,* **259,** 9790–8.
Nyunoya, H., Broglie, K.E., Widgren, E.E., and Lusty, C.J. (1985). *J. Biol. Chem.,* **260,** 9346–56.
Post, L.E., Post, D.J., and Raushel, F.M. (1990). *J. Biol. Chem.,* **265,** 7742–7.
Potter, M.D. and Powers-Lee, S.G. (1992). *J. Biol. Chem.,* **267,** 2023–31.
Powers-Lee, S.G. and Corina, K. (1986). *J. Biol. Chem.,* **261,** 15 349–52.
Powers-Lee, S.G. and Corina, K. (1987). *J. Biol. Chem.,* **262,** 9052–6.
Rodríguez-Aparicio, L.B., Guadalajara, A.M., and Rubio, V. (1989). *Biochemistry,* **28,** 3070–4.
Rubino, S.D., Nyunoya, H., and Lusty, C.J. (1986). *J. Biol. Chem.,* **261,** 11 320–7.
Rubio, V. (1986). *Biosci. Rep.,* **6,** 335–47.
Rubio, V., Britton, H.G., and Grisolia, S. (1979). *Eur. J. Biochem.,* **93,** 245–56.
Rubio, V., Britton, H.G., and Grisolía, S. (1983). *Eur. J. Biochem.,* **134,** 337–43.
Rubio, V., Britton, H.G., Rodríguez-Aparicio, L.B., and Climent, S. (1990). In *Enzymatic and model carboxylation and reduction reactions for carbon dioxide utilization* (ed. M. Aresta and J.V. Schloss), pp. 221–38. Kluwer, The Netherlands.
Rubio, V., Cervera, J., Lusty, C.J., Bendala, E., and Britton, H.G. (1991). *Biochemistry,* **30,** 1068–75.
Simmer, J.P., Kelly, R.E., Rinker, A.G., Scully, J.L., and Evans, D.R. (1990). *J. Biol. Chem.,* **265**, 10 395–402.
Souciet, J.L., Nagy, M., Le Gouar, M., Lacroute, F., and Pottier, S. (1989). *Gene,* **79,** 59–70.
Wellner, V.P., Anderson, P.M., and Meister, A. (1973). *Biochemistry,* **12,** 2061–6.
Wild, J.R., Loughrey-Chen, S.J., and Corder, T.S. (1989). *Proc. Nat. Acad. Sci. USA,* **86,** 46–50.

18

How enzymes deal with carbon dioxide and bicarbonate

Marion H. O'Leary

Introduction

Carbon dioxide is a key metabolite in living systems. Thermodynamically, it lies in a deep valley. Formation of carbon dioxide is the ultimate exothermic step in a variety of energy-producing metabolic processes. Absorption of carbon dioxide by green plants is a key endothermic step in photosynthesis.

The purpose of this article is to discuss Nature's strategies for dealing with reactions of carbon dioxide. At issue are the chemical mechanisms that are used in CO_2 uptake by carboxylases. Two principal points will be made:

1. Because of its nonpolar nature, CO_2 probably binds only weakly or not at all to most carboxylases; instead, carboxylation probably most often occurs as a bimolecular reaction between CO_2 in solution and some enzyme-bound intermediate.
2. Although some enzymes take up CO_2 from solution and some take up HCO_3^-, ultimately CO_2 appears to be the reactive species in all carboxylations.

Properties of CO_2 and HCO_3^-

Physical properties

The atmosphere currently contains about 340 ppm of CO_2. This translates into about 14 micromoles of CO_2 per litre of air near room temperature. When air reaches equilibrium with water at 15 °C, the concentration of CO_2 in solution is about the same as the concentration in air — about 14 μM (Butler 1982; Edsall and Wyman 1958; Hildebrand and Scott 1950). This solubility decreases at higher temperatures. The consequence of this for enzymes is that unless the organism has a mechanism for concentrating CO_2, the concentration of CO_2 available to the enzyme will always be low.

When the pH of that aqueous solution is near neutrality or above, CO_2 is hydrated in a reaction that may take a minute or more in the absence of carbonic anhydrase

$$CO_2 + H_2O \rightleftharpoons HCO_3^- + H^+$$

The effective pK for this interconversion is 6.3 (Butler 1982).

Thus, the total solubility of all 'CO_2' species in equilibrium with air (i.e. CO_2 plus HCO_3^- plus CO_3^{2-}) increases with increasing pH, particularly above pH 6.3. However, the concentration of CO_2 *per se* is independent of pH.

Reactivity of CO_2 and HCO_3^-

Carbon dioxide is quite reactive toward a variety of nucleophilic species. Organic chemists have long recognized this property and have made it the basis of a variety of reactions of CO_2 with organometallic reagents, enolates, etc. (Inoue and Yamazaki 1982). Bicarbonate, on the other hand, is not very reactive. Although dialkyl carbonates are reactive toward a variety of nucleophiles, monoalkyl carbonates and bicarbonate ion itself are unreactive, at least in the monoanionic form. Instead, reaction usually results in loss of CO_2. Although monoalkyl carbonates and bicarbonate ion might be expected to be reactive toward nucleophiles in acidic solution (i.e. under conditions where they are neutral, rather than negatively charged), such reactions are infrequently observed (Hegarty 1979). Certainly carbonate derivatives in which there is no negative charge (for example carbamyl phosphate) are reactive toward nucleophiles.

Binding CO_2 to enzymes

Since CO_2 is the substrate in a variety of enzymatic reactions, it is natural to assume that CO_2 forms noncovalent complexes (Michaelis complexes) with the enzymes for which it is a substrate. What interactions would be responsible for the formation of such a complex? CO_2 has no dipole moment; it is not polar. CO_2 does have a quadruple moment, and it is possible that quadrupolar interactions might be involved in the binding. The fact that CO_2 is 10–20 times more soluble in various solvents that O_2 is (Table 18.1) suggests that there may be significant solvation interactions between CO_2 and the solvent. The high solubility of CO_2 in solvents such as acetone suggests that the quadrupolar effect might be significant. However, as we will see below, little convincing evidence for CO_2 binding to enzymes exists.

Table 18.1 Solubility of CO_2 and O_2 at 25 °C*

Solvent	CO_2	O_2
Water	0.75	0.03
Benzene	2.57	0.23
Acetone	7.4	0.32
Ethanol	3.4	0.27

Source: Hildebrand and Scott (1950).

* In cc gas per g solvent at 1 atm gas.

Bicarbonate, on the other hand, is clearly polar and could easily bind to appropriately organized binding sites on a variety of enzymes. Ample evidence for HCO_3^- binding to enzymes exists.

An aside – covalent binding of CO_2 to enzymes

In the discussion to follow, we will be concerned with the question of whether CO_2 forms noncovalent complexes with enzymes. It should be noted at the outset that there are several well-known cases where CO_2 forms covalent complexes with proteins (Lorimer 1983). In haemoglobin, CO_2 decreases the affinity of the protein for O_2 by formation of a carbamate with N-terminal α-amino groups of valine residues (Roughton 1970). In ribulose bisphosphate carboxylase/oxygenase (see below), the enzyme is activated by carbamate formation with the side-chain amino group of a lysine residue (Lorimer 1983). Lorimer (1983) has suggested that carbamate formation in proteins may be more common than is generally supposed. The reaction is readily reversible, and is difficult to test for.

Identifying enzyme–CO_2 complexes

How would one ascertain the existence of a noncovalent enzyme–CO_2 complex during a carboxylation? Some of the basic issues are taken up here and then again in later sections.

The occurrence of saturation kinetics does not require the existence of a Michaelis complex. Enzymes generally operate by multistep mechanisms. When the CO_2 concentration is low, the reaction rate will be limited by CO_2 availability, whether or not there is a Michaelis complex. As the CO_2 concentration increases, the rate will increase. However, at some point the rate will begin to become limited by rates of other steps in the mechanism, and a hyperbolic dependence on CO_2 concentration will be observed even if there is no Michaelis complex.

In some cases, the kinetic patterns observed by variation of all substrate concentrations might be able to demonstrate that a complex exists. This is feasible in the case of a kinetically ordered mechanism in which CO_2 is not the last substrate to bind. However, in practice, such cases are rare (as noted below, CO_2 is generally the last substrate).

In favourable situations, carbon isotope effects may provide evidence for the formation of a Michaelis complex. For example, in a mechanism in which CO_2 is not the last substrate to bind, the observed carbon isotope effect will decrease at increasing concentrations of the last substrate. This method has been used in the case of PEP carboxykinase (see below).

Inhibitors can provide evidence for complex formation. If an inhibitor (for example COS, N_2O, NO_2^-) is competitive against CO_2, then this suggests that there is a binding site for the inhibitor, and thus, by inference, for CO_2. However, it is important to demonstrate that the inhibitor is not an alternate substrate. A number of cases are cited below.

Binding of small molecules to macromolecules can often be demonstrated by equilibrium dialysis or by gel exclusion chromatography. Neither of these methods has been successful for CO_2. This may be because such complexes do not exist or because the dissociation constants for the complexes are so high that they cannot be determined by these methods.

Magnetic resonance methods can sometimes be used to demonstrate complex formation. Most carboxylases are metalloproteins. If CO_2 (but not HCO_3^-) perturbs the metal (for example Mn^{2+}) NMR or EPR signal, that probably means that CO_2 binds. As noted below, this method has been used for PEP carboxykinase.

Ribulose bisphosphate carboxylase/oxygenase

CO_2 uptake in plants is catalysed by ribulose bisphosphate carboxylase/oxygenase (Andrews and Lorimer 1987), which catalyses the reaction

$$\text{ribulose bisphosphate} + CO_2 + H_2O \quad \rightarrow \quad 2\ \text{3-PGA}$$

In spite of a very large number of studies, there is no convincing evidence for a Michaelis complex with CO_2. The reaction mechanism is ordered, with CO_2 the last substrate (technically, H_2O is the last substrate, but its concentration cannot be varied). Oxygen is a competitive inhibitor against CO_2, but this is because oxygen is actually an alternate substrate. Earlier work (Laing and Christeller 1980) showed that COS is a competitive inhibitor against CO_2, but noncompetitive against ribulose bisphosphate. However, Lorimer and Pierce (1989) showed that COS is actually a substrate, so this experiment provides no information on the binding of CO_2. Hydrogen peroxide is also competitive against CO_2, but this is probably because there is actually a reaction with H_2O_2 (Badger *et al.* 1980).

Isotope effects have also been used to study this mechanism. Carbon isotope effects indicate that CO_2 is the innermost substrate, but give no evidence on the issue of whether there is a Michaelis complex (Roeske and O'Leary 1984). Hydrogen isotope effects have been interpreted in terms of a Theorell–Chance mechanism, and Van Dyk and Schloss (1986) suggested that there is no Michaelis complex with CO_2.

Malic enzyme

Malic enzyme catalyses the oxidative decarboxylation of malate to form pyruvate and CO_2

$$\text{malate} + \text{NADP} \quad \rightleftharpoons \quad \text{NADPH} + CO_2 + \text{pyruvate}$$

Enzyme-bound oxalacetate is an intermediate. The reaction is readily reversible. The principle of microscopic reversibility requires that the mechanism be the same in both directions, so a variety of approaches might be used to study mechanism. Isotope effects and steady-state kinetics suggest that CO_2 is the innermost substrate, and in such cases it is not possible to tell whether there is a complex (Hermes *et al.* 1982; Mallick *et al.* 1991).

Mallick *et al.* (1991) showed that NO_2^- (which is isoelectronic with CO_2) is an inhibitor of the reaction. Unfortunately, it is noncompetitive against CO_2, and thus it provides no information about a binding site for CO_2. Mallick *et al.* (1991) suggest that there may be no binding site for CO_2.

PEP carboxykinase

Phosphoenolpyruvate carboxykinase catalyses the reversible, nucleotide-dependent carboxylation of PEP (Utter and Kolenbrander 1972)

$$\text{PEP} + \text{NDP} + CO_2 \rightleftharpoons \text{OAA} + \text{NTP}$$

The substrate for the enzyme is CO_2, rather than HCO_3^-, and a divalent metal ion is required (perhaps to stabilize the enolate that is the ultimate acceptor of CO_2).

Carbon isotope effects in the carboxylation direction vary with substrate concentration (Table 18.2). This indicates that CO_2 is not the last substrate to bind. In turn, this requires that there be a binding site for CO_2.

Kinetic modification studies also suggest that there is a CO_2 binding site. Cheng and Nowak (1989) showed that PEP carboxykinase has an arginine at the active site, and CO_2 protects against arginine modification. This can most likely happen if CO_2 binds to the active site. NMR studies lead to a similar conclusion: Proton relaxation rate measurements suggest that an enzyme–Mn^{2+}–CO_2 complex exists (Hebda and Nowak 1982). Thus, in this case, several lines of evidence indicate that CO_2 forms a Michaelis complex.

Carbonic anhydrase

Carbonic anhydrase is perhaps the most studied of all enzymes for which CO_2 is a substrate

$$CO_2 + OH^- \rightleftharpoons HCO_3^-$$

The mechanism involves reaction of CO_2 with a zinc-bound OH^- (Silverman and Vincent 1983; Silverman and Lindskog 1988). The enzyme shows saturation

Table 18.2 Carbon isotope effects on PEP carboxykinase (pH 7.5, 25 °C)

[PEP] (mM)	[ADP] (mM)	KIE
22	1.3	1.024
1	1.3	1.035
22	0.06	1.040

Source: Arnelle and O'Leary (1992).

kinetics in both directions. The K_m for CO_2 is about 10 mM. However, as noted previously, this does not require that there be a binding site for CO_2. X-ray crystallography shows possible CO_2 binding sites, but does not distinguish a specific one. Molecular mechanics calculations based on the X-ray structure have identified possible binding sites (Merz 1991; Liang and Lipscomb 1990). N_2O, which is isoelectronic with CO_2, is not an inhibitor (Khalifah 1971). A tentative CO_2 binding site was identified from difference infrared spectroscopy (Riepe and Wang 1968), but this work has been questioned (Khalifah 1971).

A variety of NMR techniques are potentially capable of identifying a CO_2 binding site. Most ^{13}C NMR studies are compromised by the fact that HCO_3^- is present in high concentration and clearly binds to the metal. Whether CO_2 (present in lower concentration) also binds is problematical (Led and Neesgaard 1987; Bertini *et al.* 1979; Simonsson *et al.* 1979). It is clear from the NMR experiments that HCO_3^- is coordinated to the metal, whereas CO_2 is not. It is not certain, however, whether CO_2 binds at all.

The carbon substrate for carboxylations

There is a subclass of carboxylases that use HCO_3^- as substrate, rather than CO_2. These include phosphoenolpyruvate carboxylase and all the biotin-dependent carboxylases. Distinction between the two classes can be told by careful kinetics or by isotope studies (O'Leary and Hermes 1987). In the paragraphs below we examine the nature of the reaction with HCO_3^-.

Phosphoenolpyruvate carboxylase

PEP carboxylase catalyses the key carboxylation reaction in C_4 plants (O'Leary 1982; Andreo *et al.* 1987)

$$CH_2{=}\underset{}{\overset{O-PO_3^{2-}}{\overset{|}{C}}}-CO_2^- + HCO_3^- \longrightarrow {}^-O_2C-CH_2-\overset{O}{\overset{\|}{C}}-CO_2^- + P_i^{2-}$$

Maruyama *et al.* (1966) showed that one atom of ^{18}O is transferred from $HC^{18}O_3^-$ to P_i during the carboxylation. The reaction has a small ^{13}C isotope effect for the bicarbonate carbon (O'Leary *et al.* 1981) and shows inversion of configuration at phosphorus (Hansen and Knowles 1982). The mechanism appears to be stepwise involving rate-determining formation of carboxyphosphate, followed by carboxylation of the resulting enolate

$$PEP + HCO_3^- \longrightarrow {}^-O_2C-O-PO_3^{2-} + CH_2{=}\overset{O^-}{\overset{|}{C}}-CO_2^- \longrightarrow OAA + P_i$$

3-Fluoro-PEP is a substrate for PEP carboxylase, being carboxylated at a rate less then an order of magnitude down from the rate for PEP itself. Both the (*E*) and (*Z*) isomers of fluoro-PEP give rise to a mixture of carboxylation and hydrolysis. Use of $HC^{18}O_3^-$ gives rise to ^{18}O incorporation in substrate and about two equivalents of ^{18}O in the P_i that is formed (Janc *et al.* 1992*a*). Methyl-PEP also gives rise to this same labelling pattern (O'Laughlin 1988; Fujita *et al.* 1984).

Formate is an alternate substrate for PEP carboxylase. Labelling studies indicate that formyl phosphate is formed, but the only product is pyruvate; no formylpyruvate is formed (Janc *et al.* 1992*b*).

The most parsimonious interpretation of these results is that within the active site of PEP carboxylase, carboxyphosphate decomposes to form CO_2 and P_i; carboxylation involves reaction of enolpyruvate with CO_2

$$\mathrm{HO{-}\overset{\overset{\displaystyle O}{\|}}{C}{-}O{-}PO_3^{2-}} \longrightarrow \mathrm{O{=}C{=}O + HOPO_3^{2-}}$$

$$\mathrm{CO_2 + enolpyruvate} \longrightarrow \mathrm{OAA}$$

Isotope effects obtained with fluoro-PEP are quantitatively consistent with the free CO_2 mechanism, but not with the direct reaction of carboxyphosphate (Janc *et al.* 1992*a*). The fact that formate apparently forms formyl phosphate but is unable to formylate the enolate is consistent with the idea that carboxyphosphate (or formyl phosphate) does not react with the enolate.

Biotin-dependent enzymes

Biotin-dependent enzymes share with PEP carboxylase the use of HCO_3^-, rather than CO_2, as the one-carbon substrate. These enzymes operate by way of a carboxybiotin intermediate. As outlined below, enzyme-bound CO_2 is apparently the key carboxylating intermediate (Knowles 1989).

Biotin-dependent carboxylation of such acceptors as pyruvate, acetyl-CoA, propionyl-CoA can be shown as

$$HCO_3^- + ATP + R{-}H \rightleftharpoons R{-}CO_2^- + ADP + P_i$$

The energy derived from ATP is used to form carboxybiotin in a two-step process. The first step is reaction of ATP with HCO_3^- to form carboxyphosphate in a process reminiscent of that for PEP carboxylase

$$ATP + HCO_3^- \rightleftharpoons ADP + {}^-O_2C{-}O{-}PO_3^{2-}$$

Then this carboxyphosphate reacts with enzyme-bound biotin to form carboxybiotin

$$\text{E-biotin} + {}^{-}O_2C\text{–}O\text{–}PO_3^{2-} \rightleftharpoons \text{E-biotin–}CO_2^{-} + P_i$$

In the final step, carboxybiotin is used to carboxylate the substrate

$$\text{E-biotin} - CO_2^{-} + \text{R–H} \rightleftharpoons \text{E-biotin} + \text{R–}CO_2^{-}$$

The detailed chemistry of the two carboxyl transfer steps is controversial for many of the same reasons already cited, and a variety of mechanisms have been suggested. The large body of literature on the subject was recently summarized by Knowles (1989), who argues that both transfers involve enzyme-bound CO_2. The detailed argument is particularly persuasive because of the parsimonious nature of the mechanism, involving an alternating cycle of proton transfers and CO_2 transfer.

Conclusion

Living systems that must take up one-carbon units are provided with a choice of whether to use CO_2 or HCO_3^-. The dilemma is that whereas CO_2 is the more reactive species, HCO_3^- is the form that is easier to bind. Although earlier evidence had suggested that the mechanisms for incorporation of the one carbon unit might be fundamentally different for the two types of enzymes, it now appears that carbon dioxide is universally the one-carbon substrate for carboxylations. Enzymes that initially bind HCO_3^- ultimately convert this material to CO_2 for reaction. Enzymes which use CO_2 initially appear in most cases to react with CO_2 directly in a bimolecular reaction, rather than by forming a Michaelis complex.

Acknowledgement

Thanks are due to associates and collaborators who have participated in this work. Particular thanks are due to W.W. Cleland, who has been an invisible intellectual partner in many of these investigations. Thanks are also due to Derrick Arnelle, Scott Ausenhus, Jeff Hermes, James Janc, Janet O'Laughlin, Piotr Paneth, James Rife, and Chrissl Roeske. The research was supported by the National Institutes of Health, the National Science Foundation, and the US Department of Agriculture.

References

Andrews, T.J. and Lorimer, G.H. (1987). In *The biochemistry of plants,* (ed. M.D. Hatch and N.K. Boardman), Vol. 10, p. 131. Academic Press, New York.

Arnelle, D.A. and O'Leary, M.H. (1992). *Biochemistry,* **31,** 4363–8.

Badger, M.R., Andrews, T.J., Canvin, D.T., and Lorimer, G.H. (1980). *J. Biol. Chem.,* **255,** 7870–5.

Bertini, I., Borghi, E., and Luchinat, C. (1979). *J. Am. Chem. Soc.*, **101,** 7069–71.
Butler, J.N. (1982). *Carbon dioxide equilibria and their applications.* Addison Wesley, Reading, MA.
Cheng, K.C. and Nowak, T. (1989). *J. Biol. Chem.*, **264,** 3317–24.
Edsall, J.T. and Wyman, J. (1958). In *Biophysical chemistry,* Chap. 10. Academic Press, New York.
Fujita, N., Izui, K., Nishino, T., and Katsuki, H. (1984). *Biochemistry,* **23,** 1774–9.
Hansen, D.E. and Knowles, J.R. (1982). *J. Biol. Chem.*, **257,** 14 795–8.
Hebda, C.A. and Nowak, T. (1982). *J. Biol. Chem.*, **257,** 5515–22.
Hegarty, A.F. (1979). In *Comprehensive organic chemistry,* (ed. D.H.R. Barton), Vol. 9, p. 1967. Pergamon Press, New York.
Hermes, J.D., Roeske, C.A., O'Leary, M.H., and Cleland, W.W. (1982). *Biochemistry,* **21,** 5106–14.
Hildebrand, J.H. and Scott, R.L. (1950). *The solubility of nonelectrolytes.* Reinhold, reprinted in 1964 by Dover, p. 248. New York.
Inoue, S. and Yamazaki, N. (1982). *Organic and bio-organic chemistry of carbon dioxide.* Halsted Press, New York.
Janc, J.W., O'Leary, M.H., and Cleland, W.W. (1992*a*). *Biochemistry,* **31,** 6432–40.
Janc, J.W., Cleland, W.W., and O'Leary, M.H. (1992*b*). *Biochemistry,* **31,** 6441–6.
Khalifah, R.G. (1971). *J. Biol. Chem.*, **246,** 2561–73.
Knowles, J.R. (1989). *Ann. Rev. Biochem.*, **58,** 195–221.
Laing, W.A. and Christeller, J.T. (1980). *Arch. Biochem. Biophys.*, **202,** 592–600.
Led, J.J. and Neesgaard, E. (1987). *Biochemistry,* **26,** 183–92.
Liang, J.-Y. and Lipscomb, W.N. (1990). *Proc. Nat. Acad. Sci. USA*, **87,** 3675–9.
Lorimer, G.H. (1983). *TIBS,* **7,** 65–71.
Lorimer, G.H. and Pierce, J. (1989). *J. Biol. Chem.*, **264,** 2764–72.
Mallick, S., Harris, B.G., and Cook, P.F. (1991). *J. Biol. Chem.*, **266,** 2732–8.
Maruyama, H., Easterday, R.L., Chang, H.-C., and Lane, M.D. (1966). *J. Biol. Chem.*, **241,** 2405–12.
Merz, K.M., Jr (1991). *J. Am. Chem. Soc.*, **113,** 406–411.
O'Laughlin, J.T. (1988). Ph.D. Thesis. University of Wisconsin-Madison.
O'Leary, M.H. (1982). *Ann. Rev. Plant Physiol.*, **33,** 297–315.
O'Leary, M.H. and Hermes, J.D. (1987). *Anal. Biochem.*, **162,** 358–62.
O'Leary, M.H., Rife, J.E., and Slater, J.D. (1981). *Biochemistry,* **20,** 7308–14.
Roeske, C.A. and O'Leary, M.H. (1984). *Biochemistry,* **23,** 6275–84.
Roughton, F.J.W. (1970). *Biochem. J.*, **117,** 801–12.
Silverman, D.N. and Vincent, S.H. (1984). *CRC Crit. Rev. Biochem.*, **14,** 207–255.
Silverman, D.N. and Lindskog, S. (1988). *Acc. Chem. Res.*, **21,** 30–6.
Simonsson, I., Jonsson, B.-H., and Lindskog, S. (1979). *Eur. J. Biochem.*, **93,** 409–17.
Utter, M.F. and Kolenbrander, H.M. (1972). *Enzymes* (3rd edn) (ed. P.D. Boyer), Vol. 6, pp. 117–68. Academic Press, New York.
Van Dyk, D.E. and Schloss, J.V. (1986). *Biochemistry,* **25,** 5145–56.

19

Cyanobacterial carbon dioxide-concentrating mechanism

Murray R. Badger, G.D. Price, and J-W. Yu

Coevolution of Rubisco and CO_2 concentrating mechanisms

Photosynthesis is the process by which organisms are able to acquire inorganic carbon ($C_i = CO_2 + HCO_3^- + CO_3^{2-}$) and convert it into organic sugars which can then be used for the metabolic needs of the cell. This process of carbon fixation has developed with a number of evolutionary constraints against which, it may be argued, that organism have struggled to optimize the efficiency of this process under a range of environmental conditions. In particular, the extent to which the primary substrate for this process, CO_2, has operated as a limiting factor has changed dramatically over the time span of evolution of photosynthetic organisms. In view of this, it is not surprising that there has been a great deal of adaptation by various organisms in the mechanisms which have developed to minimize the limitation which C_i availability imposes on photosynthesis.

The primary CO_2 fixing enzyme in all photosynthetic organisms, Rubisco, appears to have arisen once in the evolution of life, presumably appearing with the chemolithotrophs, which arose some 3.5×10^9 years ago (Broda 1975). In these environments, the CO_2 level was high and the O_2 was very low. Carbon dioxide was unlikely to have presented a substrate limitation to photosynthesis, and Rubisco may have operated close to CO_2 saturation in these organisms. In addition, the deleterious effects of the oxygenase activity of Rubisco would not have been significant and photorespiration was therefore not a problem.

Eventually, however, the operation of CO_2 consuming processes, particularly Rubisco carboxylation itself, led to a decline in the level of atmospheric CO_2. Furthermore, with the advent of cyanobacteria, this was coupled to oxygenic photosynthesis and the oxygen level began to rise. The net result of these changes was that the CO_2/O_2 ratio fell dramatically and the atmospheric and aquatic environments became progressively less favourable for carboxylation by Rubisco. Eventually, the positive carbon balance of photosynthesis would have been threatened if primitive Rubiscos had kinetic properties similar to those of present-day photochemotrophs such as *R. rubrum*. These Rubiscos have low affinities for CO_2 and highly effective oxygenase activities.

Photosynthetic organisms have adapted to the changing C_i conditions of the environment in two ways. First, Rubisco itself shows some limited ability to adapt to lower CO_2/O_2 ratios, by increasing its affinity for CO_2 and decreasing the relative effectiveness of the oxygenase reaction (Badger and Andrews 1987). Second, organisms have developed mechanisms to insulate Rubisco from the external environment in compartments where the CO_2 ratio is elevated (Badger 1987).

In looking at the broad range of aerobic photosynthetic organisms, it has become clear that the majority have developed CO_2-concentrating mechanisms (CCMs). These CCMs have three essential elements which are emulated in all systems which have been examined in detailed (Fig. 19.1).

1. A process/mechanism for the capture of external C_i and its conversion to an intermediate carbon pool (either organic or inorganic carbon), using photosynthetic energy.
2. A compartment within the organism where Rubisco is co-localized with the machinery necessary to liberate CO_2 from the intermediate pool.
3. A mechanism for minimizing the leakage of the released CO_2 from the region where Rubisco is contained.

It is the balance which is struck between the rate at which CO_2 is liberated within the vicinity of Rubisco and the ease with which it leaks back to the external environment or is fixed by Rubisco, which determines the magnitude of the CO_2 elevation. Thus overall, the operation of a CCM improves both the acquisition of C_i from the external environment and the efficiency with which Rubisco performs carboxylation.

The most obvious example of organisms which have developed a CCM are C_4 higher plants, where PEP carboxylase is the primary component of C_i acquisition and fixes CO_2 into an intermediate malic acid pool. In addition, Rubisco is localized in the bundle sheath where the malic acid is decarboxylated to release CO_2 (Hatch 1987). However, similar effects have been achieved by many aquatic phototrophs. (Badger 1987). Aquatic organisms ranging from cyanobacteria and green microalgae to both marine and freshwater macrophytes, use C_i

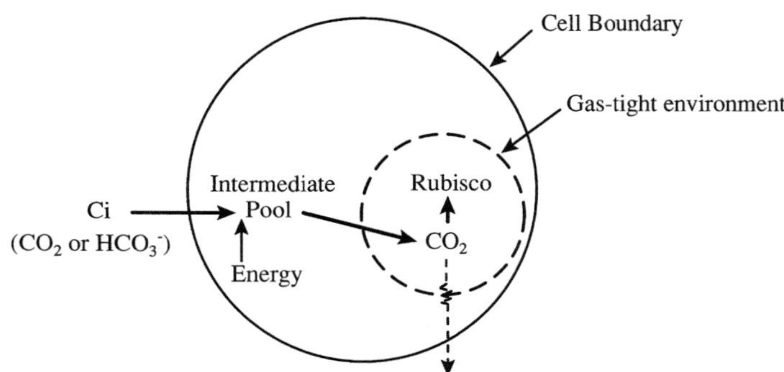

Fig. 19.1 Schematic model for the operation of a CO_2 concentrating mechanism (see text for explanation).

pumping mechanisms to elevate their intracellular [CO_2]. These pumping mechanisms vary in the extent to which CO_2 can be elevated, with cyanobacteria being the most effective, increasing internal C_i and CO_2 some 50–1000-fold over the external environment, while green algae and C_4 plants may achieve a 10–50-fold accumulation (Badger 1987).

Among the aerobic phototrophs, there is a general relationship between the kinetic properties of Rubisco and the effectiveness of the CO_2 concentrating mechanism which supports its activity. In evolutionary terms, the K_c of Rubisco increases some 10–20-fold going from C_3 plants to cyanobacteria and this is consistent with cyanobacteria having the greatest ability to concentrate CO_2. Between these extremes lie C_4 plants, aquatic macrophytes and green micro-algae, with K_cs some 2–4-fold higher than C_3 plants. All these intermediate groups may have various abilities to concentrate CO_2. The $K_{cat(c)}$ of the enzyme also increases over this evolutionary progression, but to a lesser extent (see Badger and Andrews 1987).

Photosynthesis in cyanobacteria

Cyanobacteria are prokaryotic organisms which lack organelles such as chloroplasts and mitochondria and are able to perform oxygen-evolving photosynthesis and respiration (for detailed treatments, see reviews in Packer and Glazer 1988). The photosynthetic electron-transport chain and light harvesting machinery are localized exclusively on the thylakoid membrane system. However, components of respiratory electron-transport chains are found on both the cytoplasmic and thylakoid membranes and indeed, some components of electron transport are envisaged as being shared on the thylakoid, with the cytochrome b_6/f complex acting as a branch point in electron flow. A distinct feature of cyanobacteria is the nature of their major light harvesting pigments, the phycobiliproteins. These are organized in phycobilisome structures on the surface of the thylakoid membrane and fulfil the role of the chlorophyll *a*/*b* light harvesting complex of higher plants and green algae. In addition, the thylakoids are generally arranged in concentric sheets around the perimeter of the cell and are not organized into thylakoid stacks, as is the case for the chloroplasts of green algae and higher plants.

The interior of the cell contains the genetic chromosomal material of the cell, as well as the soluble enzymes associated with cell metabolism, including photosynthesis. There are also a number of inclusion bodies which appear to serve various roles. Polyphosphate bodies, cyanophycian granules, and glycogen granules appear to function as storage reserves for phosphate, nitrogen, and organic carbohydrate respectively. In addition, the cell contains polyhedral-shaped protein bodies which contain most, if not all, of the Rubisco of the cell. These are termed carboxysomes (Codd and Marsden 1984) and, as will be described below, perform a most important role in the function of the CCM of the cell.

The membrane layers which surround the cell have many properties which are typical of gram negative bacteria. There is an inner plasma membrane, which contains, among other things, components of the respiratory electron-transport chain, and functions as the primary metabolic and ionic barrier between the cell and the external environment. Surrounding this is the outer membrane and a peptidoglycan layer, which serves to give the cell rigidity and can also participate in the selective uptake of various organic and inorganic substrates.

The acquisition of C_i by cyanobacteria

Cyanobacteria must deal with a number of critical factors in trying to optimize the efficiency with which they acquire C_i from the external environment and use it for photosynthesis. Firstly, as they live in an aquatic environment at neutral to alkaline pH, they must deal with inorganic carbon in all its forms ($CO_2 + HCO_3^-$ $+ CO_3^{2-}$). As many habitats are above pH 8, this means that while CO_2 may be in equilibrium with atmospheric CO_2 and be present at around 10 μM, HCO_3^- may be 100-fold or more higher than this. Thus HCO_3^- may in fact be the major available form of C_i. The importance of HCO_3^- is increased when one considers the diffusion of C_i species in water and the barriers posed by unstirred layers surrounding the cells. In these situations, the ability of cells to use all available C_i species, and not relying solely on CO_2, is of great importance. In addition to these extracellular factors, the efficiency of C_i fixation is dependent on the kinetic properties of Rubisco. As has been stated above, cyanobacteria posses a Rubisco which has a low affinity for CO_2 (K_c = 100–150 μM) which must be overcome before efficient capture and fixation of C_i can occur.

The solution which cyanobacteria have found to deal with these problems is the development of a CO_2 concentrating mechanism. The effect of this mechanism on the efficiency of photosynthesis can be seen in Fig. 19.2. This shows the response of photosynthesis to external C_i (pH 8.0) in a theoretical cyanobacteria operating either with or without the CCM. As can be seen, the presence of the CCM improves the affinity of photosynthesis for C_i by at least two orders of magnitude, and without its operation the cells would not be able to survive in aquatic environments.

The extent to which the CCM operates in cyanobacteria is not static. Indeed, cyanobacteria show the ability to alter their affinity for external C_i as the external medium becomes more alkaline and is depleted C_i. Cells which are grown in medium which is in equilibrium with air levels of CO_2 are normally saturated by C_i levels of around 1–2 mM at pH 8. However, when C_i is depleted to one-tenth this level, the cells increase their affinity for external C_i and may be saturated by 20–50 μM C_i. Cells are able to adapt to the levels of C_i experienced during growth by increasing the activity and affinity of the C_i transporting system at low C_i concentrations (see Badger 1987). Cells grown under limiting levels of C_i, where the CCM is maximally induced, are termed low-C_i cells, while cells

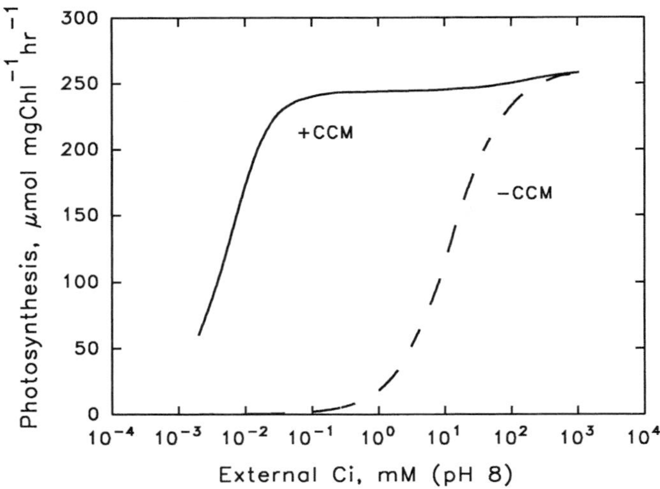

Fig. 19.2 Theoretical response of photosynthesis to external inorganic carbon (C_i) in the cyanobacterium *Synechococcus* PCC7942, both with (+CCM) and without (–CCM) the operation of the CO_2 concentrating mechanism. The response of the cells was calculated from the photosynthetic model used by Price and Badger (1989*a*) and assumes that the cells without the CCM have no active C_i transport and a carboxysome with a CO_2 conductance of 10^{-2} cm s^{-1}.

grown under nonlimiting levels of C_i and having a much reduced capacity to concentrate CO_2 are termed high-C_i cells.

Current concepts of the operation of the CCM

A current view of how the CCM achieves its function in cyanobacteria is presented in Fig. 19.3. The final elevation of CO_2 around the site of Rubisco is dependent on the three primary elements of all CCMs which are outlined in Fig. 19.1 (see Badger and Price 1992). Firstly, the cell has an active C_i uptake mechanism that is presumed to be located on the inner cell membrane. This can actively transport both CO_2 and HCO_3^- into the cell, using photosynthetic energy. The unique feature of this transport process is that regardless of whether CO_2 or HCO_3^- are the transported species, only HCO_3^- appears to be delivered to the cytosol. Thus the cell is able to accumulate an intermediate HCO_3^- pool within the cell and this may be concentrated more than 1000-fold compared to the external C_i concentration. The accumulated HCO_3^- is present in the cytosol of the cell, where there appears to be an absence of carbonic anhydrase (CA) activity, which might otherwise speed up the interconversion of HCO_3^- to CO_2. This lack of CA is crucial to preventing the leakage of accumulated C_i from the

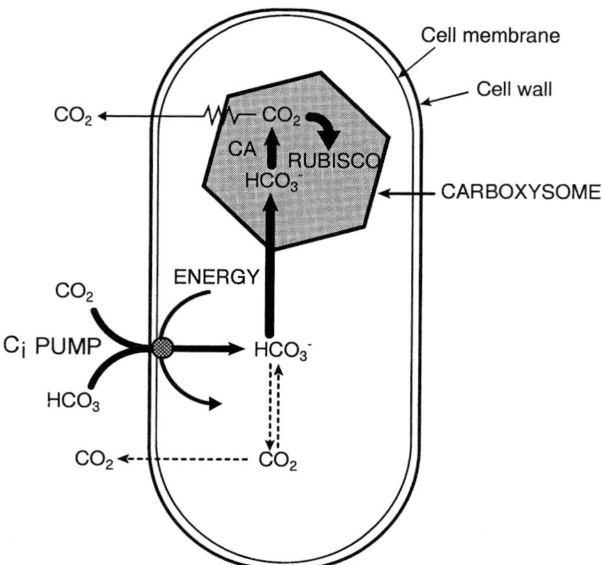

Fig. 19.3 A model showing a current view of the operation of the CCM in cyanobacteria. The thickness of the arrows denotes the relative magnitude of the fluxes of C_i species.

cell, as HCO_3^- is quite impermeable to cell membranes whereas CO_2 may diffuse quite readily back into the external solution.

Secondly, the carboxysome is the compartment where Rubisco is co-localized with the enzymatic machinery necessary to convert HCO_3^- to its primary substrate CO_2. This body contains both Rubisco and CA activity and thus HCO_3^- is envisaged to be able to diffuse into this body and be converted to CO_2 by the CA.

Thirdly, the means by which the liberated CO_2 is slowed in its diffusion from the carboxysome, to enable CO_2 levels to be elevated, is not well understood. However, suggestions thus far have been based on the notions that either the surrounding protein shell of the carboxysome may act as a diffusion barrier for CO_2 or that there is some three-dimensional arrangement of Rubisco and CA molecules within the cell which allows the Rubisco molecules themselves to act as the diffusive barrier. This will be outlined further below.

The carboxysome as the site of CO_2 elevation

A number of hypotheses for how the carboxysome might be able to function as a site for the localized elevation of CO_2 around the active site of Rubisco have been formalized by Reinhold *et al.* (1989, 1991). One model assumes that some property of the carboxysome, possibly the protein shell surrounding the Rubisco, results in a relatively low conductance to the movement of CO_2 of around 10^{-5} cm s^{-1} and a relatively high conductance to HCO_3^- of around 10^{-4} cm s^{-1}. In addition, the cell envelope has a normal conductance to CO_2 of around 10^{-2} cm s^{-1} and CA activity would be exclusively located inside the carboxysome. It was envisaged that the HCO_3^- accumulated by the C_i pump in the cytosol would penetrate into the carboxysome where CA would generate CO_2 in the immediate vicinity of Rubisco. The model predicts that CO_2 levels in the carboxysome could be efficiently elevated and maintained at a level necessary to explain the photosynthetic characteristics of cyanobacteria. A further refinement of this model has been proposed (Reinhold *et al.* 1991) in which it is postulated that the carbonic anhydrase is located in the interior of the carboxysome and surrounded by a shell of Rubisco. In this model, the CO_2 generated from CA must diffuse outwards past Rubisco sites located along the diffusion path. Using such a model, it may be unnecessary to postulate any other diffusion barrier than the Rubisco molecules themselves.

The model of Reinhold *et al.* (1989) was immediately attractive, but at its time of formulation there was no clear evidence for it. The model did, however, make several additional predictions about the operation of the CCM in cyanobacteria. These were: (a) that irrespective of the C_i species which was being taken up by the C_i pump, HCO_3^- would need to be the only species delivered to the inside of the cell; (b) that CA would be absent from the cytosol so that the slow uncatalysed conversion between HCO_3^- and CO_2 would minimize the wasteful leakage of CO_2 out of the cell; and (c) that CA, therefore be exclusively located within the carboxysome. These hypotheses were directly tested by an experiment in which a human CA gene was expressed in the cytosol of *Synechococcus* (Price and Badger 1989*b*). The induced expression of cytosolic CA resulted in an inability to accumulate C_i and such cells require a very high level of external C_i to support saturated photosynthetic rates, thus substantiating the basic predictions of the model.

The analysis of several high-CO_2 requiring (HCR) mutants has also provided strong evidence for the central involvement of carboxysomes in the CCM. These mutants appear to be able to accumulate C_i within the cells to levels which are similar to, if not higher than, wild-type cells but are impaired in the ability to use this C_i for efficient photosynthesis (see section below). It was thus obvious to suspect that some component of the carboxysome system may be impaired. Electron microscope analysis of these mutants revealed that some of them appeared to have significant alterations to the morphology of the carboxysome, although this did not disturb the pelletability of Rubisco. The genomic DNA lesions and possible open reading frames (ORFs) of these mutants have been identified and it appears that they are largely centred in regions of DNA that are clustered upstream and downstream of the Rubisco large and small subunit genes (see Fig. 19.4 and text). Indeed, a specific mutant has been constructed which has a number of these ORFs deleted (Price and Badger 1991), producing

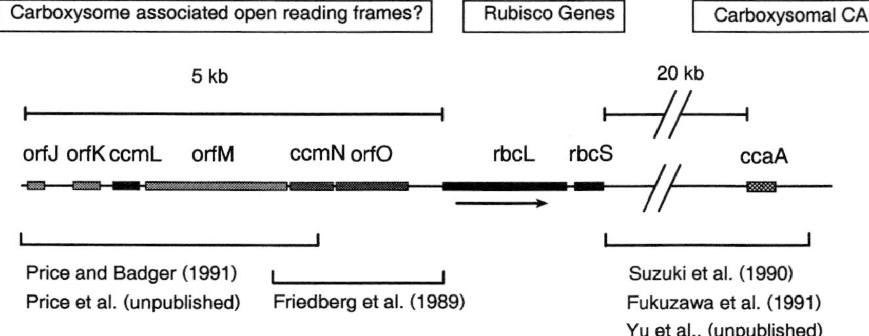

Fig. 19.4 A current view of the organization of the genes and open reading frames associated with the CCM in the cyanobacterium *Synechococcus* PCC7942 and their spatial relationship to the Rubisco large and small subunit genes (*rbc*L and *rbc*S). References listed below the map indicate the sources of the information.

a high-CO_2 requiring phenotype which has soluble Rubisco and a total absence of carboxysomal structures.

High CO_2 requiring (HCR) mutants

Rapid progress into understanding the physical, biochemical and molecular genetic basis of the CCM in cyanobacteria is currently being made through the selection and analysis of mutants which are defective in parts of this mechanism. The mutant selection procedures which have been used, so far, were adapted from mutagenic procedures that have been developed for cyanobacteria (see Golden 1988), coupled with selection for mutants which will grow at elevated CO_2 (1–5 per cent) but not in air. Kaplan and co-workers (Marcus *et al.* 1986) first applied this approach to cyanobacteria, and in the last few years a series of interesting mutants have been isolated by a number of groups (Marcus *et al.* 1986; Schwarz *et al.* 1988; Price and Badger 1989*c*; Ogawa 1990). In most cases the functional lesions have been resolved to the DNA level through the use of DNA complementation procedures (see Badger and Price 1992).

Characterization of CCM genes

Nearly all of the HCR chemical mutants have been characterized to the level of the DNA lesion, which is responsible for the phenotype. The accumulation of these various mutants is leading to an intriguing picture, which indicates that many of the genes associated with the CCM, especially those relating to carboxysome function and the utilization of internal C_i, are clustered around the

Rubisco operon. A current view of the genomic organization of this region is shown in Fig. 19.4.

The region 5′ to the *rbc*L gene contains a number of ORFs, which may be directly associated with carboxysome assembly and function. Mutations in the 300 bp *ccm*L and *ccm*N lead to altered carboxysome morphology, while a deletion of the DNA region between the *ccm*L and orfO results in a complete absence of carboxysomes. The exact role of the proteins coded for by these ORFs remains to be determined, and work is proceeding in various laboratories on their function. One hypothesis, which is attractive at present, is that some of these proteins may be coat proteins for the carboxysome. The absence of one may lead to a change in the morphology of the carboxysome, while the absence of a number could result in no carboxysomes at all. A number of other ORFs have been identified in this region by DNA sequencing, and their precise function in the CCM remains to be determined.

The region downstream of the *rbc*S gene may have a number of different functions associated with it. Approximately 20 kb downstream there is a DNA region which appears to complement two other HCR mutants. Recent physiological analysis of one of these mutants (type II, Price and Badger 1989*c*) suggests that this may code for carboxysomal CA activity (Price, Coleman, and Badger, submitted). In addition, Fukuzawa *et al.* (1991) have recently reported a DNA sequence for an ORF in this region which has significant homology with spinach chloroplast CA.

The reason for the clustering of many of the genes associated with the CCM around the Rubisco operon is unclear. It is tempting to speculate that there may be coordinated regulation of the expression of these genes, so that a functional CCM is produced and the cell can respond appropriately during periods of induction at limiting C_i. The expression of all the genes that have already been characterized, would be constitutively required if the cell is to maintain functional photosynthesis.In addition, when adapting to low C_i conditions there may be a requirement to increase the levels of many proteins required for the production of carboxysomes or C_i pumping.

The relevant ORFs, which are impaired in the C_i transport mutants isolated by Ogawa (1991), have been sequenced, but their genomic location has not been published. However, some inferences can be made on their possible function. Both ORFs code for hydrophobic polypeptides, which would suggest a membrane location. The small 240 bp ORF codes for an 80 amino acid polypeptide with little homology to any known protein. Its size suggests that it may not be part of a primary C_i transporter but part of the energization system necessary to drive transport. The 1800 bp ORF codes for a polypeptide with significant homology to a higher plant chloroplastic NADH dehydrogenase gene, the function of which is not known. This suggests that this protein may be part of a membrane bound dehydrogenase complex, which is involved in electron transport; and that this electron transport is involved in the energization of C_i transport.

Inorganic carbon transport

The C_i transport system is of central importance to the functioning of the CCM and enables cyanobacteria to actively acquire and concentrate C_i from the external environment. When this transport process is fully induced, cells possess the ability to utilize both CO_2 and HCO_3^- as substrates for transport (Badger and Andrews 1982, Miller *et al.* 1990). However, at the alkaline pHs which usually accompany these conditions, it is HCO_3^- transport that largely supports the supply of CO_2 for photosynthesis (see Miller *et al.* 1990).

The mechanism(s) by which transport occurs are still not well understood, and there has been some conjecture as to whether there are separate transport processes for both CO_2 and HCO_3^-. Two models for a single transport mechanism have been proposed, one with a central HCO_3^- transporter (Volokita *et al.* 1984) and the other with a common CO_2 transport process (Price and Badger 1989*a*). However, the existence of separate transport entities is also strongly advocated (Miller *et al.* 1990). On the whole, the present balance of evidence would support a single transporter model, in which the pump was able to use either CO_2 or HCO_3^- as substrates. This evidence includes: (1) both CO_2 and HCO_3^- transport are inhibited by the CA inhibitor ethoxyzolamide (Price and Badger 1989*a*); (2) both transport activities appear to be inhibited by the CO_2 analogue COS (Badger and Price 1990); (3) regardless of which external C_i species is taken up from the external medium, HCO_3^- appears to be the only species delivered to the interior of the cell (Price and Badger 1989*b*); (4) there is evidence for competition between CO_2 and HCO_3^- during the transport process (Volokita *et al.* 1984); (5) two high-CO_2 requiring mutants of *Synechocystis* PCC6803, which are impaired in C_i transport, show inactivation of both CO_2 and HCO_3^- transport activities (Ogawa 1990).

Arguments for the existence of multiple transport pathways for both CO_2 and HCO_3^- stem almost exclusively from studies of the differential effects of Na^+ on both CO_2 and HCO_3^- transport (for summary see Miller *et al.* 1990). It is apparent that HCO_3^- transport in some cell cultures requires mM levels of Na^+ for activity while CO_2 uptake requires only μM concentrations. In addition, for cells grown at very low C_i, where HCO_3^- is the most important substrate for C_i uptake, there appears to be a Na^+ insensitive HCO_3^- uptake. The involvement of Na^+ in transport is unknown, but it is unlikely to participate directly and stoichiometrically in the process (Miller *et al.* 1990). The most likely involvement is through its role in pH regulation through the operation of the Na^+/H^+ antiport system. The process of HCO_3^- uptake, in particular, will lead to the net appearance of OH^- ions within the cell. These must be either excreted or balanced by proton import if intracellular pH is to be stabilized during both the initial uptake of HCO_3^- and steady-state photosynthesis. Until more is understood about how Na^+ participates in the pH regulation of the cell and its role as a counterion for transport under various conditions, it would seem unwise to use the multiple effects of Na^+ on C_i transport to infer that there are multiple C_i transport mechanisms.

A possible model for transport

A possible model for operation of a single C_i transporter, located on the plasmamembrane, is shown in Fig. 19.5 but the exact details are speculative. The central transporter is proposed to have carbonic anhydrase-like properties, which allow it to accept CO_2 on the outside surface and convert it to HCO_3^-, which is then delivered to the interior of the cell. This would occur via a base catalysed reaction and could be mediated by the presence of a basic metal such as zinc at the active site. In addition, this active site could also accept HCO_3^- if it were in the protonated state. The uptake of CO_2 would be energized by the generation of the hydroxylated metal species and could be linked to an energy driven process such as an electron-transport chain in the plasmamembrane. The uptake of HCO_3^- would be facilitated by the generation of the hydrated form of the metal, and this can be envisaged to be linked to something like a plasmamembrane H^+-pumping ATPase. While a model such as this is an interesting hypothesis, it must be tested by experimental approaches.

Two other mechanisms for uptake of CO_2 which could lead to its conversion to HCO_3^- during transport have been proposed by Miller *et al.* (1991*b*). These include: (a) the formation of a carbamate, followed by an OH^- dependent decarboxylation to HCO_3^-; (b) the binding of CO_2 to a transition metal, resulting in the enhance susceptibility to nucleophilic attack by OH^-.

Effects of sulphonamide inhibitors

There is little data which will allow the transport mechanism to be clearly defined. The inhibition of transport by the CA inhibitor ethoxyzolamide (EZ)

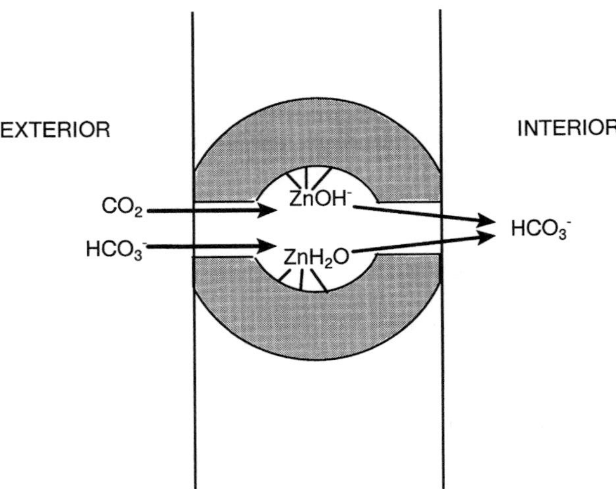

Fig. 19.5 A theoretical model for the operation of the C_i transporter in cyanobacteria.

provides the strongest support for a CA like mechanism. This inhibitor binds relatively specifically to the Zn-containing active site of the CA enzyme. The sulphonamides bind as anions, similarly to HCO_3^-, presumably with the negatively charged NH^- group liganded to the Zn (Banci *et al.* 1990). The NH^- group has the same position as the Zn-bound hydroxyl ion in the CA structure. The interactions of the rest of the inhibitor molecule with the protein depend on which substituents are situated on various locations on the aromatic ring of the inhibitor. Thus binding can be enhanced by these other interactions. The concentrations of EZ which are required to inhibit C_i transport are high (100 μM) compared with purified CA enzymes (1 nM–1 μM). However, the fact that another more hydrophilic sulphonamide, acetozolamide, is not effective as an inhibitor, (Price and Badger 1989*a*) suggests that the site of inhibition is possibly buried within a hydrophobic membrane environment. No membrane-bound CA enzymes, where the active site of the enzyme is located within the membrane, have been reported for other animal or plant-based systems, so it is difficult to predict the inhibition characteristics that would be expected for sulphonamide inhibitors.

Energization of transport

Little work has been done on understanding the mechanisms underlying the energization of C_i transport. Clearly, C_i transport is linked to photosynthetic electron transport and ATP generation, but little more has really been explored (see Badger 1987). The lack of any *in vitro* assay system for C_i transport processes, such as sealed membrane vesicles, is a significant barrier for progress in this area. Recent work has shown that there is a stimulation of oxygen photoreduction during periods of inorganic carbon transport, as distinct from photosynthetic carbon fixation, suggesting that pseudocyclic electron flow to oxygen may be important in the energization of C_i uptake (Miller *et al.* 1988, 1991*a*). In addition, chlorophyll *a* fluorescence yield is strongly quenched by the accumulation of internal inorganic carbon by the cyanobacterial cell (Miller *et al.* 1991*a*). In fact, the extent of quenching is a reasonably good indicator of the internal C_i concentration. The mechanistic basis for this quenching is unknown, but direct effects of inorganic carbon (possibly HCO_3^-) on PSII have been suggested.

Summary

Cyanobacteria have overcome the kinetic constrains of both Rubisco and the aquatic environment by developing a novel type of CO_2 concentrating mechanism. This CCM is based on the active uptake of inorganic carbon using photosynthetic energy, together with the compartmentalization of Rubisco and carbonic anhydrase within the carboxysome. Rapid progress is being made towards understanding the physiological and molecular basis of this mechanism by cyanobacteria.

References

Badger, M.R. (1987). In: *The biochemistry of plants: a comprehensive treatise,* Vol. 10, *Photosynthesis* (ed. M.D. Hatch and N.K. Boardman), pp. 219–74, Academic Press, New York.

Badger, M.R. and Andrews, T.J. (1982). *Plant Physiol.,* **70,** 517 –23.

Badger, M.R. and Andrews, T.J. (1987). In *Progress in photosynthesis research,* Vol. III (ed. J. Biggins), pp. 601–9. Martinus Nijhoff, Dordrecht, The Netherlands.

Badger, M.R. and Price, G.D. (1990). *Plant Physiol.,* **94,** 35–9.

Badger, M.R. and Price, G.D. (1992). *Physiologia Plantarum,* **84,** 606–15.

Banci, L., Bertini, I., Luchinat, C., and Moratal, J.M. (1990). In *Enzymatic and model carboxylation and reduction reactions for carbon dioxide utilization* (ed. M. Aresta and J.V. Schloss), pp. 181–98. Kluwer, Dordrecht, The Netherlands.

Broda, E. (1975). *The evolution of bioenergetic processes.* Pergamon Press, Oxford.

Codd, G.A. and Marsden, W.J.N. (1984). *Biol. Rev.,* **59,** 389–422.

Friedberg, D., Kaplan, A., Ariel, R., Kessel, M., and Seijffers, J. (1989). *J. Bacteriol.,* **171,** 6069–76.

Fukuzawa, H. and Suzuki, E. (1991). Abstract 1826, *Third international congress of plant molecular biology,* Tucson.

Golden, S. (1988). In *Methods in enzymology,* Vol. 176, *Cyanobacteria* (ed. L. Packer and A.N. Glazer), pp. 714–27. Academic Press, New York.

Hatch, M.D. (1987). *Biochem. Biophys. Acta,* **895,** 81–106.

Marcus, Y., Schartz, R., Friedberg, D., and Kaplan, A. (1986). *Plant Physiol.,* **82,** 601–12.

Miller, A.G., Espie, G.S., and Canvin, D.T. (1988). *Plant Physiol.,* **88,** 6–9.

Miller, A.G., Espie, G.S., and Canvin, D.T. (1990). *Can. J. Botany,* **68,** 1291–302.

Miller, A.G., Espie, G.S., and Canvin, D.T. (1991*a*). *Can. J. Botany,* **69,** 1151–60.

Miller, A.G., Espie, G.S., and Canvin, D.T. (1991*b*). *Can. J. Botany,* **69,** 925–35.

Ogawa, T. (1990). *Plant Physiol.,* **94,** 760–5.

Ogawa, T. (1991). *Can. J. Botany,* **69,** 951–6.

Packer, L. and Glazer, A.N. (1988). *Cyanobacteria, methods of enzymology,* Vol. 167. Academic Press, New York.

Price, G.D. and Badger, M.R. (1989*a*). *Plant Physiol.,* **89,** 37–43.

Price, G.D. and Badger, M.R. (1989*b*). *Plant Physiol.,* **91,** 505–13.

Price, G.D. and Badger, M.R. (1989*c*). *Plant Physiol.,* **91,** 514–25.

Price, G.D. and Badger, M.R. (1991). *Can. J. Botany,* **69,** 963–73.

Reinhold, L., Zviman, M., and Kaplan, A. (1989). *Plant Physiol. Biochem.,* **27,** 945–54.

Reinhold, L., Kosloff, R., and Kaplan, A. (1991). *Can. J. Botany,* **69,** 984–8.

Schwarz, R., Friedberg, D., and Kaplan A. (1988). *Plant Physiol.,* **88,** 284–8.

Suzuki, E., Fukuzawa, H., Abe, T., and Miyachi, S. (1990). In *Current research in photosynthesis,* Vol. IV, pp. 467–70. Kluwer, Dordrecht, The Netherlands.

Volokita, M., Zenvirth, D., Kaplan, A., and Reinhold, L. (1984). *Plant Physiol.,* **76,** 599–602.

20

Molecular biology and biochemistry of carbon dioxide-concentrating mechanisms in eukaryotic algae

John R. Coleman

Introduction

The ability of photosynthetic aquatic organisms to respond to fluctuations in the inorganic carbon (C_i) concentration of their environment is of considerable importance to the overall efficiency of algal carbon assimilation, and as such has been the topic of considerable research. Unlike the terrestrial environment, the supply of dissolved C_i available for photosynthesis is continually changing in response to a number of parameters such as turbulence (mixing rates with atmospheric gases), pH, and algal cell concentrations. For example, high concentrations of actively photosynthesizing algae can reduce dissolved C_i to levels well below air equilibrium and at the same time increase dissolved oxygen concentrations well above atmospheric levels. In other aquatic systems, although dissolved CO_2 levels may be in equilibrium with atmospheric concentrations, the alkaline pH of water generates a large pool of C_i in the form of HCO_3^-. It would certainly be an advantage for the algae to access this pool as a substrate for photosynthetic carbon fixation. In general, wide variation in the dissolved C_i concentration and speciation in the aquatic environment has driven the evolution of a plastic response by these algae to changing C_i concentrations.

The limiting C_i syndrome

Acclimation of changes in the dissolved C_i levels by eukaryotic algae (and cyanobacteria) has been well documented. The most prominent manifestation of the acclimation process is the change in whole cell affinity for the C_i requirements of photosynthesis (Badger 1987; Spalding 1989). When algal cells are grown at high levels of C_i (0.5–5 per cent CO_2 in air), the $K_{1/2}$ C_i requirements for whole cell photosynthesis are approximately equal to the K_m CO_2 of the principal CO_2-fixing enzyme, ribulose-1,5-bisphosphate carboxylase/oxygenase (Rubisco). This can be contrasted with the C_i requirements for photosynthesis of cells grown at inorganic carbon concentrations which are limiting for growth. These cells with exhibit $K_{1/2}$ C_i values which range from 10 to 100-fold less than the CO_2 concentration required for half saturation of purified Rubisco isolated from that same species. Data obtained from experiments with the eukaryotic green alga, *Chlamydomonas reinhardtii* provide an excellent example of this acclimation process (Table 20.1). Cells grown at high levels of CO_2 exhibit a $K_{1/2}$ for photosynthesis of approximately 200 μM HCO_3^- whereas cells grown at air levels of C_i exhibit a much lower $K_{1/2}$ of approximately 20 μM HCO_3^-. The improved photosynthetic performance at low C_i concentrations is not the result of a modification of the primary CO_2 fixing enzyme as Rubisco, isolated from *Chlamydomonas* grown at either CO_2 concentration, exhibits a K_m CO_2 of 60 μM (Berry *et al.* 1976).

Of interest to many researchers has been the ability of the algae to rapidly acclimate to sudden changes in the availability of dissolved C_i (Table 20.1). Within a matter of hours after transfer from high to limiting C_i concentrations, many algae have modified their inorganic carbon assimilation mechanisms such that they now exhibit a very high affinity for inorganic carbon. Numerous studies have now shown that it is the induction of high affinity/high capacity C_i-concentrating mechanisms (CCMs) as well as carbonic anhydrase activity, after transfer of the cells to limiting C_i conditions, that results in the development of the high affinity phenotype for photosynthetic carbon assimilation. The induction of the high affinity/high capacity CCMs may be the result of addition of a new class of C_i transporter proteins or modification of the existing constitutively expressed low affinity CCMs. It is apparent that either CO_2 or HCO_3^- can be utilized as substrates for the CCMs but there is still considerable controversy as to whether this represents the activity of one, dual-purpose transport system or the activity of two separate transporters (Sultemeyer *et al.* 1989). In the eukaryotic algae the presence of both constitutively expressed and low C_i induced carbonic anhydrase (CA) activity, and the complexity of a multicompartmented cell, have made the investigation of substrate use and membrane location of transport somewhat difficult. The impact, however, of the expression of carbonic anhydrase activity and the algal CCMs on the physiology of the cells can be readily observed. Using silicone–oil centrifugation techniques to rapidly separate the cells from the surrounding medium, it has been possible to show that the

Table 20.1 Modification of $K_{1/2}$ (C_i) for photosynthesis during acclimation of high C_i-grown *Chlamydomonas reinhardtii* to air levels of C_i

Time after transfer from high C_i (h)	$K_{1/2}$ (C_i) (μM)
0	180
0.5	100
1	73
3	45
5	28
8	25
12	22

activity of the CCMs and CA results in the accumulation of a large intracellular pool of inorganic carbon which is subsequently used by Rubisco. It is the induction of the high affinity/high capacity CCMs and CA activity during growth at limiting C_i concentrations, and the subsequence accumulation of the large intracellular C_i pool, that results in the expression of the limiting C_i phenotype in algae. Algal cells exhibiting this phenotype have a high cellular affinity for inorganic carbon, low rates of photorespiration when measured at low external C_i concentrations, low CO_2 compensation points, as well as little if any oxygen inhibition of photosynthesis. These characteristics of photosynthesis resemble those expressed by typical C_4 plants and would appear to be the product of a similar CO_2 enriched intracellular environment for Rubisco. Within the algal cell, chloroplastic CA catalysed dehydration of the intracellular HCO_3^- pool generates sufficient CO_2 for the saturation of carboxylase activity of Rubisco and repression of oxygenation. A general model describing the organization of the CCMs and the locations of carbonic anhydrase activity in the eukaryotic green alga *Chlamydomonas reinhardtii* is shown in Fig. 20.1. Both CO_2 and HCO_3^- have been proposed as transported species at the plasma membrane (Sultemeyer *et al.* 1989) and additional evidence has shown that isolated chloroplasts are capable of accumulating C_i; however, the identity of the transported species has not been determined (Moroney *et al.* 1987). CA activity has been shown to be present in three distinct compartments of eukaryotic algae, the periplasmic space, the cytosol, and the chloroplast (Coleman *et al.* 1984, 1991; Husic *et al.*

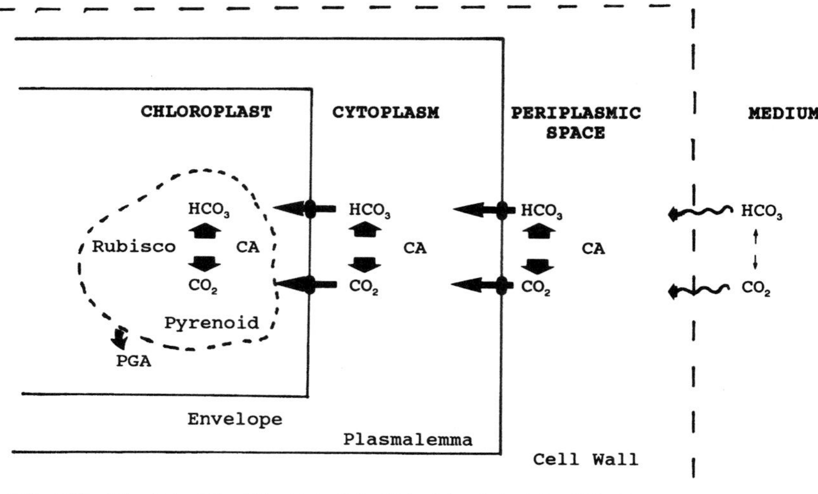

Fig. 20.1 Proposed model for the CCM of the eukaryotic alga, *Chlamydomonas reinhardtii*. Location of transport systems are indicated by the trans-membrane heavy arrows. The presence of CA activity in the three cell compartments is also indicated. PGA (phosphoglyceric acid).

1989; Sultemeyer *et al.* 1990). Its role within the periplasmic space (and possibly in the cytosol) is to maintain the equilibrium of the C_i species at the site of transport. In the absence of CA, the uncatalysed rate of C_i interconversion could be sufficiently slow as to be rate limiting for the operation of either a CO_2 or HCO_3^- transporter. The principal role of the chloroplastic CA would be to maintain the rate of supply of CO_2 for the efficient operation of Rubisco.

Polypeptide and mutant analyses

Although much is known about the physiology and photosynthetic performance of algae during acclimation to limiting C_i, only recently has some progress been made in the molecular and biochemical analyses of the acclimation response. The strategies that have been employed in the search for proteinaceous components of algal CCMs have been: (1) *in vivo* labelling of polypeptides synthesized during acclimation to limiting C_i; and (2) biochemical and molecular analyses of mutants that have an obligate growth requirement of high CO_2 concentrations.

In vivo labelling studies of algae — primarily *Chlamydomonas* — have resulted in the identification of a number of polypeptides which are synthesized in response to limiting CO_2 concentrations. In the addition to the well characterized periplasmic CA, at least six other polypeptides are synthesized during acclimation to low C_i concentrations (Bailly and Coleman 1988; Manuel and Moroney 1988; Spalding and Jeffrey 1989). These include membrane proteins (nonthylakoid) or 35, 36, 21, and 19 kDa which are expressed and accumulate during the acclimation period. An additional two soluble polypeptides of 44–50 kDa are transiently synthesized but do not appear to accumulate during acclimation. It has also been shown that the synthesis of both the large and small subunits of Rubisco is transiently inhibited during acclimation (Coleman and Grossman 1984) and that the regulation of expression of the small subunit is occurring at the level of translation (Spalding *et al.* 1991). In contrast to translational control of Rubisco, expression of the 36 kDa polypeptide appears to be regulated by an increase in mRNA abundance that parallels the acclimation to low C_i concentrations (Geraghty *et al.* 1990). It is interesting to note that a polypeptide of the same molecular mass has been localized to the *Chlamydomonas* chloroplast membrane and that these chloroplasts when isolated have some capacity for C_i transport and accumulation (Mason *et al.* 1990; Moroney *et al.* 1987). These data are suggestive of a role for this induced 36 kDa polypeptide in the algal CCMs.

Carbonic anhydrase activity in many green algae has been shown to increase dramatically after the transfer of cells from high to limiting C_i concentrations (Fig. 20.2). In *Chlamydomonas,* much of the CA activity is found within the periplasmic space and it is this low C_i induced protein that has been the most extensively characterized. It is synthesized as a 42–44 kDa precursor on 80 S ribosomes and then processed to yield a glycosylated 35–37 kDa monomer and a

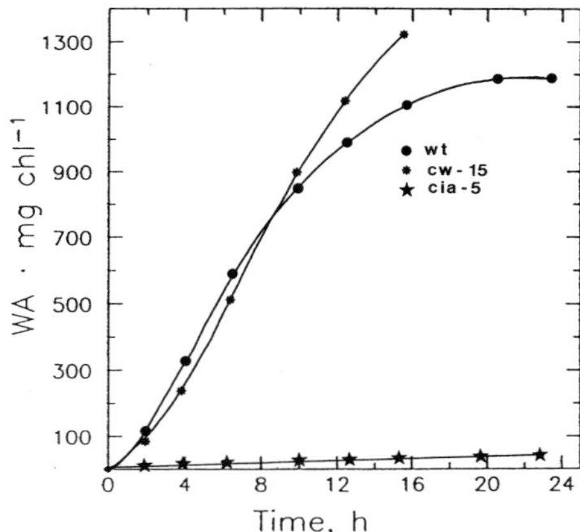

Fig. 20.2 Induction of CA activity in *Chlmaydomonas reinhardtii* cell lines. Wild-type (●), cell wall-less mutant (✱), and acclimation deficient mutant (★) cultures were grown at 1 per cent CO_2 in air and then transferred to air levels of CO_2 at time 0. Wild-type cells exhibit a saturation curve for CA activity as the protein accumulates in the periplasmic space and synthesis is reduced. The cell wall-less mutant over-expresses CA activity as it is unable to accumulate the CA protein within the periplasmic space. The acclimation deficient mutant (cia-5) does not respond to the change in extracellular C_i and continues to express only basal levels of CA activity.

smaller subunit of 4 kDa (Coleman and Grossman 1984; Toguri *et al.* 1989; Fukuzawa *et al.* 1990*a*). The holoenzyme is thought to be a heterotetramer composed of two large and two small subunits joined by disulphide bonds (Kamo *et al.* 1990). Expression of the low C_i induced periplasmic CA is regulated by transcription, with large increases in CA mRNA abundance upon transfer of the cells from high to limiting CO_2 concentrations (Fig. 20.3) (Bailly and Coleman 1989; Fukuzawa *et al.* 1990*a*). Both transcription and translation have been shown to be light dependent and can be blocked by the addition of DCMU in the light at limiting C_i concentrations (Fukuzawa *et al.* 1990*a*). Most recently it has been shown that two periplasmic CA genes are present in the *Chlamydomonas* genome (Fukuzawa *et al.* 1990*b*). These genes, designated CAH1 and CAH2, are very similar in sequence but exhibit significant differences with respect to regulation of expression. Transcript analysis has shown that CAH1 codes for the more abundant polypeptide expressed only under low C_i conditions whereas CAH2 codes for low abundance protein that is expressed only at high levels of CO_2 (Fujiwara *et al.* 1990). Presumably, gene duplication and alteration of promoter structures resulted in the abundant synthesis of the protein coded for by the low C_i inducible CAH1 gene. With the characterization of the CA genomic sequences and the development of *Chlamydomonas* transformation and expression system it will be possible to identify promoter elements which are

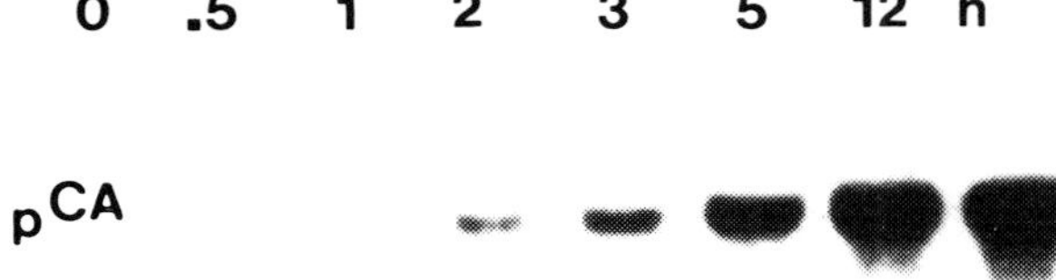

Fig. 20.3 Time course of periplasmic CA transcript accumulation in *Chlamydomonas reinhardtii*. Cells were transferred from high to limiting concentrations of CO_2 for varying periods of time after which total RNA was extracted, electrophoresed, and transferred to nitrocellulose. The RNA blot was then probed with a cDNA clone of the CAH1 gene and the extent of hybridization determined by autoradiography.

responsive to control by environmental CO_2 concentrations. The rapid and high level of transcription of the CAH1 gene will be useful as a model system for the isolation of *cis*- and *trans*-acting elements.

The generation of high CO_2 requiring mutants of *Chlamydomonas* has provided some information on the number of components of the algal CCMs. Unlike studies with cyanobacterial mutants however, the inability to perform complementation assays with algal mutants has meant that specific DNA fragments containing CCM genes have not been isolated. The physiological and biochemical studies of specific cell lines have described a number of categories of mutations that will generate a high CO_2 requiring phenotype. These include cell lines with presumed defects in C_i transport, in utilization of the intracellular C_i pool, in the photorespiratory pathway, and in the ability to induce acclimation (Moroney *et al.* 1989; Suzuki and Spalding 1989; Suzuki *et al.* 1990). From these studies it is obvious that a number of different loci are required for the expression of functional CCMs and that at least partial expression of the CCMs is constitutive and required for efficient photosynthesis even at high C_i concentrations.

The pyrenoid

The pyrenoid, an inclusion body of the algal chloroplast, has been shown to contain the majority of the cellular Rubisco and as such is presumably the site of CO_2 fixation in *Chlamydomonas* as well as in many other algal species. As a homologue of the algal pyrenoid, the Rubisco-containing carboxysome, has been shown to be an integral component of the cyanobacterial CCMs (Price and Badger 1991), and it has been proposed that the algal pyrenoid is also involved in the eukaryotic CCMs. In an extension of the cyanobacterial carboxysome model, the presence of CA activity within the Rubisco aggregate could result in localized CO_2 generation and a reduction in the passive leak of CO_2 from the chloroplast. Although the pyrenoid is not bounded by a protein membrane as is the cyanobacterial carboxysome, it is partially enclosed within a starch sheath. In addition to the presence of biochemically active Rubisco, the enzyme Rubisco activase has been shown to be localized in the algal pyrenoid (McKay and Gibbs 1991). Dansylamide fluorescence of isolated pyrenoids from *Chlamydomonas* suggests that CA activity is also present within these structures (Kuchitsu *et al.* 1991). Additional evidence for the involvement of the pyrenoid in the algal CCMs includes the observed elaboration of this structure during acclimation to limiting C_i concentrations (Miyachi *et al.* 1986). An important step in our understanding of the role of the pyrenoid would be the isolation of mutants unable to form these structures or cell lines which are deficient in the various polypeptide components of this chloroplast inclusion.

In conclusion, although our understanding of how eukaryotic algae acclimate to external C_i concentrations has increased significantly in the past five years, there are still a number of areas that require more research. For example, the environmental or metabolic signal that the alga detects and then uses to elicit the induction of the low C_i response has not been determined. It is possible that a detected change in the concentration of, or flux through, a metabolite pool in the photorespiratory pathway; or a change in the ratio of Rubisco carboxylase/oxygenase activity could be responsible for activating the acclimation response. An alternative model would be that the C_i transport systems themselves could be used as sensors of the extracellular C_i environment and that a reduced flux through constitutively expressed low affinity CCMs is a signal for the expression of the high affinity C_i transport systems and the additional components of the acclimation response. In the future, the application of molecular biology techniques, including newly developed procedures for algal transformation, will certainly speed the molecular dissection of the low C_i acclimation response. Presumably the data generated in these studies will provide the needed information on some of the more elusive aspects of this process, such as organization and role of the pyrenoid, energization and localization of the transport systems, as well as the identification of the inducing signal for acclimation.

References

Badger, M.R. (1987). In *The biochemistry of plants: a comprehensive review*, Vol. 10 (ed. M.D. Hatch and N.K. Boardman), pp. 219–74. Academic Press, New York.

Berry, J., Boynton, J., Kaplan, A., and Badger, M.R. (1976). *Carnegie Inst. Washington Yearbook*, **76**, 423–32.

Bailly, J. and Coleman, J.R. (1988). *Plant Physiol.*, **87**, 833–40.

Coleman, J.R. and Grossman, A.R. (1984). *Proc. Nat. Acad. Sci. USA*, **81**, 6049–53.

Coleman, J.R., Rotatore, C., Williams, T., and Colman, B. (1991). *Plant Physiol.*, **95**, 331–4.

Fujiwara, S., Fukuzawa, H., Tachiki, A., and Miyachi, S. (1990). *Proc. Nat. Acad. Sci. USA*, **87**, 9779–83.

Fukuzawa, H., Fujiwara, S., Yamamoto, Y., Dionisio-Sese, M.L., and Miyachi, S. (1990*a*). *Proc. Nat. Acad. Sci. USA*, **87**, 4383–7.

Fukuzawa, H., Fujiwara, S., Tachiki, A., and Miyachi, S. (1990*b*). *Nucleic Acids Res.*, **18**, 6441–2.

Geraghty, A.M., Anderson, J.C., and Spalding, M.H. (1990). *Plant Physiol.*, **93**, 116–21.

Husic, H.D., Kitayama, M., Togasaki, R.K., Moroney, J.V., Morris, K.L., and Tolbert, N.E. (1989). *Plant Physiol.*, **84,** 904–9.

Kamo, T., Shimogawara, K., Fukuzawa, H., Muto, S., and Miyachi, S. (1990). *Eur. J. Biochem.*, **192**, 557–62.

Kuchitsu, K., Tsuzuki, M., and Miyachi, S. (1991). *Can. J. Botany*, **69**, 1062–9.

McKay, R.M.L. and Gibbs, S.P. (1991). *Can. J. Botany*, **69**, 1040–52.

Manuel, L.J. and Moroney, J.V. (1988). *Plant Physiol.*, **88**, 491–6.

Mason, C.B., Manuel, L.J., and Moroney, J.V. (1990). *Plant Physiol.*, **93**, 833–6.

Miyachi, S., Tsuzuki, M., Maruyama, I., Gantar, M., and Miyach, S. (1986). *J. Phycol.*, **22**, 313–19.

Moroney, J.V., Kitayama, M., Togasaki, R.K., and Tolbert, N.E. (1987). *Plant Physiol.*, **83**, 460–3.

Moroney, J.V., Husic, H.D., Tolbert, N.E., Kitayama, M., Manuel, L.J., and Togasaki, R.K. (1989). *Plant Physiol.*, **89**, 897–903.

Price, G.D. and Badger, M.R. (1991). *Can. J. Botany*, **69**, 963–73.

Spalding, M.H. (1989). *Aquatic Botany*, **34**, 181–209.

Spalding, M.H. and Jeffrey, M. (1989). *Plant Physiol.*, **89**, 133–7.

Spalding, M.H., Winder, T.L., Anderson, J.C., Geraghty, A.M., and Marek, L.F. (1991). *Can. J. Botany*, **69**, 1008–16.

Sultemeyer, D.F., Miller, A.G., Espie, G.D., Fock, H.P., and Canvin, D.T. (1989). *Plant Physiol.*, **89**, 1213–19.

Sultemeyer, D.F., Fock, H.P., and Canvin, D.T. (1990). *Plant Physiol.*, **94**, 1250–7.

Suzuki, K. and Spalding, M.H. (1989). *Plant Physiol.*, **90**, 1195–200.

Suzuki, K., Marek, L.F., and Spalding, M.H. (1990). *Plant Physiol.*, **93**, 231–7.

Toguri, T., Muto, S., Mihara, S., and Miyachi, S. (1989). *Plant Cell Physiol.*, **30**, 533–9.

Index